World Regional Geography

World Regional Geography

VOLUME 1

Omer Lasme

Kruger Brentt
Publishers
2025

Kruger Brentt Publishers UK. LTD.
Company Number 9728962

Regd. Office: 68 St Margarets Road, Edgware, Middlesex HA8 9UU

ISBN: 978-1-78715-405-6 (Vol. 1)

For information on all our publications visit our website at http://krugerbrentt.com/

PREFACE

"World Regional Geography" offers a comprehensive exploration of the diverse landscapes, cultures, and environments that shape the world's regions. This book serves as an essential resource for students, educators, researchers, and anyone interested in understanding the complex interactions between human societies and their geographical contexts.

Geography is the study of the Earth's surface and the processes that shape it, including human activities, physical landscapes, and environmental dynamics. World regional geography provides a framework for understanding the spatial patterns, cultural diversity, and economic disparities that characterize different parts of the globe.

As editors of this volume, we have endeavored to compile a diverse array of perspectives, methodologies, and case studies from leading experts and practitioners in the field of world regional geography. Through a blend of theoretical foundations, empirical analysis, and real-world examples, this book covers a wide range of topics, including the physical geography of regions, cultural landscapes, economic activities, political systems, and environmental challenges.

Our overarching goal in presenting this book is to provide readers with a comprehensive understanding of the world's regions and the complex interactions between human societies and their environments. By offering insights into the geographical diversity, cultural richness, and interconnectedness of the world's regions, we aim to foster a deeper appreciation and understanding of global issues and challenges.

We extend our sincere gratitude to all the contributors who have generously shared their expertise, experiences, and insights in the field of world regional geography. It is our hope that this book will serve as a valuable reference and guide for students, educators, researchers, and anyone interested in exploring the dynamic and interconnected nature of the world we inhabit.

Omer Lasme

CONTENTS

VOLUME–2

Contents

1 Chapter INTRODUCTION TO THE WORLD

1.1: INTRODUCTION: AN OVERVIEW

Welcome to a world brimming with wonder and possibility, where each corner holds stories waiting to be uncovered and adventures longing to be embarked upon. Picture vibrant landscapes that stretch as far as the eye can see, from lush, green forests teeming with life to majestic mountains that touch the sky. Here, cultures intertwine, creating a rich tapestry of traditions, languages, and cuisines that invite exploration and connection. Imagine bustling cities alive with energy, where history meets modernity, and quiet villages that offer a glimpse into a simpler, slower way of life. This world is not just a backdrop; it's a living entity, pulsating with the dreams, struggles, and triumphs of its inhabitants. As we embark on this journey together, prepare to be inspired by the resilience of the human spirit, the beauty of nature, and the endless possibilities that await at every turn. Welcome to a world where every moment is an opportunity to learn, grow, and discover the extraordinary in the ordinary.

Geography

Geography is the spatial examination of the Earth's surface, derived from the Greek terms "geo," meaning "Earth," and "graphein," meaning "to write." Geographers examine the Earth's physical attributes, its populations and civilizations, climatic occurrences, and the Earth's position in the universe.

Geography analyzes the spatial links among all physical and cultural phenomena globally. Geographers examine the alterations in the Earth's climate and landscapes resulting from cultural impact.

The term geography was first utilized by Eratosthenes of Cyrene, a Greek scholar who lived from 276 to 194 BCE in present-day Libya, North Africa. He developed one of the earliest systems of longitude and latitude and computed the Earth's diameter. Moreover, he produced one of the earliest globe maps with the knowledge accessible during that period. Simultaneously, numerous ancient civilizations in China, southern Asia, Polynesia, and the Arabian Peninsula devised maps and navigation techniques employed in geography and cartography.

Geography is divided into two primary domains: physical geography and human geography. These two primary domains are analogous in that they both employ a spatial viewpoint and encompass the examination of place as well as the comparative analysis of one location with another.

Physical geography is the geographical examination of natural phenomena that constitute the environment, including rivers, mountains, landforms, weather, climate, soils, vegetation, and other physical features of the Earth's surface. Physical geography emphasizes geography as a branch of earth science. It emphasizes the primary physical components of the Earth—the lithosphere (surface layer), the atmosphere (air), the hydrosphere (water), and the biosphere (living organisms)—and the interrelationships among these components.

The major forms of study within physical geography include the following:

- Geomorphology (the study of the earth's surface features)
- Glaciology (the study of glaciers)
- Coastal geography (the study of the coastal regions)
- Climatology (the study of climates and climate change)
- Biogeography (the study of the geographic patterns of species distribution)

Some physical geographers study the earth's place in the solar system. Others are environmental geographers, part of an emerging field that studies the spatial aspects and cultural perceptions of the natural environment. Environmental geography requires an understanding of both physical and human geography, as well as an understanding of how humans conceptualize their environment and the physical landscape.

Physical landscape is the term used to describe the natural terrain at any one place on the planet. The natural forces of erosion, weather, tectonic plate action, and water have formed the earth's physical features. Many US state and national parks attempt to preserve unique physical landscapes for the public to enjoy, such as Yellowstone, Yosemite, and the Grand Canyon.

Human geography is the study of human activity and its relationship to the earth's surface. Human geographers examine the spatial distribution of human populations, religions, languages, ethnicities, political systems, economics, urban dynamics, and other components of human activity. They study patterns of interaction between human cultures and various environments and focus on the causes and consequences of human settlement and distribution over the landscape. While the economic and cultural aspects of humanity are primary focuses of human geography, these aspects cannot be understood without describing the landscape on which economic and cultural activities take place.

The cultural landscape is the term used to describe those parts of the earth's surface that have been altered or created by humans. For example, the urban cultural landscape of a city may include buildings, streets, signs, parking lots, or vehicles, while the rural cultural landscape may include fields, orchards, fences, barns, or farmsteads. Cultural forces unique to a given place—such as religion, language, ethnicity, customs, or heritage—influence the cultural landscape of that place at a given time. The colors, sizes, and shapes of the cultural landscape usually symbolize some level of significance regarding societal norms. Spatial dynamics assist in identifying and evaluating cultural differences between places.

Traditionally, the field of cartography, or map making, has been a vital discipline for geographers. While cartography continues to be an extremely important part of geography, geographers also look at **spatial** (space) and **temporal** (time) relationships between many types of data, including physical landscape types, economies, and human activity.

Geography also examines the relationships between and the processes of humans and their physical and cultural environments. Because maps are powerful graphic tools that allow us to illustrate relationships and processes at work in the world, cartography and geographic information systems have become important in modern sciences. Maps are the most common method of illustrating different spatial qualities, and geographers create and use maps to communicate spatial data about the earth's surface.

Geospatial techniques are tools used by geographers to illustrate, manage, and manipulate spatial data.

Cartography is the art and science of making maps, which illustrate data in a spatial form and are invaluable in understanding what is going on at a given place at a given time.

Making maps and verifying a location have become more exact with the development of the global positioning system (GPS). A GPS unit can receive signals from orbiting satellites and calculate an exact location in latitude and longitude, which is helpful for determining where one is located on the earth or for verifying a point on a map. GPS units are standard equipment for many transportation systems and have found their way into products such as cell phones, handheld computers, fish finders, and other mobile equipment. GPS technology is widely implemented in the transport of people, goods, and services around the world.

Remote sensing technology acquires data about the earth's surface through aerial photographs taken from airplanes or images created from satellites orbiting the earth. Remotely sensed images allow geographers to identify, understand, or explain a particular landscape or determine the land use of a place. These images can serve as important components in the cartographic (map-making) process. These technologies provide the means to examine and analyze changes on the earth's surface caused by natural or human forces. Google Earth is an excellent example of a computer tool that illustrates remotely sensed images of locations on the earth.

Figure 1.1: Low Elevation Air Photo of Cultural Landscape in Morehead, Kentucky. Photo by R. Berglee – CC BY-NC-SA.

Cultural Landscape in Morehead, Kentucky. Photo by R. Berglee – CC BY-NC-SA.

Geographic information science (GIS), often referred to as geographic information systems, uses a computer program to assimilate and manage many layers of map data, which then provide specific information about a given place. GIS data are usually in digital form and arranged in layers. The GIS computer program can sort or analyze layers of data to illustrate a specific feature or activity. GIS programs are used in a wide range of applications, from determining the habitat range of a particular species of bird to mapping the hometowns of university students.

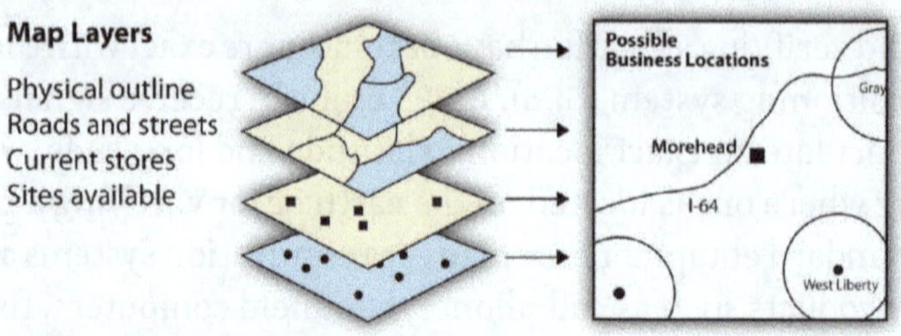

Figure 1.2: Illustration of Layers in a GIS Process

GIS specialists often create and analyze geographical information for government agencies or private businesses. They use computer programs to take raw data to develop the information these organizations need for making vital decisions. For example, in business applications, GIS can be used to determine a favorable location for a retail store based on the analysis of spatial data layers such as population distribution, highway or street arrangements, and the locations of similar stores or competitive establishments. GIS can integrate a number of maps into one to help analysts understand a place in relation to their own specific needs.

GIS also focuses on storing information about the earth (both cultural and natural) in computer databases that can be retrieved and displayed in the form of specialized maps for specific purposes or analyses. GIS specialists require knowledge about computer and database systems. Over the last two decades, GIS has revolutionized the field of cartography: nearly all cartography is now done with the assistance of GIS software. Additionally, analysis of various cultural and natural phenomena through the use of GIS software and specialized maps is an important part of urban planning and other social and physical sciences. GIS can also refer to techniques used to represent, analyze, and predict spatial relationships between different phenomena.

Geography encompasses a significantly wider scope than is commonly perceived. Many individuals perceive area studies as synonymous with the entirety of geography. Geography is the examination of the Earth and the impact of human action on its transformation.

Geography encompasses disciplines that extend well beyond merely comprehending the configuration of the earth's landforms. Physical geography encompasses all of the Earth's physical systems.

Human geography encompasses the examination of human culture, spatial relationships, interactions between humans and their environment, as well as various other research domains within the diverse subspecialties of geography. Students aspiring to pursue a career in geography should acquire geospatial techniques and develop proficiency in GIS and remote sensing, as these domains have witnessed the most significant growth in employment prospects during recent decades.

The Earth and Graticule Location

The initial stage in recognizing an area or location on Earth is to comprehend its relative and absolute locations. Relative location refers to a position on the Earth's surface in relation to other locations, including factors such as transportation accessibility and topography. The relative location facilitates the comparison of the benefits of one site against another. Absolute location denotes a specific point on the Earth's surface, independent of its relation to other locations. Absolute location is essential to cartography and to human activities that necessitate a standardized way for defining a location or point.

Similar to the geometric principle that a circle or sphere comprises 360 degrees, the Earth is similarly divided into 360 degrees, which are quantified using a grid system known as the graticule. Lines of latitude and longitude provide any absolute point on Earth with a distinct address of degrees north or south and east or west, enabling geographers to precisely identify, measure, and analyze spatial activity.

Geographers and cartographers delineate locations on Earth via a network of imaginary lines that encircle the planet. The two principal lines are the equator and the prime meridian. The systems of longitude and latitude are established from these lines, enabling precise location identification anywhere on the earth. The route is longest when traversed in an east-west orientation. At the equator, the sun is positioned directly overhead at midday during the equinoxes in March and September.

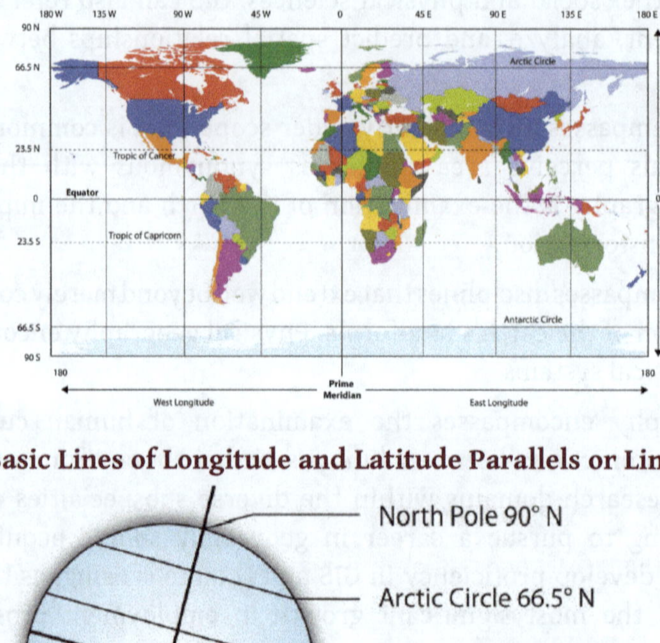

Figure 1.3 Basic Lines of Longitude and Latitude Parallels or Lines of Latitude

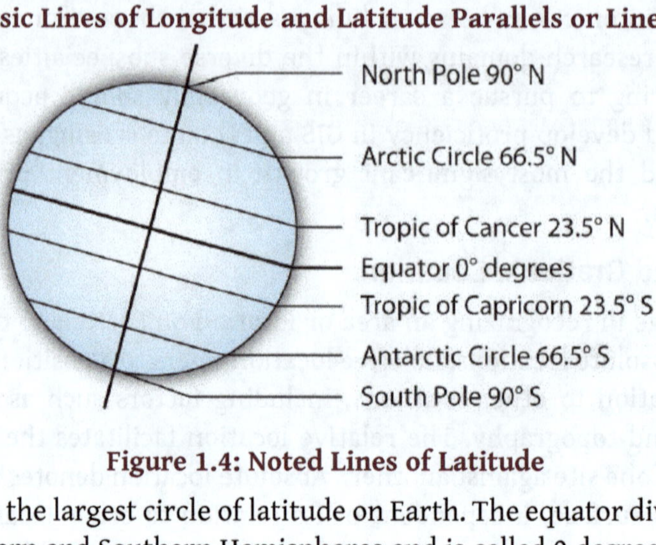

Figure 1.4: Noted Lines of Latitude

The **equator** is the largest circle of latitude on Earth. The equator divides the earth into the Northern and Southern Hemispheres and is called 0 degrees latitude. The other lines of latitude are numbered from 0 to 90 degrees going toward each of the poles. The lines north of the equator toward the North Pole are north latitude, and each of the numbers is followed by the letter "N." The lines south of the equator toward the South Pole are south latitude, and each of the numbers is followed by the letter "S." The equator (0 latitude) is the only line of latitude without any letter following the number. Notice that all lines of latitude are parallel to the equator (they are often called parallels) and that the North Pole equals 90 degrees N and the South Pole equals 90 degrees S. Noted parallels include both the Tropic of Cancer and the Tropic of Capricorn, which are 23.5 degrees from the equator. At 66.5 degrees from the equator are the Arctic Circle and the Antarctic Circle near the North and South Pole, respectively.

Meridians or Lines of Longitude

The **prime meridian** sits at 0 degrees longitude and divides the earth into the Eastern and Western Hemispheres. The prime meridian is defined as an imaginary line that runs through the Royal Observatory in Greenwich, England, a suburb of London. The Eastern Hemisphere includes the continents of Europe, Asia, and Australia, while the Western Hemisphere includes North and South America. All meridians (lines of longitude) east of the prime meridian (0 and 180) are numbered from 1 to 180 degrees east (E); the lines west of the prime meridian (0 and 180) are numbered from 1 to 180 degrees west (W). The 0 and 180 lines do not have a letter attached to them. The meridian at 180 degrees is called the **International Date Line**. The International Date Line (180 degrees longitude) is opposite the prime meridian and indicates the start of each day (Monday, Tuesday, etc.). Each day officially starts at 12:01 a.m., at the International Date Line. Do not confuse the International Date Line with the prime meridian (0 longitude). The actual International Date Line does not follow the 180-degree meridian exactly. A number of alterations have been made to the International Date Line to accommodate political agreements to include an island or country on one side of the line or another.

Climate and Latitude

The earth is tilted on its axis 23.5 degrees. As it rotates around the sun, the tilt of the earth's axis provides different climatic seasons because of the variations in the angle of direct sunlight on the planet. Places receiving more direct sunlight experience a warmer climate. Elsewhere, the increased angle of incoming solar radiation near the earth's poles results in more reflected sunlight and thus a cooler climate. The Northern Hemisphere experiences winter when sunlight is reflected off the earth's surface and less of the sun's energy is absorbed because of a sharper angle from the sun.

The **Tropic of Cancer** is the parallel at 23.5 degrees north of the equator, which is the most northerly place on Earth, receiving direct sunlight during the Northern

Hemisphere's summer. Remember that the earth is tilted 23.5 degrees, which accounts for seasonal variations in climate. The **Tropic of Capricorn** is the parallel at 23.5 degrees south of the equator and is the most southerly location on Earth, receiving direct sunlight during the Southern Hemisphere's summer.

The tropics (Cancer and Capricorn) are the two imaginary lines directly above which the sun shines on the two solstices, which occur on or near June 20 or 21 (summer solstice in the Northern Hemisphere) and December 21 or 22 (winter solstice in the Northern Hemisphere). The sun is directly above the Tropic of Cancer at noon on June 20 or 21, marking the beginning of summer in the Northern Hemisphere and the beginning of winter in the Southern Hemisphere. The sun is directly above the Tropic of Capricorn at noon on December 21 or 22, marking the beginning of winter in the Northern Hemisphere and the beginning of summer in the Southern Hemisphere.

Solstices are the extreme ends of the seasons, when the line of direct sunlight is either the farthest north or the farthest south that it ever goes. The region between the Tropics of Cancer and Capricorn is known as the tropics. This area does not experience dramatic seasonal changes because the amount of direct sunlight received does not vary widely. The higher latitudes (north of the Tropic of Cancer and south of the Tropic of Capricorn) experience significant seasonal variation in climate.

Figure 1.5: Road Sign South of Dakhla, Western Sahara (Claimed by Morocco), Marking the Tropic of Cancer This sign was placed in this desert location by the Budapest-Bamako rally participants. The non-English portion is in Hungarian because of the European participants in the race. Wikimedia Commons– public domain.

The **Arctic Circle** is a line of latitude at 66.5 degrees north. It is the farthest point north that receives sunlight during its winter season (90 N - 23.5 = 66.5 N). During winter, the North Pole is away from the sun and does not receive much sunlight. At times, it is dark for most of the twenty-four-hour day. During the Northern Hemisphere's summer, the North Pole faces more toward the sun and may receive sunlight for longer portions of the twenty-four-hour day. The **Antarctic Circle** is the corresponding line of latitude at 66.5 degrees south. It is the farthest location south that receives sunlight during the winter season in the Southern Hemisphere (90 S - 23.5 = 66.5 S). When it is winter in the north, it is summer in the south.

The Arctic and Antarctic Circles mark the extremities (southern and northern, respectively) of the polar day (twenty-four-hour sunlit day) and the polar night (twenty-four-hour sunless night). North of the Arctic Circle, the sun is above the horizon for twenty-four continuous hours at least once per year and below the horizon for twenty-four continuous hours at least once per year. This is true also near the Antarctic Circle, but it occurs south of the Antarctic Circle, toward the South Pole.

Equinoxes, when the line of direct sunlight hits the equator and days and nights are of equal length, occur in the spring and fall on or around March 20 or 21 and September 22 or 23.

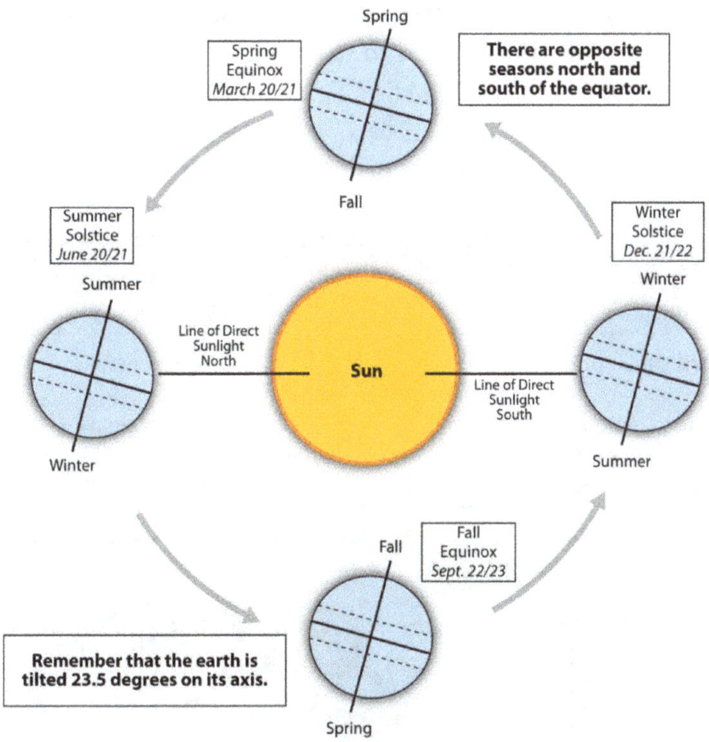

Figure 1.6: Graphic of the Four Seasons

Photo by R. Berglee – CC BY-NC-SA.

Time Zones

Universal Time (UT), Coordinated Universal Time (UTC), **Greenwich Mean Time** (GMT), or Zulu Time (Z): all four terms can be defined as local time at 0 degrees longitude, which is the prime meridian (location of Greenwich, England).

This is the same time under which many military operations, international radio broadcasts, and air traffic control systems operate worldwide. UTC is set in zero- to twenty-four-hour time periods, as opposed to two twelve-hour time periods (a.m. and p.m.). The designations of a.m. and p.m. are relative to the central meridian: a.m. refers to *ante meridiem*, or "before noon," and p.m. refers to *post meridiem*, or "after noon." UT, UTC, GMT, and Z all refer to the same twenty-four-hour time system that assists in unifying a common time in regard to global operations. For example, all air flights use the twenty-four-hour time system so the pilots can coordinate flights across time zones and around the world.

The earth rotates on its axis once every twenty-four hours at the rate of 15 degrees per hour (15 × 24 = 360). **Time zones** are established roughly every 15 degrees longitude so that local times correspond to similar hours of day and night. With this system, the sun is generally overhead at noon in every time zone that follows the 15-degree-wide system. The continental United States has four main time zones (see Table 1.1 "Four Main Time Zones in the Continental United States and Their Central Meridians" and Figure 1.7 "Major Time Zones of the World").

Table 1.1 Four Main Time Zones in the Continental United States and Their Central Meridians

USA Time Zones	Central Meridian
Eastern standard time zone	75 degrees W
Central standard time zone	90 degrees W
Mountain standard time zone	105 degrees W
Pacific standard time zone	120 degrees W

Figure 1.7 Major Time Zones of the World

The twenty-four times zones are based on the prime meridian in regard to Universal Coordinated Time (UTC), Greenwich Mean Time (GMT), or Zulu Time (Z), which all operate on the twenty-four-hour time clock. Local time zones are either plus or minus determined by the distance from the prime meridian.

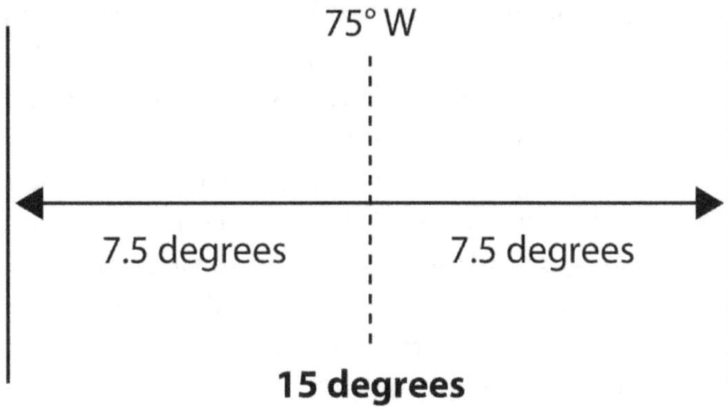

Figure 1.8: Diagram Illustrating the Width of a Time Zone

In this diagram, 75 W is the central meridian for the eastern standard time zone in the United States.

The eastern standard time zone is five hours earlier than the time at the prime meridian (UTC) because it is about 75 degrees west of 0 degrees (5 × 15 = 75). For example, if it is noon in London, then it is 7 a.m. in New York. If it is 1 p.m. in New York, it is 10 a.m. in San Francisco, which is three times zones to the west. Since there are twenty-four hours in a day, there are twenty-four time zones on Earth. Each time zone is 15 degrees wide.

A problem with the 15-degree time zones is that the zones do not necessarily follow state, regional, or local boundaries. The result is that time zones are seldom exactly 15 degrees wide and usually have varied boundary lines. In the United States, the boundaries between the different time zones are inconsistent with the lines of longitude; in some cases, time zones zigzag to follow state lines or to keep cities within a single time zone. Other countries address the problem differently. China, for example, is as large in land area as the United States yet operates on only one time zone for the entire country.

Regions in Geography: Basic Unit of Study in Geography

A region is a basic unit of study in geography—a unit of space characterized by a feature such as a common government, language, political situation, or landform. A region can be a formal country governed by political boundaries, such as France or Canada; a region can be defined by a landform, such as the drainage basin of all the water that flows into the Mississippi River; and a region can even be defined by the area served by a shopping mall. Cultural regions can be defined by similarities in human activities, traditions, or cultural attributes. Geographers use the regional unit to map features of particular interest, and data can be compared between regions to help understand trends, identify patterns, or assist in explaining a particular phenomenon.

Regions are traditionally defined by internal characteristics that provide a sense of place. Their boundaries vary with the type of region, whether it is formal, functional, or vernacular; each type has its own meaning and defined purpose. A formal region has a governmental, administrative, or political boundary and can have political as well as geographic boundaries that are not open to dispute or debate. Formal boundaries can separate states, provinces, or countries from one another. Physical regions can be included within formal boundaries, such as the Rocky Mountains or New England. An official boundary, such as the boundary of a national park, can be considered a formal boundary. School districts, cities, and county governments have formal boundaries.

Natural physical geographic features have a huge influence on where political boundaries of formal regions are set. If you look at a world map, you will recognize that many political boundaries are natural features, such as rivers, mountain ranges, and large lakes. For example, between the United States and Mexico, the Rio Grande makes up a portion of the border. Likewise, between Canada and the United States, a major part of the eastern border is along the Saint Lawrence Seaway and the Great Lakes. Alpine mountain ranges in Europe create borders, such as the boundary between Switzerland and Italy.

While geographic features can serve as convenient formal borders, political disputes will often flare up in adjacent areas, particularly if valuable natural or cultural resources are found within the geographic features. Oil drilling near the coast of a sovereign country, for example, can cause a dispute between countries about which one has dominion over the oil resources. The exploitation of offshore fisheries can also be disputed. A Neolithic mummy of a man who died in 3300 BCE caused tension between Italy and Switzerland: the body was originally taken to Innsbruck, Switzerland, but when it was determined that the body was found about 90 meters (180 feet) inside the border of Italy, Italian officials laid claim to the body.

Functional regions have boundaries related to a practical function within a given area. When the function of an area ends, the functional region ends and its boundaries cease to exist. For example, a functional region can be defined by a newspaper service or delivery area. If the newspaper goes bankrupt, the functional region no longer exists. Church parishes, shopping malls, and business service areas are other examples of functional regions. They function to serve a region and may have established boundaries for limits of the area to which they will provide service. An example of a common service area—that is, a functional region—is the region to which a local pizza shop will deliver.

Vernacular regions have loosely defined boundaries based on people's perceptions or thoughts.

Vernacular regions can be fluid—that is, different people may have different opinions about the limits of the regions.

Vernacular regions include concepts such as the region called the "Middle East." Many people have a rough idea of the Middle East's location but do not know precisely which countries make up the Middle East. Also, in the United States, the terms *Midwest* or *South* have many variations. Each individual might have a different idea about the location of the boundaries of the South or the Midwest. Whether the state of Kentucky belongs in the Midwest or in the South might be a matter of individual perception. Similarly, various regions of the United States have been referred to as the Rust Belt, Sun Belt, or Bible Belt without a clear definition of their boundaries. The limit of a vernacular area is more a matter of perception than of any formally agreed-upon criteria. Nevertheless, most people would recognize the general area being discussed when using one of the vernacular terms in a conversation.

Employing a State as a Benchmark for Comparison

When comparing one formal political region to another, use a familiar country, state, province, or political entity as a reference might be advantageous. Regardless of your location, you can investigate the statistical data for a formal region with which you are acquainted to establish a common reference point.

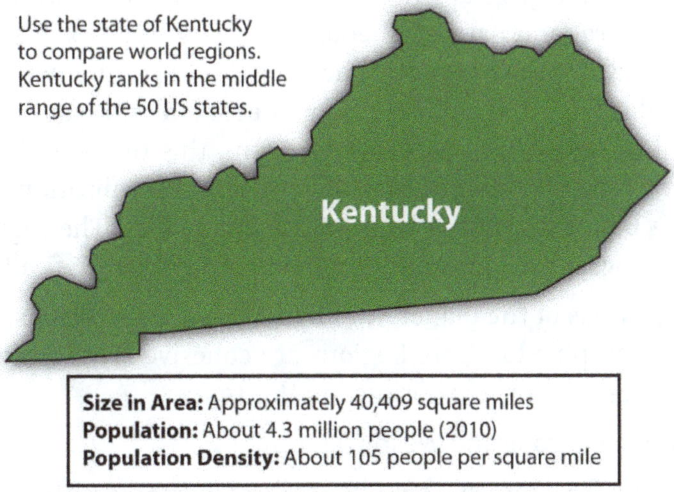

Figure 1.9 The state of Kentucky can be used as a comparison guide for understanding other formal political regions around the world.

Kentucky serves as an exemplar for comparing formal political regions. Kentucky is positioned near the median among the fifty US states, with a population of 4.3

million individuals. Kentucky ranks within the median range of the fifty states in terms of total physical area. The state's physical area of 40,409 square miles ranks thirty-seventh in size among the United States. Kentucky is smaller in land area than the western states but greater than numerous eastern states. Kentucky encompasses a portion of the rural outlying region of Appalachia, while also featuring cosmopolitan urban hubs like Lexington and Louisville. Kentucky also adjoins the metropolitan area of Cincinnati. The state's rural peripheral regions are characterized by agriculture and mining activities. The urban core regions host industrial and service centers. Additional US states may also serve as exemplars. Recognizing a state's geographical characteristics offers readers, both domestically and internationally, a benchmark for geographic comparison.

World Regional Geography

World regional geography examines distinct global areas in relation to the broader world. Comparison factors encompass both the physical and cultural landscapes. The primary inquiry is, Who resides there? What is the nature of their lives? What is their occupation? Significant physical elements may encompass location, climatic classification, and topography. Human variables encompass cultural traditions, ethnicity, language, religion, economy, and politics.

World regional geography examines regions of varying dimensions across the Earth's terrain, aiming to comprehend the distinctive characteristics of these regions in relation to their natural and cultural features. Spatial studies significantly contribute to regional geography. The scientific method may concentrate on the spread of cultural and natural phenomena across regions defined by diverse environmental and cultural elements. The emphasis is on the spatial relationships among several disciplines, including regional economics, resource management, regional planning, and landscape ecology. Again, this textbook takes a regional approach with a focus on themes that illustrate the globalization process, which in turn helps us better understand our global community. The regions studied in world regional geography can be combined into larger portions called realms.

Realms are large areas of the planet, usually with multiple regions, that share the same general geographic location. Regions are cohesive areas within each realm. The following eleven realms are outlined in this text:

1. Europe (Eastern Europe and Western Europe)
2. The Russian Realm (Russian republic of the former Soviet Union)
3. North America (United States and Canada)
4. Middle America (Caribbean, Mexico, Central America)
5. South America

6. North Africa, the Middle East and central Asia
7. Subsaharan Africa (Africa south of the Sahara Desert)
8. Southern Asia (India and its neighbors)
9. Eastern Asia (China, Mongolia, Japan, and the Koreas)
10. Southeast Asia (mainland region and the islands region)
11. Australia and the Pacific (including New Zealand)

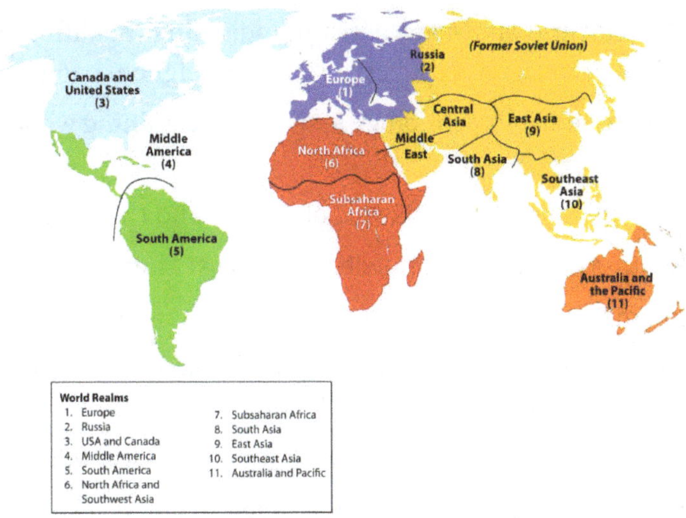

Figure 1.10 Major World Realms

Key Takeaways

- Geography is the spatial study of the earth's surface. The discipline of geography bridges the social sciences with the physical sciences. The two main branches of geography include physical geography and human geography. GIS, GPS, and remote sensing are tools that geographers use to study the spatial nature of physical and human landscapes.
- A grid system called the graticule divides the earth by lines of latitude and longitude that allow for the identification of absolute location on the earth's surface through geometric coordinates measured in degrees. There are twenty-four time zones that are set at 15-degree intervals each and organize time intervals around the world.
- The tilt of the earth's axis at 23.5 degrees helps create the earth's seasonal transitions by either absorbing or reflecting the sun's energy. The line of direct sunlight always hits the earth between 23.5 degrees north (Tropic of Cancer) and 23.5 degrees south (Tropic of Capricorn), depending on the time of year.

- A region is the basic unit of study in geography. Three main types of boundaries define a region: formal, functional, and vernacular. World regional geography is the study of a particular group of world regions or realms as each compares with the rest of the world.

Discussion and Study Questions

1. How does the discipline of geography provide a bridge between the social sciences and the physical sciences?
2. How does the cultural landscape assist in indicating the differences between a wealthy neighborhood and a poverty-stricken neighborhood?
3. How can remote sensing technology assist in determining what people do for a living?
4. What is the significance of the Tropic of Cancer and the Tropic of Capricorn?
5. What occupations depend on knowledge of the seasons for their success?
6. If it is 4 p.m. in San Francisco, what time is it in London, England?
7. How would GIS, GPS, or remote sensing technology be used to evaluate the destruction caused by a tornado in Oklahoma?
8. How is the cultural landscape influenced by the physical landscape?
9. Can you list a formal region, a functional region, and a vernacular region that would include where you live?
10. What methods, topics, or procedures would be helpful to include in the study of world geography?

Geography Exercise

Identify the following key places on a map:

- Arctic Circle
- Antarctic Circle
- Equator
- International Date Line
- North Pole
- Prime meridian
- Tropic of Cancer
- Tropic of Capricorn
- South Pole

Activities

1. Use Google Earth to locate your current school or residence.
2. Draw a map of your home state or province and include lines of latitude and longitude.
3. Compile the statistical data on your home state, province, or territory to use in comparing formal political regions.

1.2: THE INTERACTION BETWEEN THE ENVIRONMENT AND HUMAN ACTIVITY

The interaction between the environment and human activity is a complex and dynamic relationship that shapes both natural ecosystems and human societies. Human activities, such as agriculture, urbanization, and industrialization, significantly impact the environment by altering landscapes, depleting resources, and contributing to pollution and climate change. Conversely, the environment influences human behavior and development, as communities adapt to natural conditions, resource availability, and climate variations. This interplay highlights the importance of sustainable practices that balance human needs with environmental preservation, ensuring that future generations can thrive in a healthy and stable ecosystem. Recognizing and addressing these interconnections is crucial for promoting resilience and sustainability in an increasingly interconnected world.

Figure 1.11: Glaciers in Glacier National Park, Montana Mountainous type H climates vary with elevation, with warmer temperatures at the base and colder temperatures at higher elevations. Photo by R. Berglee – CC BY-NC-SA.

Climate and Human Habitation

The Earth's capacity to accept and absorb sunlight is a fundamental determinant of its environment and significantly influences human populations. Antarctica lacks significant towns or human settlements due to its extreme cold; the majority of sunlight reaching Antarctica is reflected off the Earth at that latitude, attributable to the tilt of the Earth's axis and the consequent angle of incoming solar energy. Addressing the fundamental inquiries regarding the primary habitats of humanity on Earth and the rationale behind their settlement patterns necessitates an understanding of climate. Moderate class C climates often offer the most favorable conditions for human existence.

The area between the Tropic of Cancer and the Tropic of Capricorn receives the most direct sunlight annually, making it conducive to plant and animal life, contingent upon sufficient moisture or precipitation. Humans have inhabited the tropics for an extended period, even during the era when ice sheets enveloped portions of the midlatitudes. The issue with tropical regions is that the soils typically possess worse quality, and the nutrients have been depleted by leaching. Currently, the distribution of human population on Earth is primarily influenced by two factors: temperate climates and availability of water resources.

Over 70 percent of the Earth's surface is comprised of water. Less than 3 percent of the Earth's water is fresh, with the majority of this fresh water contained in ice caps in the North and South Poles. This results in less than 1 percent of the world's freshwater available for human utilization, typically found in lakes, rivers, streams, or groundwater and subterranean aquifers. Climate significantly influences human habitation, as precipitation is essential for agriculture, livestock rearing, and providing fresh water to urban populations.

Several geographers have developed categories to identify climate types. **Climate** can be defined as a long-term average weather pattern evident in a particular region of the world. **Weather** is a term usually used to define conditions on a short-term or even daily basis. The two main elements in climate conditions are temperature and precipitation. For the purposes of this overview of world geography, the various climate types have been broken down into six basic types—A, B, C, D, E, and H—after the Köppen-Geiger classification system. Type H climates are actually a subset of the type E climate category.

- Type A: Tropical or equatorial climates
- Type B: Dry or arid climates
- Type C: Moderate or temperate climates
- Type D: Cold or continental climates

- Type E: Polar or extreme climates
- Type H: (Unclassified) highland climates

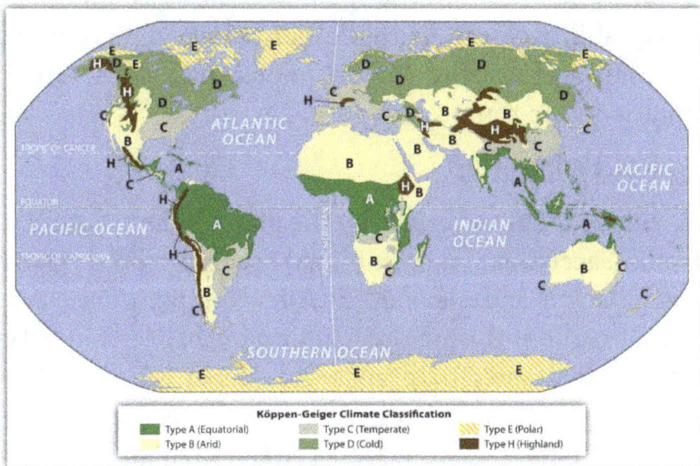

Figure 1.12 Basic Climate Regions Based on the Köppen-Geiger Classification System (Local Conditions May Vary Widely)

Type A: Tropical or Equatorial Climates

Figure 1.13 Tropical type A climates experience warm temperatures year round and a higher annual rainfall, such as this location in coastal Belize.

Photo by R. Berglee – CC BY-NC-SA.

The humid tropical climate type A, commonly located in the tropics, is characterized by consistently warm temperatures and elevated precipitation levels, predominantly as rainfall. Type A climates possess multiple subcategories that reflect the variability of annual rainfall distribution. Certain type A climates exhibit a dry season and a rainy season (monsoon), whilst others experience uniform precipitation year-round.

Type B: Dry or Arid Climates

The arid type B climate is characterized by the planet's desert areas. Temperatures may reach extremes, accompanied by minimal precipitation. Type B climate zones exhibit less precipitation and elevated daytime temperatures, accompanied by colder conditions at night or during the winter months. Terrain in type B climates may vary from sandy deserts to prairie grasses or steppes. Type B climates exhibit a lower density of arboreal vegetation compared to the majority of other climatic zones. A direct correlation exists between highlands and type B climates in numerous locations globally. The precipitation shadow effect occurs when one side of a mountain range experiences significant rainfall, while the opposite side is characterized by dry or arid conditions. This phenomena is observable in regions with sufficient height to impede the movement of precipitation-laden clouds.

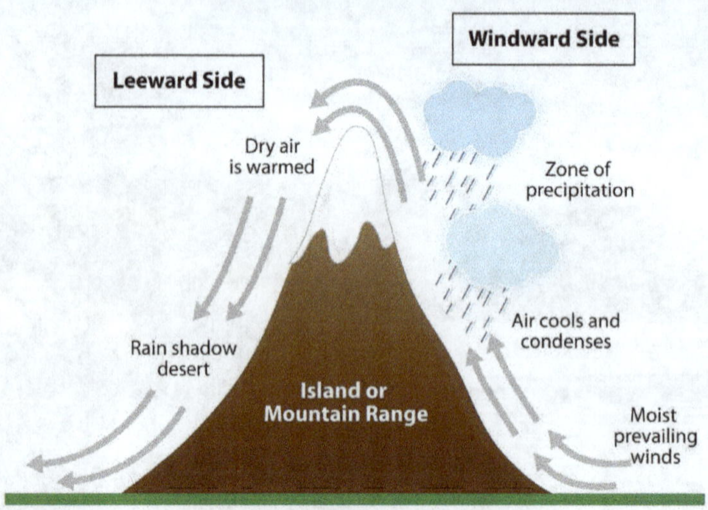

Figure 1.14: Rain Shadow Effect

Rain shadows occur when prevailing winds laden with moisture ascend rapidly along a hillside, causing the air to cool and condense, resulting in precipitation as rain or snow. Upon reaching the mountain summit, the air mass experiences a significant reduction in moisture content. The desiccated air descends the opposite slope of the mountain range, where its temperature rises. The warm, dry air descending from the mountains persistently extracts moisture from the terrain, leading to desert or arid climatic conditions.

Figure 1.15: Dry or arid type B climates such as central Arizona have fewer trees than other climates. The saguaro cactus can withstand long periods with little precipitation.

Kauai, a Hawaiian island, exemplifies the rain shadow effect in an extreme manner. The island's windward side experiences precipitation levels exceeding those of nearly any other location on Earth, with annual rainfall reaching up to 460 inches (nearly 40 feet). Nevertheless, only a portion of the island receives that quantity of precipitation. The elevation of the mountains induces a rain shadow effect on the arid leeward side, resulting in semidesert conditions and type B climates.

Death Valley in California is a consequence of the rain shadow phenomenon. Minimal precipitation occurs in Death Valley as moisture in the prevailing winds is deposited on the western slopes of the adjacent mountain ranges. The entirety of Nevada is arid due to the rain shadow effect. Precipitation from the Pacific Ocean descends upon the coastal mountains and the Sierra Nevada in California. The mountains are sufficiently elevated to cast a shadow over that portion of Nevada, while the basin and ranges exacerbate the local rain shadow effect.

The Himalayas exemplify mountains that produce the rain shadow effect. A majority of western China experiences type B climates due to the rain shadow effect created by elevated mountains that obstruct rain clouds from reaching the area. The southern Himalayas experience substantial precipitation due to monsoon rains originating from the Indian Ocean, whereas western China is predominantly arid. It is sparsely populated in contrast to the densely populated areas of China. Located to the east, where precipitation is abundant.

Type C: Moderate or Temperate Climates

Figure 1.16: Moderate type C climates are good for agriculture. For example, in Appalachia, tobacco and hardwood forests grow well.

Type C climates, characterized by moderate temperature and precipitation, are the most conducive to human habitation, supporting the highest population densities globally. Type C climates predominantly occur in the midlatitudes adjacent to the tropics. Seasonal variations are significant, characterized by separate winters and summers. Winters are characterized by cool to frigid temperatures, whereas summers are typically mild. Precipitation fluctuates from minimal to substantial, based upon geographic location. C climates predominate in the southeastern and western coastal regions of the United States.

Type C climates, while not the most prevalent globally, have drawn the largest human populations. The attractiveness is partly due to the prevalence of woods, agricultural land, and freshwater in type C locations. The primary population centers of the planet are located in type C climates. With a global population over seven billion and continuing to increase, humans have inhabited the majority of places characterized by type C climates and are now encroaching upon areas with A, B, or D climate classifications.

Figure 1.17: Colder type D climates are prominent in states such as North Dakota, which has the coldest annual winter temperatures of the continental forty-eight states.

Type D: Cold or Continental Climates

Type D climate regions are often found in the interiors of continents away from the moderating influence of large bodies of water. They are often farther north than type C regions, resulting in colder winters. Seasonal variations exist, with cool to hot summers and cold winters. Precipitation is usually in the form of rain in summer and snow in winter. Regions with type D climates can be found in the Great Lakes region of the United States, much of Canada, and a large portion of Russia.

Type E: Polar or Extreme Climates

Type E is an extreme climate type found in the polar regions near or to the north of the Arctic Circle and near or to the south of the Antarctic Circle. Regions with type E climates are cold with permanent ice or permafrost year round. Vegetation is minimal, and there are no trees. Temperatures may warm slightly during the short summer months but rarely rise above 50 degrees.

Type H: Highland Climates

Type H highland climates are usually listed as a subcategory of type E climates. Mountain ranges can create a variety of climate types because of the change in elevation from the base of the range to the summit. Different climate types can be found on the same mountain at different elevations. Type H climates designate highlands or mountain terrain. Variations in climate exist on most mountain ranges. Climates at the base of mountains will vary depending on whether the mountains are found in the tropics or in the higher latitudes. For example, high mountains near the equator may have a type A climate at their base and a type E climate at their summit with various type C and type D climates between them. Type H climates are found where elevation differences are profound enough to provide different climate zones. Higher elevation relief can reach above the tree line and have permanent snow cover at the summit. The term *relief* is used in geography to indicate elevations of a land surface. Elevation zones with permanent ice or snow can resemble a type E polar climate.

Deforestation

The planet's growing population has increased demands on natural resources, including forest products. Humans have been using trees for firewood, building homes, and making tools for millennia. Trees are a renewable resource, but deforestation occurs when they are removed faster than they can be replenished. Most people in rural areas in developing countries rely on firewood to cook their food. Many of these areas are experiencing a fast decline in the number of trees available. People living in mainly type B climates may not have access to a lot of trees to start with; therefore, when trees are cut down for firewood or for building materials, deforestation occurs. In the tropical areas, it is common for hardwood trees to be cut down for lumber to gain income or to clear the land for other agricultural purposes, such as cattle ranching. Countries that lack opportunities and advantages look to exploit their natural resources—in this case, trees—for either subsistence agriculture or economic gain.

Deforestation has increased across the globe with the rapid rise in worldwide population.

During the Industrial Revolution, European countries chopped down their forests at a rapid rate. Much of the British Isles was forested at one point, but today few forests remain on the British Isles, and they are typically protected.

Colonialism brought the Europeans to the Americas. The United States, in its early development, pushed west from the original thirteen colonies, and many old growth forests were cut down in the process. As railroad tracks were laid down and pioneer development pushed west into the Great Plains, where there were few trees, the great cutover occurred in the eastern and central forests—*cutover* is a term indicating the systematic deforestation of the eastern and central forests. Michigan and Wisconsin saw their trees removed in a systematic deforestation.

Figure 1.18: Forests are removed for timber, and burning the excess then clears the land for other purposes.

Photo by R. Berglee – CC BY-NC-SA.

Some areas were allowed to grow back, but many other areas were turned into farmland. Few old-growth forests remain in the United States. Today there are conflicts over how the timber industry is handling the forests in places such as the Pacific Northwest region of the United States.

Countries that are better off economically no longer have to cut down their own trees but can afford to substitute other resources or import lumber from other places. Developing regions of the world in Latin America, Africa, and parts of Asia are experiencing serious problems with deforestation.

Deforestation is widespread: Residents of Haiti have cut down about 99 percent of the country's forests; most of the wood has been used as fuel to cook food. People in Afghanistan have cut down about 70 percent of their forests. Nigeria has lost about 80 percent of its old-growth forests since 1990. Ethiopia has lost up to 98 percent of its forested acreage, and the Philippines has lost about 80 percent of its forests.

Brazil's Amazon basin has undergone many projects that have driven deforestation. For example, about half the state of Rondônia in western Brazil has been deforested since 1990. The countries of Central America have lost about half their original forests, and deforestation continues on a systematic basis. Tropical regions of Southeast Asia and Africa are being exploited for their timber at unsustainable rates, causing deforestation that the next generation will have to deal with. India, with over a billion people, still has a high demand for firewood and building materials; their forests are declining faster than they can be replanted. China, with its billion plus population, has been attempting to address its deforestation problems by implementing a massive replanting program and conservation measures. Other countries are starting to adopt similar measures.

Figure 1.19: A man in Malawi carries firewood for cooking and heating purposes.

David Stanley – Girls Carrying Firewood – CC BY 2.0.

Tropical rain forests only make up about 5 percent of the earth's surface but contain up to 50 percent of the earth's biodiversity. These forests are cut down for a variety of reasons. Norman Meyers, a British environmentalist, estimated that about 5 percent of deforestation in tropical regions is caused by the push for cattle production. Nineteen percent of these forests are cut down by the timber industry, 22 percent are cut down for the expansion of plantation agriculture, and 54 percent are removed due to slash-and-burn farming. Most tropical rain forests are located in the Amazon basin of South America, in central Africa, and in Southeast Asia. All these areas are looking for advantages and opportunities to boost their economies; unfortunately, they often target their tropical rain forests as a revenue source.

Problems with Deforestation

Deforestation results in more than merely the depletion of trees for fuel, construction materials, paper goods, or industrial use. A similar concern in the context of deforestation is soil erosion. In the absence of trees to stabilize the soil during intense rainfall, erosion occurs, rendering the land unproductive. In tropical regions, soils frequently exhibit degradation and nitrogen deficiency. The majority of nutrients in tropical regions are found in decomposing matter at the base of trees, which reintroduces energy into the environment. After the trees are eliminated, there is minimal restoration of this energy source. Soil erosion in tropical regions hinders the regeneration of forests after deforestation.

Landslides can exacerbate the issue of soil erosion. Following substantial precipitation, fully saturated hillsides may descend, resulting in significant structural damage to edifices, residences, and agricultural land. Tree roots stabilize hillsides, so mitigating the risk of landslides.

Forests significantly contribute to the hydrological cycle. Trees extract moisture from the earth through their roots and release it into the atmosphere via transpiration through their leaves. Moisture in the atmosphere collects into clouds, condenses, and falls back to Earth. Trees not only retain water, but the organic stuff at their base also retains water, making it accessible to the broader ecosystem, perhaps mitigating water runoff. Forest canopies distribute water during precipitation and generate an additional layer of moisture on their foliage and branches, which is either utilized by other species or evaporates back into the atmosphere.

Deforestation eradicates the function of forests within the hydrological cycle.

Forest habitats support a vast array of creatures. Tropical rain forests are one of the most vibrant ecosystems on the planet. Their rich biodiversity may offer insights into unexplored solutions for the future. Plants and organisms in these habitats may hold the key to medical or biological breakthroughs, but wildlife and vegetation will be lost as deforestation eliminates their habitat and accelerates the extinction of endangered species.

Figure 1.20: Lumber Mill Processing Hardwood Timber. Photo by R. Berglee – CC BY-NC-SA.

Trees and plants extract carbon dioxide from the atmosphere and sequester it inside their structures via photosynthesis. Carbon dioxide is a significant greenhouse gas involved in the climate change phenomenon. Carbon dioxide and analogous gases diminish the escape of long-wave radiation (heat) from the Earth's atmosphere, leading to elevated planetary temperatures. Increased carbon dioxide emissions into the atmosphere result in climate change. Deforestation leads to a reduction in the absorption of carbon dioxide from the atmosphere, hence exacerbating climate change. Slash-and-burn agricultural practices that incinerate forests emit carbon from vegetation directly into the atmosphere, exacerbating climate change.

Climate Change

Climate change has been a persistent phenomenon in the planet's evolution. The rise in environmental temperature has garnered the greatest attention in recent years. Concerns have been expressed over the pace and magnitude of global climate change. Comprehending the dynamics of temperature escalation can aid in elucidating its correlation with human activity.

The atmosphere is the gaseous envelope around the Earth, delineating the boundary between its surface and outer space. The atmosphere is composed of a mixture of gases: nitrogen (77 percent), oxygen (21 percent), and trace elements (1 percent) such as argon, helium, carbon dioxide, and water vapor. A minimal concentration of carbon dioxide is essential for regulating the Earth's temperature. The atmosphere spans over three hundred miles above the Earth's surface, with its lower strata constituting the planet's climate system. The lowest layer is termed the troposphere and is accountable for the conditions that facilitate life on the planet's surface.

Since the 1960s, researchers have expressed alarm regarding the levels of carbon dioxide, methane, nitrous oxide, and chlorofluorocarbons in the atmosphere. Greenhouse gases can capture heat energy released from the Earth's surface, thereby elevating global temperatures and contributing to climate change. Since the Industrial Revolution, human activities, like as the combustion of fossil fuels and extensive deforestation, have augmented the concentration of heat-trapping greenhouse gases in the atmosphere. Carbon dioxide and analogous gases function like to greenhouse glass panels, permitting shortwave solar energy to penetrate but obstructing the escape of long-wave thermal radiation into space.

An elevation in carbon dioxide and greenhouse gas concentrations in the atmosphere typically results in a rise in global temperatures, subsequently leading to alterations in weather patterns across different regions of the Earth. Variations in temperature can impact precipitation patterns and modify weather conditions, thereby affecting agricultural yields and influencing energy requirements, so contributing to economic instability. Climate alterations also affect the environmental conditions for creatures adapted to particular habitat ranges. Climate change can alter an organism's habitable zone, thereby affecting entire ecosystems.

Deforestation and the combustion of fossil fuels can exacerbate climate change. Fossil fuels, including coal, oil, and natural gas, are formed when deceased flora and fauna undergo prolonged disintegration under pressure, preserving their carbon content. Combustion of fossil fuels emits carbon into the atmosphere. The escalating demand for energy and timber due to human activities will persist in exacerbating climate change unless viable alternatives are identified. The rise in temperatures may lead to the melting of ice caps, thereby elevating sea levels and affecting global human activities. Chapter 13 on Antarctica contains additional information regarding climate change.

Tectonic Plates

The movement of tectonic plates is another aspect of the earth's dynamics that affects human activity. The earth's crust, which is between 10 and 125 miles thick, is not one big solid chunk but rather a series of plates that cover a molten iron core at the center of the planet. The plates that cover the earth's surface slowly shift and move. Plates can slide away from each other or they can collide, and they can slide parallel to each other in opposite directions. When two plates collide and one plate slides under an adjacent plate, the process is called **subduction.** Movement or shift where two plates meet can cause earthquakes and is usually associated with volcanic activity.

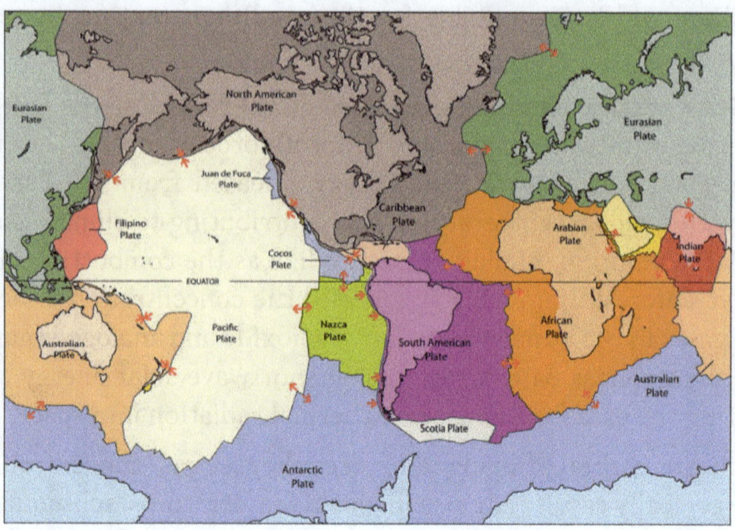

Figure 1.21: General Pattern of Tectonic Plates. Image courtesy of USGS.

Mountain chains, such as the Himalayas, are a direct result of two plates colliding. The collision pushes up the earth into a mountain chain, either by direct pressure or by volcanic activity. Plates can move up to an inch a year in active regions. Driven by the earth's internal heat, these plates have created the planet's mountain landscapes. Earthquakes and volcanic action along plate boundaries (called **faults**) continue to affect human activity and can cause serious economic damage to a community. Plate boundaries can be found near many natural edges of continents. The continuous action of the plates causes serious earthquakes and volcanic eruptions that can devastate human activities. Earthquakes near the sea often trigger tsunamis that can create enormous waves that bring destruction to coastal regions in their path. The earthquake off the east coast of Japan in 2011 created a tsunami that brought additional destruction to nuclear energy facilities, exposing parts of Japan and the rest of the world to radiation.

Key Takeaways

- Human activity on the planet correlates with the type of climate and terrain that presents itself to humans in the form of natural resources or habitability. Six basic climate zones (A, B, C, D, E, and H) describe the earth's climate types. Temperature and precipitation are the two main variables that create a climate zone and its corresponding environmental attributes.
- The earth's crust consists of a number of separate plates that move, creating earthquakes and volcanic activity. Most mountain ranges on Earth are a product of tectonic plate activity.
- Removing trees faster than they can grow back is called deforestation. Humans are cutting down the forests in many areas at an unsustainable rate.

Deforestation can result in soil erosion, changes in weather patterns, and the loss of habitats. Trees are being cut down for firewood, building materials, or profit.

- Mountains or high elevation relief can restrict the passage of rain clouds and cause the clouds to lose their precipitation as the air mass increases in elevation. The other side of the mountain or range does not receive any precipitation and is reduced to a more arid or drier region, creating desert conditions.
- Climate change is a phenomenon whereby gases such as carbon dioxide and methane increase in the troposphere and restrict long-wave radiation from escaping the planet, which can result in warmer temperatures on Earth. Trees remove carbon dioxide from the atmosphere, which may reduce climate change.

Discussion and Study Questions

1. What climate type do you live in?
2. What are the main attributes of each climate type?
3. What two main climatic qualities do humans gravitate toward?
4. What is the difference between weather and climate?
5. How can the main causes of deforestation be alleviated or diminished?
6. What are tectonic plates, and how do they help shape the planet?
7. How would tectonic plate movement affect human activities?
8. Who is responsible for addressing the problems caused by climate change? What can you do about it?
9. What causes a rain shadow effect? Name some examples of this phenomenon.
10. Which regions of the planet are being affected the most by deforestation?

Activities

1. Research and determine when the last three earthquakes nearest where you live occurred. What damage did they cause?
2. Chart the annual average temperature for where you live for the past one hundred years to illustrate any trend in climate change.
3. Determine the nearest location to where you live that is impacted by a rain shadow effect.

1.3: POPULATION AND CULTURE

Population and culture are intricately linked elements that shape societies across the globe. The dynamics of population—its size, distribution, and demographics—greatly influence cultural practices, beliefs, and identities. As populations grow and migrate, they bring diverse traditions, languages, and worldviews, enriching the cultural tapestry of regions. Conversely, cultural factors can impact population trends, affecting birth rates, migration patterns, and community cohesion. Understanding this interplay is essential for grasping how societies evolve, adapt, and interact in an increasingly interconnected world.

Demographic Transition

Demography is the study of how human populations change over time and space. It is a branch of human geography related to population geography, which is the examination of the spatial distribution of human populations. Geographers study how populations grow and migrate, how people are distributed around the world, and how these distributions change over time.

For most of human history, relatively few people lived on Earth, and world population grew slowly. Only about five hundred million people lived on the entire planet in 1650 (that's less than half India's population in 2000). Things changed dramatically during Europe's Industrial Revolution in the late 1700s and into the 1800s, when declining death rates due to improved nutrition and sanitation allowed more people to survive to adulthood and reproduce. The population of Europe grew rapidly. However, by the middle of the twentieth century, birth rates in developed countries declined, as children had become an economic liability rather than an economic asset to families. Fewer families worked in agriculture, more families lived in urban areas, and women delayed the age of marriage to pursue education, resulting in a decline in family size and a slowing of population growth. In some countries (e.g., Russia and Japan), population is actually in decline, and the average age in developed countries has been rising for decades. The process just described is called the **demographic transition**.

At the beginning of the twentieth century, the world's population was about 1.6 billion. One hundred years later, there were roughly six billion people in the world, and as of 2011, the number was approaching seven billion. This rapid growth occurred as the demographic transition spread from developed countries to the rest of the world. During the twentieth century, death rates due to disease and malnutrition decreased in nearly every corner of the globe. In developing countries with agricultural societies, however, birth rates remained high. Low death rates and high birth rates resulted in rapid population growth. Meanwhile, birth rates—and family size—have also been declining in most developing countries as people leave agricultural professions and move to urban areas. This means that population growth rates—while still higher in the developing world than in the developed world—are declining. Although the exact figures are unknown, demographers expect the world's population to stabilize by 2100 and then decline somewhat.

In 2010, the world's population was growing by about eighty million per year, a growth rate found almost exclusively in developing countries, as populations are stable or in decline in places such as Europe and North America. World population increase is pronounced on the continent of Asia: China and India are the most populous countries in the world, each with more than a billion people, and Pakistan is an emerging population giant with a high rate of population growth. The continent of Africa has the highest fertility rates in the world, with countries such as Nigeria—Africa's most populous and the world's eighth most populous country—growing rapidly each year. The most striking paradox within population studies is that while there has been marked decline in fertility (a declining family size) in developing countries, the world's population will grow substantially by 2030 because of the compounding effect of the large number of people already in the world—that is, even though population growth rates are in decline in many countries, the population is still growing. A small growth rate on a large base population still results in the birth of many millions of people.

Earth's human population is growing at the rate of about 1.4 percent per year. If the current growth rate continues, the human population will double in about fifty years to more than twelve billion. The current population increase remains at about eighty million per year. A change in the growth rate will change the doubling time. Between 2010 and 2050, world population growth will be generated exclusively in developing countries.

The three largest population clusters in the world are the regions of eastern China, south Asia, and Europe. Southeast Asia also has large population clusters. Additional large population centers exist in various countries with high urbanization. An example is the urbanized region between Boston and Washington, DC, which includes New York City, Philadelphia, Baltimore, and neighboring metropolitan areas, resulting in a region often called a *megalopolis*. The coastal country of Nigeria in West Africa or the island of Java in Indonesia are good examples of large population clusters centered in the tropics.

Social dynamics and geography will determine where the new additions to the human family will live. Providing food, energy, and materials for these additional humans will tax many countries of the world, and poverty, malnutrition, and disease are expected to increase in regions with poor sanitation, limited clean water, and lack of economic resources. In 2010, more than two billion people (one-third of the planet's population) lived in abject poverty and earned less than the equivalent of two US dollars per day. The carrying capacity of the planet is not and cannot be known. How many humans can the earth sustain in an indefinite manner? There is the possibility that we have already reached the threshold of its carrying capacity.

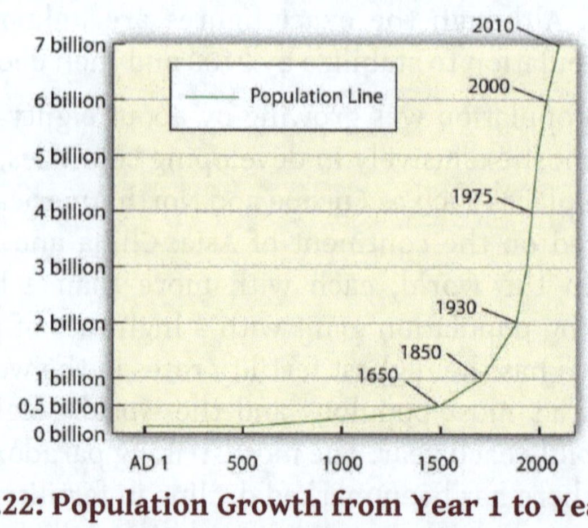

Figure 1.22: Population Growth from Year 1 to Year 2010 AD

Human population will continue to grow until it either crashes due to the depletion of resources or stabilizes at a sustainable carrying capacity. Population growth exacts a toll on the earth as more people use more environmental resources. The areas most immediately affected by increased populations include forests (a fuel resource and a source of building material), fresh water supplies, and agricultural soils. These systems get overtaxed, and their depletion has serious consequences.

Type C climates, which are moderate and temperate, are usually the most productive and are already vulnerable to serious deforestation, water pollution, and soil erosion. Maintaining adequate food supplies will be critical to supporting a sustainable carrying capacity. The ability to transport food supplies quickly and safely is a major component of the ability to manage the conservation of resources.

Deforestation by humans using wood for cooking fuel is already a serious concern in the arid and dry type B climates.

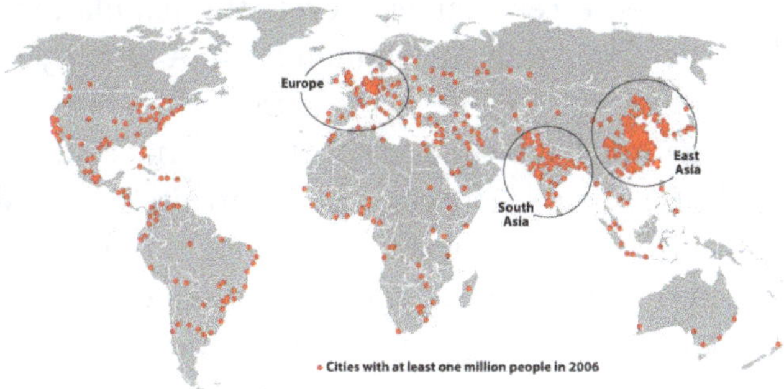

Figure 1.23: The three main human population clusters on the planet are eastern Asia, southern Asia, and Europe. Most of these regions with high population densities are in type C climates. Wikimedia Commons – CC BY-SA 3.0.

Urbanization and Family Size

As countries move from an agricultural to an industrial economy, there is a major shift in population from rural to urban settings. The

Industrial Revolution of the nineteenth century ushered in major technological developments and changes in labor practices, which encouraged migration from the farm to the city. Because of increased mechanization, fewer farm workers are needed to produce larger agricultural yields. At the same time, factories in urban areas have a great need for industrial workers. This shift continued into the information age of the late twentieth century and continues in many parts of the developing world in the current century.

Figure 1.24: Rural-to-Urban Shift. Wikimedia Commons – CC BY-SA 2.0.

A basic principle of population growth that addresses this rural-to-urban shift states that as countries industrialize and urbanize, family size typically decreases and incomes traditionally increase. Though this may not be true in all cases, it is a general principle that is consistent across cultural lines. Agricultural regions generally have a larger average family size than that of their city counterparts. Fertility rate is the average number of children a woman in a particular country has in her lifetime, whether or not they all live to adulthood. If a fertility rate for a given country is less than 2.1—the replacement level—the population of that country is in decline, unless there is significant immigration. A fertility rate greater than 2.1 indicates that the country's population is increasing. Some children will never reach reproductive age nor have children of their own, so the replacement rate has to be slightly greater than 2. The concept of fertility rate is slightly different from the term family size, which indicates the number of living children raised by a parent or parents in the same household. In this textbook, family size is used to illustrate the concept of population growth and decline.

Population Demands

A country's demographic statistics can be illustrated graphically by a population pyramid. A population pyramid is essentially two bar graphs that depict male and female age cohorts either in absolute size or as a percentage of the total population. Male cohorts are typically shown on the left side of the pyramid, and females are on the right side.

The shape of a country's population pyramid tells a story about the history of its population growth. For example, a high-growth-rate country has a pyramid that is narrow at the top and wide at the bottom, showing that every year more children have been born than the year before. As family size decreases and women in a society have fewer children, the shape of the pyramid changes. A population pyramid for a postindustrialized country that has negative growth would be narrower at the bottom than in the middle, indicating that there are fewer children than middle-aged people. Four basic shapes indicate the general trends in population growth:

1. Rapidly expanding
2. Expanding
3. Stationary
4. Contracting

These shapes also illustrate the percentage of a population under the age of fifteen or over the age of sixty-five, which are standard indicators of population growth. Many postindustrial countries have a negative population growth rate. Their population pyramids are narrow at the bottom, indicating an urbanized population with small family sizes.

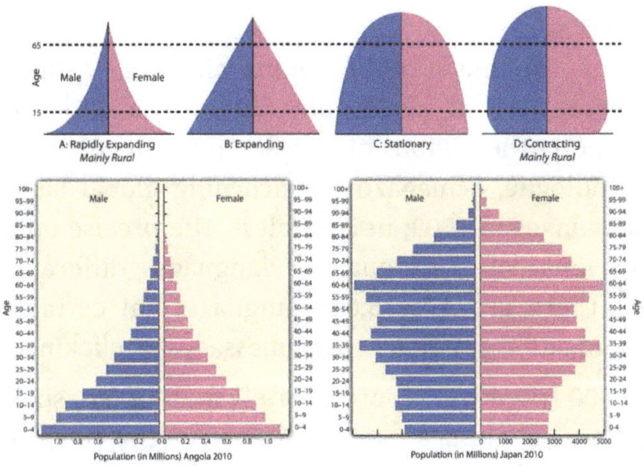

Figure 1.25: Population Pyramids Angola had a fertility rate of 5.6 in 2011. Japan had a fertility rate of 1.4 in 2011. Data courtesy of US Census Bureau International Programs.

Culture and Ethnicity

The term culture is often difficult to differentiate from the term ethnicity. In this textbook, *ethnicity* indicates traits people are born with, including genetic backgrounds, physical features, or birthplaces. People have little choice in matters of ethnicity. The term *culture* indicates what people learn after they are born, including language, religion, and customs or traditions. Individuals can change matters of culture by individual choice after they are born. These two terms help us identify human patterns and understand a country's driving forces.

The terms *culture* and *ethnicity* might also be confused in the issue of ethnic cleansing, which refers to the forced removal of a people from their homeland by a stronger force of a different people.

Ethnic cleansing might truly indicate two distinct ethnic groups: one driving the other out of their homeland and taking it over. On the other hand, ethnic cleansing might also be technically cultural cleansing if both the aggressor and the group driven out are of the same ethnic stock but hold different cultural values, such as religion or language. The term *ethnic cleansing* has been used to describe either case.

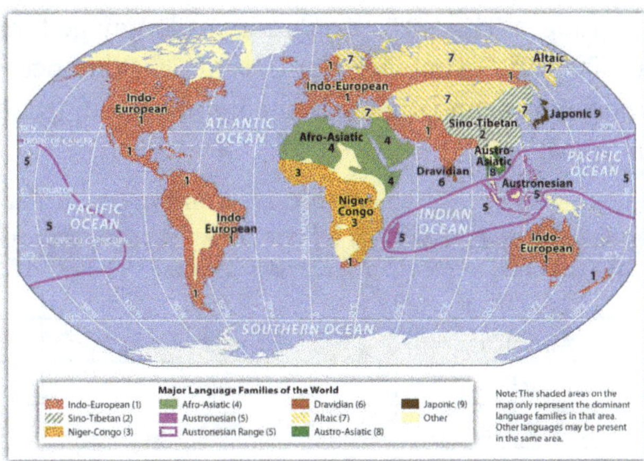

Figure 1.26 Major Language Families of the World

Global Languages

Language serves as the medium of communication within human culture, encapsulating the entirety of philosophy, literature, and the arts' diversity. Every individual among the billions of people on Earth communicates in at least one language. Ethnologue, a magazine concerning global languages, estimated 6,909 surviving languages in 2009; nevertheless, the precise number may remain indeterminate. Various data sets quantify languages differently; however, the consensus is that there are over 6,000 languages. In certain global cultures, individuals communicate through whistled messages or clicking sounds.

Among the over 6,000 languages, approximately a dozen are spoken by more than one hundred million individuals apiece. These are the predominant languages utilized in the most populous nations worldwide. Nevertheless, the overwhelming majority of the world's languages are utilized by a comparatively limited population. Many languages lack a written form and are spoken by diminishing populations. Language specialists project that as much as fifty percent of the world's extant languages may become extinct by the conclusion of the twenty-first century due to globalization. New languages emerge when populations exist in isolation; nevertheless, in the contemporary period of heightened global interaction, languages are being forsaken as their speakers transition to more pragmatic languages.

Nine predominant language families exist globally. Every language within a linguistic family have a shared ancestral language. An illustration of a language family is the Indo-European family, which comprises several branches of language groups derived from a common ancestor: Proto-Indo-European, likely spoken approximately six thousand years ago. As populations dispersed from their ancestral homeland, their language evolved and diverged into numerous new languages. The three predominant language families of the Indo-European family utilized in Europe are the Germanic, Romance, and Slavic languages. Additional Indo-European languages encompass Hindi (used in India) and Persian (employed in Iran).

Table 1.2 Languages of the Continents

World Area	Number of Languages	Percentage of All Languages
Africa	2,110	30.5
Americas	993	14.4
Asia	2,322	33.6
Europe	234	3.4
Pacific	1,250	18.1
Totals	6,909	100.0

Table 1.3 Thirteen Major Languages of the World

Language	First Language (Millions of Speakers)	Second Language (Millions of Speakers)	Total Speakers (Millions)
Mandarin	845	180	1,025
Hindi/Urdu	242	224	466
Arabic (All)	206	246	452
English	340	110	450
Spanish	329	53	382
Russian	144	106	250
Bengali	181	69	250
Portuguese	178	42	220
Indonesian	23	140	163
German	95	28	123
Japanese	122	1	123
French	65	55	120
Punjabi	109	6	115
Languages with more than one hundred million speakers (speakers given in millions)			
These thirteen languages are spoken by more than four billion people, or about 60 percent of the current world population in 2009.			

Source: M. Paul Lewis, ed., *Ethnologue: Languages of the World*, 16th ed. (Dallas: SIL International, 2009), accessed September 13, 2011, http://www.ethnologue.com.

Table 1.4 Nine Major Language Families

1. Indo-European languages
2. Sino-Tibetan languages
3. Niger-Congo languages
4. Afro-Asiatic languages
5. Austronesian languages
6. Dravidian languages
7. Altaic languages
8. Austro-Asiatic languages
9. Japonic languages
Note: There are more than nine language families, but these are the dominant ones with over one hundred million speakers each.

Source: M. Paul Lewis, ed., *Ethnologue: Languages of the World*, 16th ed. (Dallas: SIL International, 2009), accessed September 13, 2011, http://www.ethnologue.com.

Language Characteristics

The following terms are used to describe language characteristics.

- **accent.** An accent is the pronunciation of words within a language that is different from that used by a different group of the same language. For example, people in Mississippi pronounce words differently from people in North Dakota, but the differences are less severe than dialects.
- **creole.** Similar to pidgin, a creole language arises from contact between two other languages and has features of both. However, Creole is a pidgin that becomes a primary language spoken by people at home. Creole languages are often developed in colonial settings as a dialect of the colonial language (usually French or English). For example, in the former French colony of Haiti, a French-based creole language was developed that is spoken by people at home, while French is typically used for professional purposes.
- **dead language.** A dead language is one that is no longer used for local communication. For example, Latin is no longer used by local people to communicate but is still used by the Roman Catholic Church in some of its services.
- **dialect.** A dialect is a regional variety of a language that uses different grammar or pronunciation. Examples include American English versus British English. Linguists suggest that there are three main dialects of the English language in the United States: a Southern dialect, a midland dialect, and a Northern dialect. Television and public media communication has brought a focus on more uniform speech patterns that have diminished the differences between these three dialects.
- **isolated language.** An isolated language is one not connected to any other language on Earth. For example, Basque is not connected to any other language and is only spoken in the region of the Pyrenees between Spain and France.
- **lingua franca.** A lingua franca is a second language used for commercial purposes with others outside a language group but not used in personal lives. For example, Swahili is used by millions in Africa for doing business with people outside their own group but is not used to communicate within local communities.

- **official language.** The official language is the language that is on record by a country to be used for all its official government purposes. For example, in India the official language is Hindi, though in many places the lingua franca is English and several local languages may be spoken.
- **pidgin.** A pidgin is a simplified, created language used to communicate between two or more groups that do not have a language in common. For example, Residents of New Guinea mix English words with their own language to create a new language that can bridge speakers of different local language groups. Though the words are in English, the grammar and sentence structure is mixed up according to local vocabulary. There are many English-based pidgin languages around the world.
- **slang.** Slang is the local use of informal words or phrases that are not part of the official language. For example, a lot of musicians use slang in their lyrics.

Religions of the World

Religious geography is the study of the distribution of religions and their relationship to their place of origin. Religious geographers recognize three main types of religions: **universal** (or universalizing), **ethnic** (or cultural), and **tribal** (or traditional) religions. Universal religions include Christianity, Islam, and various forms of Buddhism. These religions attempt to gain worldwide acceptance and appeal to all types of people, and they actively look for new members, or converts. Ethnic religions appeal to a single ethnic group or culture. These religions do not actively seek out converts. Broader ethnic religions include Judaism, Shintoism, Hinduism, and Chinese religions that embrace Confucianism and Taoism. Finally, traditional religions involve the belief in some form of supernatural power that people can appeal to for help, including ancestor worship and the belief in spirits that live in various aspects of nature, such as trees, mountaintops, and streams (this is often called **animism**). Subsaharan Africa is home to many traditional religions.

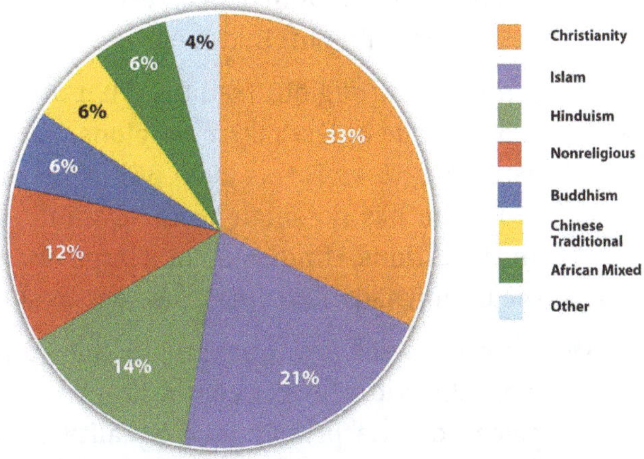

Figure 1.27: Major Religions of the World and Their Respective Percentage of the World Population

Although the world's primary religions are listed here, many other religions are practiced around the world, as well as many variations of the religions outlined here. The top four religions by population are Christianity, Islam, Hinduism, and Buddhism. Because the official doctrine of Communism was nonreligious or atheist, there are actually many more followers of Buddhism in China than demographic listings indicate. The percentage of the world's population that follows Buddhism is probably much higher than the 6 percent often listed for this religion.

- Christianity and Islam originated out of Judaism in the eastern Mediterranean and the Arabian Peninsula. Both are monotheistic religions that look to the Jewish patriarch Abraham as a founding personage. Christianity, based on the life and teachings of Jesus Christ, who lived in Palestine in the first century CE, spread rapidly through the Roman Empire. Islam is based on the teachings of Muhammad, a seventh-century religious and political figure who lived on the Arabian Peninsula. Islam spread rapidly across North Africa, east across southern Asia, and north to Europe in the centuries after Muhammad's death.
- Buddhism is a religion or way of life based on the teachings and life of Siddhartha Gautama, who lived in what is now India/Nepal around the fifth century BCE. There are three main branches of Buddhism: southern or Theravada Buddhism, eastern or Mahayana Buddhism, and northern or Vajrayana (Tibetan) Buddhism.
- Hinduism, a religious tradition that originated on the Indian subcontinent, is one of the oldest major religions still practiced in the world, and it may date back to as far as 2000 BCE or earlier. Unlike other world religions, Hinduism has no single founder and is a conglomerate of diverse beliefs and traditions. Hinduism has a large body of scripture, including the Vedas, the Upanishads, and epic tales such as the Mahabharata and the Ramayana.
- Sikhism, a religion founded in the Punjab region of southern Asia, is a monotheistic religion centered on justice and faith. High importance is placed on the principle of equality between all people. The writings of former gurus are the basis for the religion.
- Judaism is the religion of the Jewish people, whose traditions and ethics are embodied in the Jewish religious texts, the Tanakh, and the Talmud. According to Jewish tradition, Judaism began with the covenant between God and Abraham around 2000 BCE.
- Shintoism is a major ethnic religion of Japan focused on the worship of kami, which are spirits of places, things, and processes.
- Confucianism and Taoism are ethnic Chinese religions based on morality and the teachings of religious scholars such as Confucius.

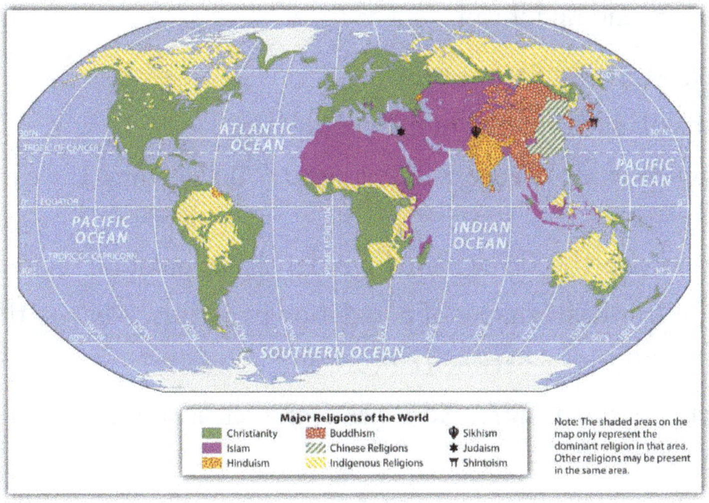

Figure 1.28 Major Religions of the World. Wikimedia Commons – CC BY-SA 3.0.

Key Takeaways

- The human population was approaching seven billion in 2011 and is increasing rapidly, mainly in developing regions of Asia and Africa. No one can agree on the earth's carrying capacity for our human population, but unless the growth rate changes, the human population will double in about fifty years.
- Since the

Industrial Revolution, humans have been moving from rural areas to urban areas. Workers were needed in the factories and fewer workers were needed on the farms because of improved technology. This trend is still happening in many rural areas of developing countries.

Population pyramids are one method of illustrating demographic data for a country to show if the population is declining or increasing and at what rate.

- Though often interchangeable in general terms, for the purpose of geography in this textbook, ethnicity is what you are born with and culture is what you learn after you are born.
- There are about six thousand languages in the world today, with about thirteen of them spoken by over one hundred million people or more. Of the main language families, nine include at least 1 percent or more of the human population.
- There are thousands of religions or variants of them in the world. Religious geographers recognize three main types of religions: universal, ethnic, and traditional. The four main religions of the world are Christianity, Islam,

Hinduism, and Buddhism.

Discussion and Study Questions

1. What is the planet's current human population?
2. What happens when overpopulation occurs? Do we know the earth's carrying capacity?
3. Where are the earth's three main high-density population regions?
4. Outline the four basic shapes of population pyramids. What do they indicate?
5. What is the difference between culture and ethnicity?
6. What is ethnic cleansing? Where has ethnic cleansing occurred in the world during your lifetime?
7. Approximately how many languages are there in the world? Which continent has the most?
8. What are the five most widely spoken languages in the world? Where would each be mainly spoken?
9. What are the four major world religions by percentage of followers?
10. What is the difference between an ethnic, universal, or a traditional religion? Give an example of each.

Activities

1. Create a population pyramid for your state, province, or country using the population data available at http://www.census.gov.
2. Compare population pyramids for countries from each world region found at this site: www.census.gov/population/international/data/idb/informationGateway.php.
3. Learn five common phrases in a language you do not speak.

1.4: GLOBALIZATION AND DEVELOPMENT

Globalization and development are intricately linked, as the former influences economic growth, cultural exchange, and technological advancement across the globe. By facilitating trade and investment, globalization allows developing nations to access larger markets, attract foreign capital, and adopt new technologies. This can lead to improved infrastructure, job creation, and enhanced living standards. However, the benefits of globalization are not evenly distributed; some regions and communities may face challenges such as job displacement, cultural homogenization, and environmental degradation. Balancing the positive aspects of globalization with the need for sustainable development is crucial for ensuring that its benefits reach all segments of society and contribute to a more equitable world.

Figure 1.29: Cultural Landscapes Representing the Urban Core Region of Los Angeles and the Peripheral Regions of Rural Montana The tall structures shown in Montana are grain elevators used to store grain of local farmers before loading it into railroad cars to be shipped to market. R. Berglee – CC BY-NC-SA.

Globalization

US President George H. W. Bush gave a speech during his presidency (1988–92) indicating the real possibility of a New World Order. The speech, given during the Persian Gulf War when Iraq occupied Kuwait, was quite surprising to many, and it created questions about what he meant. Geographers and professionals in other disciplines understand that the world is not static. Cultural forces continue to act on human activities as globalization creates new alliances and global networks. The goal is to understand globalization and to make sense of what is happening. The better we understand the world and human dynamics, the better we will be prepared to address the changes that are occurring. Geography provides a means to spatially examine these changes.

Globalization is a process with a long history. People have been exploring, migrating, and trading with each other throughout human history, and these activities have created interactive networks connecting the different parts of the planet and producing dependent economic relationships. In modern times, globalization can be recognized by noting iconic global corporations, such as Walmart, McDonald's, or Toyota, that trade across international borders and integrate labor and resources from different countries to sell a product or service in the global marketplace. In a number of countries, people have protested against the building of new Walmarts or McDonalds', and such protests exemplify concerns about globalization and the growing expansion of dominant global economic units into local communities. These ubiquitous corporations represent corporate interests that are primarily concerned with company profits. Global corporations tend to view countries or communities as either markets for their products or sources of labor or raw materials.

Globalization can seriously impact local communities for better or for worse, depending on local circumstances. The main force that encourages globalization is economic activity based on technological advancements. Cultural and societal changes often occur as a consequence and are no less significant.

Political geography examines geography's influence on political systems and globalization, which are related to issues concerning the development of borders and the distribution of government types. One of the most significant events in political geography was colonialism, which is connected to the development of capitalism. Eric Wolf, in his book *Europe and the People without History*, details the expansion of European powers throughout the world to control both human and natural resources as well as expand a country's world power and promote Christianity. The British parliamentary system of government was exported to various areas of the British Empire. Now the parliamentary system is used throughout the world in former British colonies and throughout the British Commonwealth. Additionally, colonialism helped spread the English language throughout the world, for example, to India and the United States.

European colonialism was an early wave of globalization that changed the planet and shaped most of the world's current political borders. This early wave of global conquest was fueled by the Industrial Revolution.

Colonialism transferred technology, food products, and ideas around the globe in merchant ships that centered on the European power bases of the colonial empires of Europe—mainly Britain, Spain, France, Portugal, and the Netherlands. When the United States became independent of these European colonial powers, it began to extend its power and influence around the world. Thus the first major wave of globalization was a result of European colonialism.

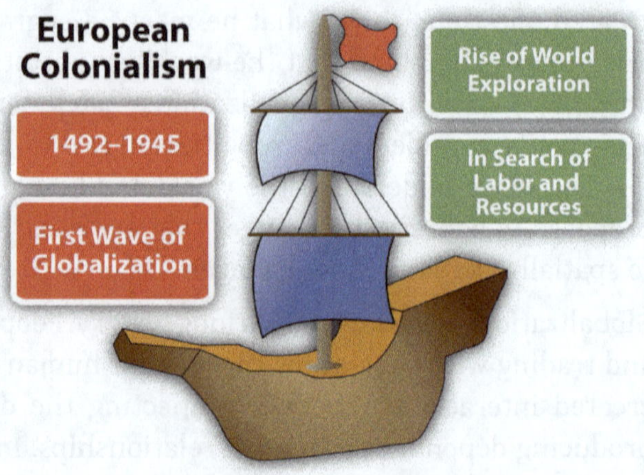

Figure 1.30: European colonialism changed many things. The expansion of European empires was driven by the search for profits from resources and labor in the acquisition of new lands. European colonialism had a significant impact on people and cultures.

The space race and the information revolution in the latter part of the twentieth century sparked a second significant wave of globalization. The space race was a rivalry between the United States and the Soviet Union to advance space-related technologies, such as satellites, and to achieve lunar landing. The conclusion of the

Cold War, marked by the dissolution of the Soviet Union in 1991, coincided with technological developments in computing that propelled the second significant wave of contemporary globalization. Technological advancements and corporate endeavors have catalyzed a surge of globalization that is influencing the economics of nations globally. European colonialism involved the physical conquest of territory and peoples by the mother country, resulting in colonies governed by the European colonizer's administration. Great Britain was the most fervent colonizer and accumulated substantial wealth through its colonial holdings. A distinction between European colonialism and contemporary globalization is that internationally engaged multinational businesses do not aspire to possess the country or govern directly. Corporations are indifferent to the type of government in power or the identity of the country's leaders, provided they can function and generate profit. This neocolonialism, like to European colonialism, perpetuates the exploitation of natural resources, labor, and markets for economic gain. Critics assert that corporate colonialism is only a legal means of exploitation, while proponents argue it represents the most efficient allocation of labor and resources to provide the lowest-cost products globally.

Figure 1.31: After the turn of the twenty-first century, Walmart became the world's largest corporation. Its global expansion and economic relationships are good examples of the activities of corporate colonialism.Walmart – Walmart store exterior – CC BY 2.0.

Examples of corporate colonialism can be seen in the trade relationships between the United States and places such as Mexico and China. US corporations move their manufacturing plants to Mexico to earn more profits by exploiting cheap labor. The corporations do not take over Mexico politically; they exploit it economically.

The many US corporations that have started manufacturing their products in China do not attempt to overthrow the Communist Chinese government; they want to exploit the cheap Chinese labor pool and open up markets to sell products to Chinese consumers. Desire for profits drives corporate colonialism.

Opportunity and Advantage

Considering the drive of individuals to increase their opportunities or advantages is one way to understand our world. People who have access to opportunities and advantages can generally achieve a higher standard of living than people without. This can be true for individuals, global corporations, or whole countries. Countries that have an opportunity or an advantage over others can achieve a higher standard of living for their people, and countries without such opportunities or advantages will struggle in a global economy. A country's opportunities and advantages can be determined by various factors, such as the amounts of available natural or human resources, arable land for farming, forests for timber, and freshwater for fishing or a specific location that provides a greater access to the world markets. Examples of human resources are a large labor pool or a high percentage of educated professionals. Individuals seek to gain opportunities or advantages by increasing their education, learning new skills that can translate into higher earning power, or migrating to a place with improved employment prospects.

Rural-to-urban shift occurs when people move from rural agricultural areas to the cities for employment or in search of a better life and is an example of a migration pattern based on people seeking greater opportunities or advantages. When people migrate from a poorer country to a postindustrial country, they are seeking opportunities or advantages in life. The "have" countries are those with opportunities and advantages; the "have-not" countries are those with fewer opportunities and advantages for their people or for their country's future. Migration patterns around the world usually shift people from places without resources to places with resources. These migration patterns, which are evident in rural-to-urban shift and periphery-to-core migration, allow individuals to seek a greater opportunities or advantages for the future.

Core-Periphery Spatial Relationship

Economic conditions vary across the globe. There are wealthy countries and there are poor countries, and the determination of which countries are wealthy and which countries are poor has generally been determined by the availability of economic opportunities and advantages. There are three **core** areas of wealthy industrialized countries, all of which are found in the Northern Hemisphere: North America, Western Europe, and eastern Asia. The main market centers of these regions are New York City, London, and Tokyo. These three core areas and their prosperous neighbors make up the centers of economic activity that drive the global economy. Other wealthy countries can be found dispersed in regions with large amounts of natural resources, such as the Middle East, or places of strategic location, such as Singapore. The world's poorer countries make up the **peripheral** countries. A few countries share qualities of both and may be called semiperipheries.

The periphery countries and the core countries each have unique characteristics. Peripheral locations are providers of raw materials and agricultural products. In the periphery, more people earn their living in occupations related to securing resources: farming, mining, or harvesting forest products. For the workers in these occupations, the profits tend to be marginal with fewer opportunities to advance. In the periphery, there is a condition known as brain drain, which describes a loss of educated or professional individuals. Young people leave the peripheral areas for the cities to earn an education or to find more advantageous employment. Few of these individuals return to the periphery to share their knowledge or success with their former community.

Brain drain also happens on an international level—that is, students from periphery countries might go to college in core countries, such as the United States or countries in Europe. Many international college graduates do not return to their poorer countries of origin but instead choose to stay in the core country because of the employment opportunities. This is especially true in the medical field. There is little political power in the periphery; centers of political power are almost always located in the core areas or at least dominated by the core cities. The core areas pull in people, skills, and wealth from the periphery. Lack of opportunities in the periphery pushes people to relocate to the core.

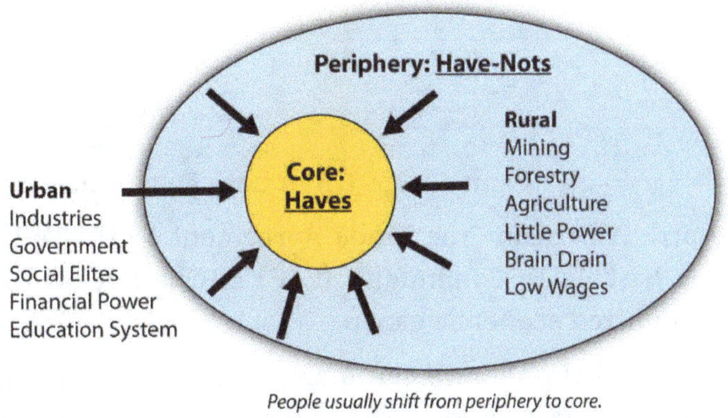

Figure 1.32: The Core-Periphery Spatial Relationship

Power, wealth, and opportunity have traditionally been centered in the core areas of the world. These locations are urbanized and industrialized and hold immense economic and political power. Ideas, technology, and cultural activity thrive in these core areas. Political power is held in the hands of movers and shakers who inhabit the core. The core depends on the periphery for raw materials, food, and cheap labor, and the periphery depends on the core for manufactured goods, services, and governmental support.

The core-periphery spatial relationship can be viewed on various levels. On a local level, one can select eastern Kentucky, with the city of Morehead as an example. Morehead has a population of about ten thousand people and is the only significant town in its county. Morehead, with a university, regional hospital and retail services, serves as a core hub for the surrounding periphery. The **hinterland** of Morehead has an economy based on agriculture, coal mining, and timber, which is typical of a peripheral region. The city of Morehead has the political, economic, and educational power that serves the people of its local area.

If we move up a level, we can understand that entire regions of the United States can be identified as peripheral areas: the agricultural Midwest, rural Appalachia, and the mountain ranges and basins of the western United States. The large metropolitan areas of the East and West Coasts and the Industrial Belt act as the core areas. Los Angeles and New York City anchor each coast, and cities such as Chicago, St. Louis, Denver, and Indianapolis represent the heartland. All the other large cities in the United States act as core areas for their surrounding peripheral hinterlands. Southern cities such as Atlanta, Memphis, Dallas, or Phoenix act as core centers of commerce for the South in a region known as the Sun Belt.

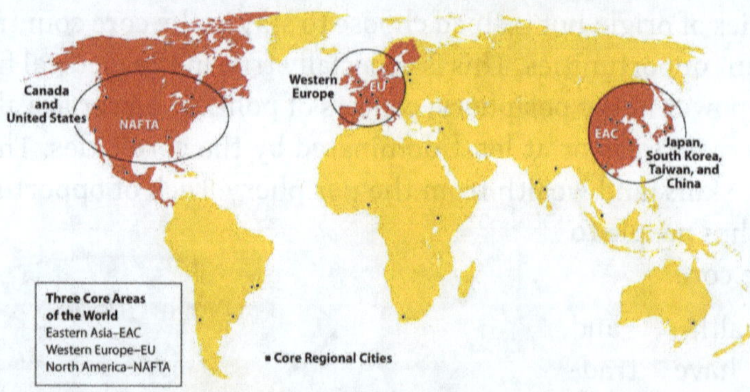

Figure 1.33: Three Core Economic Areas of the World:

North American Free Trade Agreement (NAFTA); the European Union (EU); and the East Asian Community (EAC) EAC is not an "official" organization but is a recognized economic group.

On a global scale, we can understand why North America, Western Europe, and eastern Asia represent the three main economic core areas of the world. They all possess the most advanced technology and the greatest economic resources. Core regions control the corporate markets that energize and fuel global activity. Peripheral regions include portions of Africa, Asia, Latin America, and all the other places that primarily make their living from local resources and support the economic core. These peripheral regions may include key port cities. A semiperiphery would be a transitional area between the core and the periphery, which could include countries such as Russia, India, or Brazil that are not exactly in the core and not really in the periphery but might have qualities of both. World migration patterns follow the core-periphery spatial relationship in that people and wealth usually shift from the peripheral rural regions to the urban core regions. The "have" countries of the world are in the core regions, while the "have-not" countries are most likely in the peripheral regions.

National Income Methods

It is easier to understand why people move from rural to urban, from periphery to core, from Mexico to the United States when one begins to understand the global economy. Economic conditions are connected to how countries gain national income, opportunities, and advantages. One way of gaining wealth is simply by taking someone else's wealth. This method has been common practice throughout human history: a group of armed individuals attacks another group and takes their possessions or resources. This is regularly practiced through warfare. Unfortunately, this pillage-and-plunder type of activity has been a standard way of gaining wealth throughout human history. The taking of resources by force or by war is frowned upon today by the global economic community, though it still occurs. The art of piracy, for example, is still practiced on the high seas in various places around the globe, particularly off the coast of Somalia.

The main methods countries use to gain national income are based on sustainable national income models and value-added principles. The traditional three areas of agriculture, extraction/mining, and manufacturing are a result of primary and secondary economic activities. Natural resources, agriculture, and manufacturing have been traditionally targeted as the means to gain national income. Postindustrial activities in the service sector would include tertiary and quaternary economic activities, which make up a large part of a nation's economy but might not hold the same value-added quotient for national income as the traditional three areas.

Agriculture is the method of growing crops or trees or raising livestock that provides food and some raw materials. The excess is usually sold for profits. This is a renewable method of gaining wealth, as long as conditions are favorable. Profits for agricultural products might be low because of global competition. Countries with minerals, oil, or other natural resources can earn income from the extraction and sale of those items. Saudi Arabia and other countries with abundant petroleum reserves can gain wealth by selling that resource to other countries. Since these resources are not renewable, once the minerals or oil run out, the country must turn to other activities to gain national income.

Places around the world have sometimes been named after the methods used to gain wealth. For example, the Gold Coast of western Africa received its label because of the abundance of gold in the region. The term *breadbasket* often refers to a region with abundant agricultural surpluses. Another example is the Champagne region of France, which has become synonymous with the beverage made from the grapes grown there. Banana republics earned their name because their large fruit plantations were the main income source for the large corporations that operated them. Places such as Copper Canyon and Silver City are examples of towns, cities, or regions named after the natural resources found there. The United States had its Manufacturing Belt, referring to the region from Boston to St. Louis, which was the core industrial region that generated wealth through heavy manufacturing for the greater part of nineteenth and twentieth centuries

Manufacturing has offered the industrialized world the opportunity for the greatest value-added profits. From the beginning of the Industrial Revolution, solid profits have been made by turning raw materials into useable products that can be mass-produced and sold in high quantities. The core areas of the world have all made enormous wealth from manufacturing profits. Today, information technology and high-tech manufactured products generate substantial wealth. The Microsoft Corporation manufactures information by placing data in the form of computer programs on inexpensive disks or in digital files that are then sold at a profit to a world computer market. The geographic region of California south of San Francisco was labeled Silicon Valley after this type of information-generating activity.

For a country to gain national wealth, income must be brought in from outside the country or be generated from within. Other than the big-three methods of agriculture, extraction/mining, and manufacturing, there are additional ways a country can gain wealth, such as through tourism and services. Postindustrial economic activities (the service sector) contribute to a large percentage of employment opportunities in industrialized countries and generate a large percentage of their national economies. However, the service sector does not have the high value-added profits traditionally provided by the manufacturing sector. In the gaming industry, for example, casinos do not have a large wealth-generating potential for the country unless they can either attract gamblers from other countries or provide for other wealth-generating services. Gambling shifts wealth from one person to another and provides little national income. The activity of gambling does, however, generate service jobs for the gaming industry.

Figure 1.34 Major Methods of Gaining National Income Based on Sustainable National Income Models and Value-Added Principles

Methods of Gaining National Income Pillage and Plunder (Unnaceptable) Discouraged by Civilized Nations

1. Grow it: Renewable with Low Profits (Agriculture, Forestry, etc.)
2. Extract It: Nonrenewable and Selective (Ores, Oil, and Minerals)

3. Manufacture It: High Value-Added Profits (Mass-Produced Products)

Postindustrial Services (Mixed Profits) Communications, Financial/Credit, Information Data, Travel and Tourism **Requirement for Success** Good Education System, Highly Skilled Workforce, Quality Infrastructure

National Debt

Countries with few opportunities to gain wealth to support their governments often borrow money to provide services for their people.

National debt is a major problem for national governments. National income can be consolidated into the hands of a minority of the population at the top of the socioeconomic strata. These social elites have the ability to dominate the politics of their countries or regions. The elites may hold most of a country's wealth, while at the same time their government might not always have enough revenues to pay for public services. To pay for public services, the government might need to borrow money, which then increases that country's national debt. The government could have a high national debt even when the country is home to a large number of wealthy citizens or a growing economy. Taxes are a standard method for governments to collect revenue. If economic conditions decline, the amount of taxes collected can also decline, which could leave the government in a shortfall. Again, the government might borrow money to continue operating and to provide the same level of services. Political corruption and the mismanagement of funds can also cause a country's government to lack revenues to pay for the services it needs to provide its citizens.

Figure 1.35: Inflationary Banknote from Zimbabwe—Considered the Largest Denomination Banknote Ever Printed for Legal Tender

National debt, defined as the total amount of money a government owes, is a growing concern across the globe. Many governments have problems paying their national debt or even the interest on their national debt. Governments whose debt has surpassed their ability to pay have often inflated their currency to increase the amount of money in circulation, a practice that can lead to hyperinflation and eventually the collapse of the government's currency, which could have serious negative effects on the country's economy. In contrast to the national debt, the term **budget deficit** refers to the annual cycle of accounting of a government's excess spending over the amount of revenues it takes in during a given fiscal year.

Development and Demographics

The Industrial Revolution, which prompted the shift in population from rural to urban, also encouraged market economies, which have evolved into modern consumer societies. Various theories and models have been developed over the years to help explain these changes. For example, in 1929, the American demographer Warren Thompson developed the demographic transition model (DTM) to explain population growth based on an interpretation of demographic history. A revised version of Thomson's model outlines five stages of demographic transition, from traditional rural societies to modern urban societies:

- Stage 1: High birth and death rates; rural preindustrial society
- Stage 2: Declining death rate; developing country
- Stage 3: Declining birth rate; high urbanization rate
- Stage 4: Low birth and death rates; stabilized population
- Stage 5: Declining population; urban postindustrial society

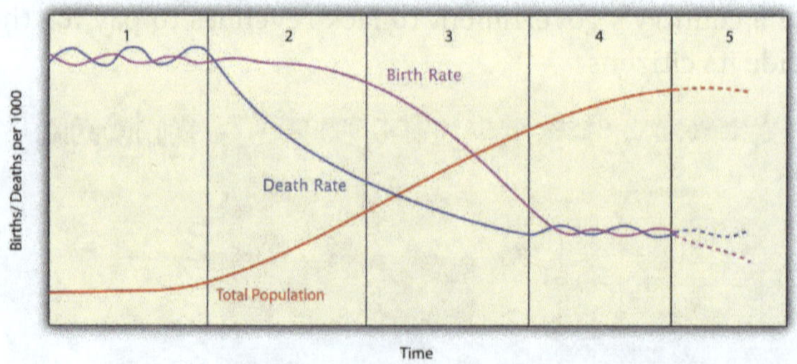

Figure 1.36: The Five Stages of the Demographic Transition Model

In the 1960s, economist Walt Rostow outlined a pattern of economic development that has become one model for growth in a global economy. Rostow's model outlined the five stages of growth in the economic modernization of a country:

- Stage 1: Traditional society
- Stage 2: Preconditions for take-off
- Stage 3: Take-off
- Stage 4: Drive to maturity
- Stage 5: Age of high mass consumption

The human development index (HDI) was developed in 1990 and used by the United Nations Development Program to measure a standard of human development, which refers to the widening opportunities available to individuals for education, health care, income, and employment. The HDI incorporates variables such as standards of living, literacy rate, and life expectancy to indicate a measure of well-being or the quality of life for a specific country. The HDI is used as an indicator of a country's economic and technological development.

The basic principles of Rostow's DTM and the HDI can be illustrated in a general index for understanding development. Two variables of the DTM's correlating five stages are family size and economic income, which can illustrate the population pattern or development potential. All countries of the world are at one of the five stages. The general index addresses how population growth rates relate to rural-to-urban shift, which has traditionally been a result of industrialization. By tracking both family size and economic conditions, a pattern of population growth and economic development can be illustrated and more clearly understood.

Fertility rate is often defined as the number of children born to a woman in her lifetime, regardless of whether they all live to adulthood. Fertility rate may (or may not) vary from **family size**, which is an indication of the number of living children raised by a parent or parents in the same household. A high infant mortality rate may account for a fertility rate that is greater than family size. To simplify the understanding, family size is used in this textbook to illustrate economic dynamics. Though the statistical data may vary slightly between the two terms, this should not present any problem in understanding basic patterns of development.

As a general trend, when a country experiences increasing levels of industrial activity and greater urban growth, the outcome is usually a higher standard of living for its people. Additionally, rural-to-urban shift takes place, driven by the pull of opportunities and advantages in the industrializing and urbanizing areas. Though there are exceptions, a decrease in family size usually coincides with a higher level of urbanization. There are cases in which only core regions within a country transition through the five development stages without the peripheral regions experiencing the coinciding levels of economic benefit. The five stages of the index of economic development assist in illustrating these general patterns.

Stage 1 indicates traditional rural societies, which are usually based on agriculture and not as dependent on the outside world. Stage 1 families are larger, their income levels are low, and their advantages and economic development opportunities are low. Health care, education, and social services are in short supply or nonexistent. High birth and death rates maintain a high fertility rate/family size and a low population-growth status. Populations in stage 1 development have a stationary population pyramid. Though there may be regions of the world that exhibit stage

1 development patterns, few if any entire countries fall into this category as of the year 2000.

Stage 2 countries experience high population growth rates because family size remains high but modern medicine or improved nutrition allows people to live longer, which lowers the death rate. Population is exploding in countries in stage 2. During this stage, young people from rural areas often migrate to the cities looking for employment. Rural stage 2 regions are starting to urbanize and integrate their economic activities with the outside world. Regions in stage 2 often have a surplus of cheap labor. Income levels remain low and family size continues to be large. Countries in this stage often have a rapidly expanding population pyramid.

Societies that have made business connections that provide for manufacturing of products, industrial activities, or an increased service sector might progress to stage 3, the rural-to-urban shift stage. These regions are experiencing a high rate of rural-to-urban shift in their populations. These regions are often targeted by multinational corporations for their labor supply, and as people migrate from rural areas to the cities looking for employment, urban populations grow and core or central cities experience high rates of self-constructed housing (slums). Income levels start to increase and family size starts to drop significantly. Stage 3 countries have an expanding population pyramid.

Societies that have urbanized and industrialized and are members of the global marketplace might enter stage 4. Members of an urban workforce assist in building a networked economy.

Family size is lower as urban women enter the workforce and have fewer children. Health care, education, and social services become increasingly available, and income levels continue to rise. In stage 4, there is typically a high level of growth in the industrial and service sectors with a great need for infrastructure in the form of transportation, housing, and human services. Countries in stage 4 development have populations that resemble a stationary population pyramid.

As incomes increase and family size decreases, a consumer society emerges, creating stage 5, where high mass consumption can drive the economy. Many countries in stage 5 can eventually experience a negative population growth rate in which the fertility rate (family size) is below replacement levels (statistically around 2.1 children). With a low number of young people entering the workforce, stage 5 regions become an attractive magnet for people looking for opportunities and advantages in the job market. Illegal immigration might become an issue. Europe and the United States are now experiencing this condition. Japan has the same low family size, but because of their island location and strict laws, they have a different set of illegal immigration issues compared with Europe or the United States. Populations in stage 5 development have a contracting population pyramid.

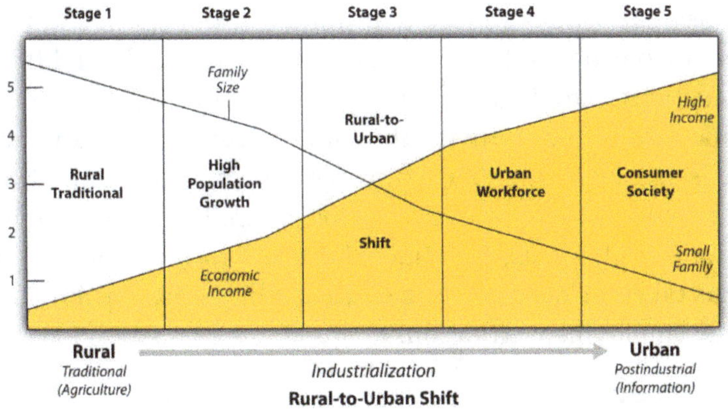

The formula states that as countries urbanize and industrialize, family size will naturally go down and incomes traditionally go up.

Figure 1.37: Index of Economic Development

The four basic shapes of population pyramids can parallel the various stages of a country's economic development. Many of the concepts used in this textbook are interrelated. The various methods, models, or theories used in geography are often used by other disciplines as well. Understanding one concept usually assists the reader in learning about the other concepts and how they apply to different geographic locations. In this case, the stages of economic development and the population pyramids illustrate the contrast between rural and urban societies and the changes in family size during the industrialization process.

Geography of Opportunity

Highly industrialized countries in Western Europe, eastern Asia, or North America can offer more economic opportunities compared with developing countries. The push-pull factors that push people out of poorer countries and pull them to an industrialized country are strong. Portions of the population of countries in the earlier stages of the index of economic development often migrate to countries in the latter stages of the index looking for work and other opportunities. This transition has created a dichotomy between people who have opportunities and advantages and those who do not. People who do not have opportunities and advantages often want to move to places that do have them so they can work to attain greater economic security for themselves and their families. The haves and have-nots are general categories of economic status and not necessarily cultural values.

Countries in stage 4 or 5 of the index of economic development are often attractive places for those seeking greater economic opportunities or advantages. Populations in these stages generally have fewer children, so the demand for entry-level workers is often higher. Immigrants with fewer skills take the entry-level jobs to enter the economic workforce. An established country with a long-standing history and culture does not always welcome an influx of new immigrants. The arrival of an immigrant labor pool often includes individuals who hold different cultural traditions or customs than those of the mainstream society. Social tensions arise if different ethnic groups are vying for the same cultural spaces and opportunities. One example is the large number of people entering the United States across its southern border. Europe is experiencing a similar immigration issue, with immigrants from North Africa, the Middle East, and former colonies. Japan, on the other hand, has taken pride in holding on to its ubiquitous Japanese culture but is facing the same employment situation.

Opportunities in Geography

The following information on careers in geography is from the website of the Association of American Geographers (AAG), which is a resource for those interested in pursuing employment in the field of geography (www.aag.org).

Many occupations require knowledge of and skills in geography. Geographers work in many different areas, such as environmental management, education, disaster response, city and county planning, community development, and more. Geography is an interdisciplinary field that offers diverse career opportunities.

Many geographers pursue rewarding careers in business; local, state, or federal government agencies; nonprofit organizations; and schools. Geographers with graduate (master's and doctorate) degrees may become educators in higher education (community colleges and universities).

Using information from the US Department of Labor, the AAG has compiled data on a broad list of occupations related to geography. Using this online career database, you can explore the diverse career opportunities available to geographers, as well as retrieve data on salaries, projected growth, demand for key skills, and much more to guide your career planning. Occupations can be found at the following site: www.aag.org/cs/salary_data_and_trends/salary_data_and_trends_overview.

Key Takeaways

- The search for cheap labor and resources drives the need for profits by multinational corporations that fuel the global economy. This activity is creating a second major wave of globalization, often referred to as neocolonialism or corporate colonialism.
- The concepts of opportunity and advantage provide a means to understand the attractiveness or unattractiveness of a place to immigrants or economic activities. Opportunities and advantages drive rural-to-urban shift, migration, and movement of corporate activity.

- Core areas are usually urban with high levels of industrial and economic development. Peripheral areas are typically suppliers of food and raw materials used in the core. Political and economic power is held in the core, while the periphery suffers from lower incomes and brain drain.
- National income methods are based on standard economic practices and value-added principles. Agricultural activities are renewable, but extractive activities are not. Manufacturing has historically provided the highest value-added profits and has been the main means of the core economic regions of the world to gain income. The service sector provides a high number of jobs but might not contribute to national income at the same levels as agriculture, extraction activities, or manufacturing.
- Development and population models can help one understand a country's socioeconomic dynamics.

Family size and economic income are two indicators that can be tracked to assist in understanding the industrialization or urbanization levels of a country or demographic region.

- Globalization has prompted a greater understanding of how opportunities and advantages relate to haves and have-nots. Whether it is individuals or countries, some have greater levels of opportunity and advantage than others. Human migration patterns usually coincide with the push-pull forces of opportunity or advantage levels.
- Many of the concepts used to understand the dynamics of a region or place are related. The four main types of population pyramids can help illustrate the various stages of a country's socioeconomic situation as illustrated in the index of economic development.

Discussion and Study Questions

1. What prompted the first major wave of rural-to-urban migration in modern history?
2. What human economic activities would you find in a core region? What human economic activities would you encounter in the periphery?
3. What is meant by the terms *opportunity* and *advantage*? How can they help us understand world geography?
4. What are the three main traditional methods through which a country can gain national income? Of the three, which has traditionally provided the highest value-added profits?
5. What are the three main economic core areas of the world that have dominated the global economy?

6. What are the main elements of business that global corporations seek in terms of a geographic location to make the highest profits?
7. What is the main human migration pattern between the core and the periphery?
8. What does each stage of the index of economic development represent or illustrate?
9. Would minimum wage tend to go up or down for countries moving from stage 4 to stage 5 of the index of economic development?
10. How is it that countries with strong or stable economies run up a high national debt?

Activities

1. Compare the main opportunities and advantages for the place you live with another place either in the same country or in another country with a similar population pyramid profile.
2. Determine the main economic activities for your city or country and check if they coincide with any of the methods listed in the explanation in Section 1.1.4 "World Regional Geography" on how a country gains national income.
3. Outline the main migration pattern for people moving into or out of your current location.

1.5 CONCISE SUMMARY

Geography is the spatial analysis of the Earth's surface. The dynamic field of geography connects the social sciences and the physical sciences. The spatial aspect of geography is exemplified by the production of maps, which serve as a vital method for conveying information. The two primary domains of the discipline are human geography and physical geography. Geographers utilize GPS, GIS, and remote sensing as tools to spatially analyze a site or the physical and cultural landscape.

Geographers segment the Earth into a geometric grid to offer locational references for sites on its surface. Human activity is structured around the grid system to facilitate time zones, navigation, and the structuring of communication systems. This grid system of longitude and latitude enables the monitoring of climate and seasonal variations.

Regions constitute the fundamental units of geographic analysis. World geography categorizes the globe into regions known as realms, which serve as frameworks for comparative analyses of human and physical landscapes and activities. Climate areas or zones facilitate comprehension of the Earth's natural circumstances. Historically, human activity has been significantly influenced by climatic variations.

Type C conditions typically draw substantial human populations.

The interplay between the environment and human activity constitutes a significant aspect of geography. The movement of tectonic plates occasionally induces earthquakes and volcanic activity, impacting human endeavors. The rain shadow effect can influence human habitation and lifestyle. Human activity exacerbates environmental issues, like deforestation, which affects ecosystems through habitat loss, soil erosion, and potential climate change.The human population persists in its growth. The Earth's carrying capacity for humans is contentious, although the influence of humanity on the world is unequivocally significant. Certain circumstances and conditions promote high population growth, whilst others inhibit it. Migration, rural-to-urban transition, and urbanization are associated with demographic changes. With the rise in population, the quantity of languages diminishes, exemplifying the effects of globalization.

European colonization initiated a significant wave of globalization that persisted until around the onset of World War II. Following the conclusion of the Cold War, a second wave of globalization was propelled by the information age and the advent of novel communication and transportation technology. Concepts such as labor and resources, opportunity and advantage, or the affluent and the impoverished can elucidate the dynamics of globalization.

The core-periphery spatial relationship elucidates the organization of human economic activity around an urban core or a rural periphery. The methods employed by nations to generate national revenue provide insight into the categorization and explanation of economic activity. The economic development index is a framework that delineates the steps a nation may undergo to achieve a postindustrial development status. All these notions, models, and theories serve as instruments for comprehending the human behaviors that constitute aspects of the globalization process.

2 Chapter

EUROPE

2.1: INTRODUCTION: AN OVERVIEW

Europe's historical trajectory, from the Roman Empire to the European Union (EU), serves as a paradigmatic case in regional geography. The continent grapples with the cultural dynamics that both unify and divide it, ranging from historic empires to varied nation-states and a multinational union. The significant influence of European colonialism on the world since the Industrial Revolution persists to this day. The rural-to-urban migration initiated by the Industrial Revolution initially affected Europe

Figure 2.1 Map of Europe

and continues to influence developing nations. Comprehending the geographic geography of Europe is crucial for grasping our globe. This concise overview of fundamental topics will impart a significant lesson in globalization, which influences every individual on Earth. The concepts and principles applicable to Europe are also relevant to other nations and regions.

Location and Climate

Europe is a northern continent. All the British Isles, for example, fall above the fiftieth parallel. If we compare Europe's position on a globe with that of the contiguous United States, we see that much of Europe lies north of the United States. Paris, France, is at about the same latitude as Fargo, North Dakota. Athens, Greece, is at about the same latitude as St. Louis, Missouri. Europe's northern position affects its growing seasons and people's moods, and it should be taken into consideration as an important influence in the evolution of the European character. Europe is also surrounded by bodies of water: the Atlantic Ocean borders Europe on the west, the Arctic Ocean borders Europe to the north, and many seas surround the various peninsulas and coastal regions.

The oceans exert significant influence on the world's climates. The oceans collect and store vast amounts of solar energy, particularly around the equator, and transport that heat with their currents. Ocean currents can move water for thousands of miles from one temperature zone to another. Because oceans can absorb so much heat, maritime climates are often milder than continental ones, with smaller temperature variations from day to night as well as from

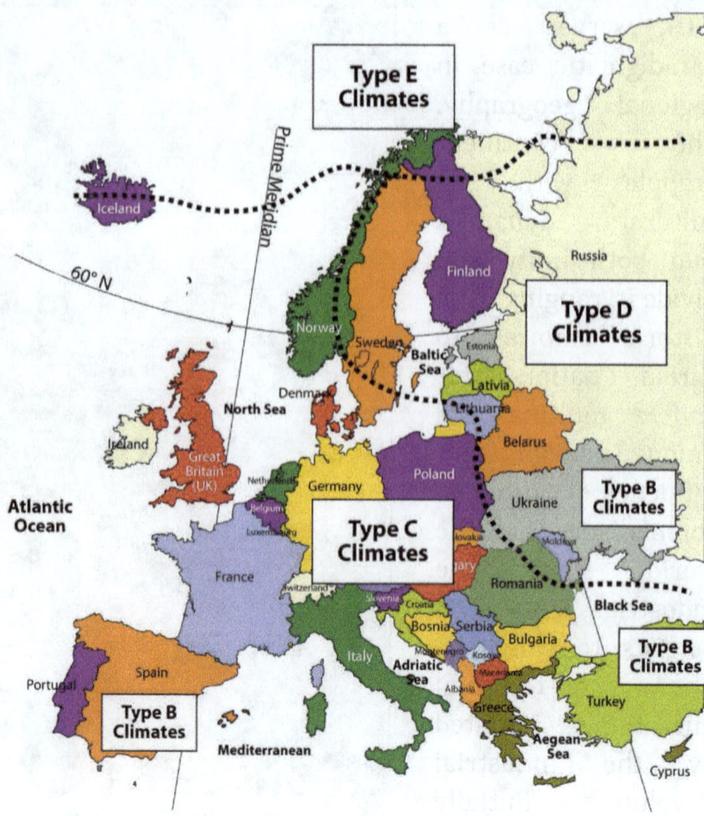

Figure 2.2: The Dominant Climate Types of Europe

winter to summer. This influences not only temperature but also precipitation patterns over wide regions of Europe and the rest of the world. Water moderates coast environments in a number of ways. Water heats and cools more slowly than land. This heat inertia allows coastal communities to have climates that tend to be more moderate than one might imagine for places so far north. Interior Europe does not benefit from coastal waters and can have winters as cold as those found within the upper midwestern United States.

The Gulf Stream is perhaps the most important current for Western Europe's climate and is responsible for producing a temperate climate for a northern latitude location. Most of Western Europe has a moderate type C climate. The Gulf Stream originates in the Gulf of Mexico, where the waters are warmed. This powerful current follows the Eastern Seaboard of the United States before crossing the Atlantic Ocean for Europe. The Gulf Stream's most dramatic effect can be found in the western coastal islands of Scotland, which has a mild enough climate to support some forms of tropical flora, even though it is a degree of latitude as far north as Hudson Bay, Canada.

The coast of Norway provides another example. While most of Norway's coastal area lies within the Arctic region, it remains free of ice and snow throughout the winter. People living farther inland and closer to Eastern Europe and Russia encounter the colder type D climates. Colder air sweeps down from the Arctic north or from eastern Siberia and provides colder winters in this eastern region. The Mediterranean Sea moderates the temperature to the south, providing a type C climate around its shores. Type C climates meet up with type E climates at or near the Arctic Circle in Norway and in Iceland.

Four Main European Landforms

Europe has four main landforms, many islands and peninsulas, and various climate types. The four main landforms include the **Alpine region**, **Central Uplands**, **Northern Lowlands**, and **Western Highlands**. Each represents a different physical part of Europe. The wide-ranging physical environment has provided Europe with an abundance of biodiversity. Biodiversity refers to diversity of the number of species in an ecosystem and the quantity of members in each species. The physical environment also provides natural resources and raw materials for human activities. Europe's moderate climates and favorable relative location are supported by its access to the many rivers and seas. These advantageous developmental factors supported the development of the Industrial Revolution in Europe, which gave rise to highly technical and urban societies. Europe has emerged as one of the core economic centers of the global economy. Associated with the urbanization of Europe are high human population densities that have placed a strain on the natural environment. As result, there has been significant deforestation and the loss of natural habitat, which has in turn has decreased the realm's level of biodiversity.

Rivers are abundant in Europe and have provided adequate transportation for travel and trade throughout its history. Most of Europe is accessible by water transport either via the many rivers or along the extensive coastlines of the peninsulas and islands. Two main rivers divide Europe: the **Danube** and the **Rhine**. Both have their origins in the region of southern Germany on or near the border with Switzerland. The Rhine River flows north and empties into the North Sea in Rotterdam, Holland, one of the world's busiest ports. The Danube flows east through various major European cities, such as Vienna, Budapest, and Belgrade before emptying into the **Black Sea**.

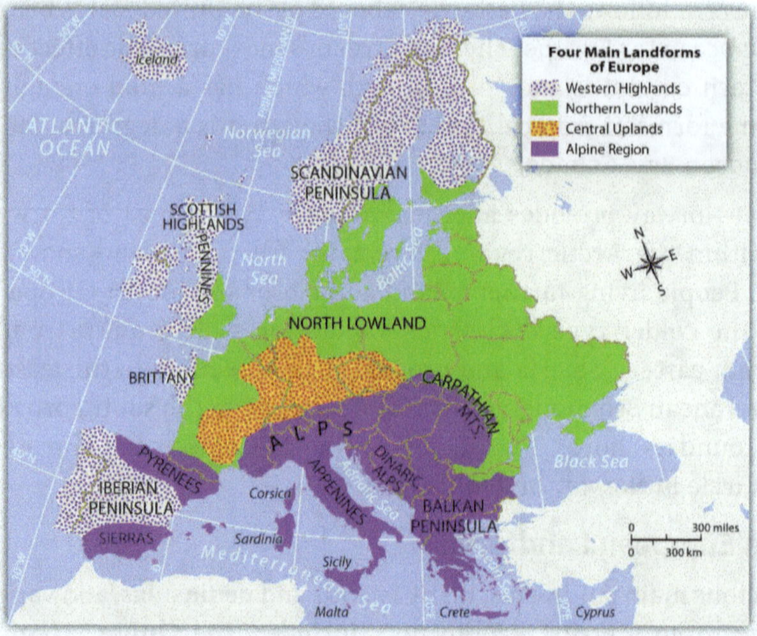

Figure 2.3: Four Main Landforms of Europe: Western Highlands, Northern Lowlands, Central Uplands, and the Alpine Region

Alpine Region

The **High Alps**, which range from eastern France to Slovenia, are central to the **Alpine region**. Included in the **Alpine Range** are the **Pyrenees**, located on the border between France and Spain; the **Apennines**, running the length of Italy; the **Carpathians**, looping around Romania from Slovakia; and finally, the shorter **Dinaric Alps** in former Yugoslavia. Mountains usually provide minerals and ores that were placed there when the earth's internal processes created the mountains. Mountains also isolate people by acting as a dividing range that can separate people into cultural groups.

Figure 2.4: The Alpine Region: Eiger, Mönch, and Jungfrau from Männlichen—Swiss Alps

The Alpine region encircles the Mediterranean coastlines, which have more temperate type C climates that are particularly warm with hot, dry summers and cool, wet winters. This climate type allows for the cultivation of food products such as olives, citrus fruit, figs, apricots, and grapes. Evergreen scrub oaks and other drought-resistant shrubs are common in the Mediterranean region.

The Central Uplands

The region bordering the main Alps to the north, which includes a large portion of southern Germany extending eastward, is known as the **Central Uplands**. These foothills to the Alps are excellent sources of raw materials such as forest products and coal, which are valuable resources for industrial activities. The Central Uplands are also good locations for dairy farming and cattle raising. This middle portion of the continent has a mixed deciduous-coniferous forest, and the vegetation includes oak, elm, and maple trees intermingled with pine and fir trees. There are four distinct seasons in this region with moderate amounts of precipitation year round.

Northern Lowlands

Figure 2.5: The Western Highlands meet the lowlands in central Scotland.

Photo by R. Berglee – CC BY-NC-SA.

Similar to the breadbasket of the midwestern United States, Europe's **Northern Lowlands** possess excellent farmland. Major agricultural operations here provide for a large European population. The land is flat to rolling with relatively good soils. The Northern Lowlands are a great plain that extends across northern Europe from southern France, north through Germany, and then all the way to the Ural Mountains of Russia. This area is typified by prairies and areas of tall grasses and is mostly used as farmland. The lowlands area also contains bogs, heaths, and lakes. The eastern part of this great plain around Ukraine is characterized by a steppe biome. It is a flat and relatively dry region with short grasses and is generally an agricultural region. This eastern area has great swings in temperature, both from day to night and from summer to winter. Winter temperatures in the eastern steppe can drop to below -40 °F, with summer temperatures reaching as high as 105 °F. This is similar to the steppes of eastern Montana or western North Dakota in the United States.

Western Highlands

On the western edges of the European continent arise short rugged mountains called highlands that extend throughout Norway, parts of Britain, and portions of the **Iberian Peninsula** of Portugal and Spain. These Western Highlands hold sparser populations and are less attractive to large farming operations. Agriculture is usually limited to grazing livestock or farming in the valleys and meadows. The **Scottish Highlands** are noted for their wool products and Highland cattle. In England, the central chain of highlands called the **Pennines** proved valuable during the Industrial Revolution because they enabled hydropower and, later, coal mining. Coal mining was prominent in the highland regions of Wales. In the far northern regions of Scandinavia, tundra environments prevail. In this coldest and driest biome, permafrost dominates the landscape, and the land becomes soggy for brief periods during the few weeks of summer. The flora consists primarily of lichens, mosses, low shrubs, and wildflowers.

Natural Resources in Europe

The physical landforms of Europe provide a diversity of geographic opportunities that have catapulted Europe through the

Industrial Revolution and into the information age. With an abundance of natural resources, European countries have gained wealth from the land and leveraged their geographic location to develop a powerhouse of economic activity for the global marketplace. Europe has placed a strong focus on manufacturing activity to take advantage of its natural resources. The highly urbanized society has struggled to find a balance between modernization and environmental concerns. Industrial activities have contributed to the degradation of the environment and the demise of a number of species.

Different regions of Europe are blessed with fresh water supplies, good soils, and various minerals. Chief among the mineral deposits in Europe is iron ore, which can be found in Sweden, France, and Ukraine. Other minerals exist in smaller quantities, including copper, lead, bauxite, manganese, nickel, gold, silver, potash, clay, gypsum, dolomite, and salt. Extraction activities have supported the continent's industrialization.

The ready access to vast areas of the Atlantic Ocean and a number of major seas, lakes, and rivers has elevated fish to an important natural resource in Europe. The seas around Europe provide about 10 percent of the world's fish catches. Mirroring the situation around the world, European fishing activity increased as population increased. Europeans are becoming increasingly aware of the effects of overfishing. Stocks of Atlantic cod and Atlantic mackerel are considered to be at risk because of the twin threats of overfishing and changes in the environment that are affecting natural mortality and slowing spawning. Stocks of eastern North Atlantic bluefin tuna are also attracting attention for the same reason. Other species of fish in the North Atlantic and the Baltic Sea are considered overexploited. Changes to the fishery are a major concern for strictly ecological reasons and also because fish comprise such a significant portion of the European economy.

The countries of Europe, with the formation of the EU, began to work toward shared research and policies with respect to the fishery. The Common Fisheries Policy was drafted and includes strict and extensive rules and guidelines for fishing, particularly of cod. The rules are so strict and thus so controversial that Norway and Iceland decided to opt out of the EU rather than agree to abide by these rules. The effect of these rules is incompletely known, but the cod stocks in the North Sea do not seem to be rebounding as quickly or substantially as expected.

Forest covers more than 40 percent of the continent's land area, with the majority on the Russian side. Forests exist primarily in the less populous Nordic and Baltic countries and in Central Europe. About half the forest land in Europe is privately owned. Interestingly, there are several different forms of private ownership, including large family holdings, holdings by forest industries, and small to very small holdings by thousands of individuals. Most of the forested land is managed, and about 85 percent of the forests produce exploitable resources. The percentage of forested land in Europe is rebounding because of an extensive tree-planting initiative since 2000. Pollution has caused great harm to the forests of Poland, the Czech Republic, and eastern Germany, and acid rain and air pollution have seriously harmed forests such as the **Black Forest** of southern Germany, which has also experienced heavy logging activity.

Soil resources are of critical importance. Soil is necessary for growing food, supporting livestock, supporting other natural resources such as the forests, and supplying groundwater. Soil resources are only just beginning to receive attention in Europe. Problems to be addressed include loss of topsoil from development and building activities, older-style agricultural practices, mining, contamination from industries and other sources, and acidification as a result of acid rain.

Coal, now substantially depleted, is abundant in several areas of Great Britain, including some offshore areas, as well as in the industrial centers of Germany and in Ukraine. Other coals deposits are found in Belgium, France, Spain, the Czech Republic, Poland, Slovakia, and Russia. The burning of coal has produced high levels of air pollution.

Acid rain has been a major concern in the northern countries, where wind currents carry pollutants north into Scandinavia from the industrial regions of Central Europe. In Scandinavia, acid rain has diminished fish populations in many of the lakes. Forest health is also being challenged, which is diminishing the economic conditions of regions that depend on forests for their economic survival.

Petroleum and natural gas deposits exist underneath the North Sea and were first tapped in the 1970s. Five European countries have rights to these resources, including Norway, the United Kingdom, Denmark, the Netherlands, and Germany, with Norway holding the bulk of the rights. The governments of these five nations agree that, although tapped only decades ago, half the North Sea oil reserves have been consumed.

Before the extraction of petroleum products from the North Sea, Russia and the former Soviet Union's other republics supplied petroleum to Europe. These areas still have a number of active extraction operations. Hydroelectric power has been important in Europe as well. With both coal and oil resources largely depleted and the desire to avoid the environmental damage caused by dams, the European Energy Commission is devoting substantial energy and resources to encouraging use of renewable resources such as wind and solar energy. In March 2007, European leaders agreed that a binding target of 20 percent of all energy must be from alternative sources by 2020. Also, 10 percent of the transportation fuels used by EU members must be sustainable biofuels.

Key Takeaways

- The Gulf Stream provides a moderate type C climate for much of Western Europe. Eastern Europe can experience colder type D climates.
- Europe has four main physical landforms that provide a diversity of natural resources. The North European Lowland holds the majority of its agricultural potential.
- An increase in population has also increased the demand on the environment. Various environmental concerns are becoming more evident.

Acid rain from industrialization has caused extensive damage to forests and fish populations in northern Europe. Atlantic fisheries are also experiencing a decline in production.

Discussion and Study Questions

1. Identify and locate the four main physical landforms of Europe. What are the main features of each?
2. How do these landforms and natural resources provide wealth, opportunities and advantages to the European community?
3. Why is there a higher concentration of acid rain in northern Europe? How is acid rain affecting the environment?
4. What two rivers act to naturally divide Europe? Which rivers flow through Paris and London?
5. Track the flow of each of the two main rivers and identify the major cities that each flows through.
6. Where are the major mountain ranges of the Alpine region?
7. How does the Alpine region contribute to the economies of the countries of Europe?
8. Why do the countries of Western Europe have more moderate type C climates even though they are in the northern latitudes?
9. How does agricultural production vary with physical landforms or climate changes?
10. How are Europeans addressing the decline in the North Atlantic fishing industry?

Geography Exercise

Identify the following key places on a map:

- Alpine region
- Alps
- Apennines
- Black Forest
- Black Sea
- Carpathians
- Central Uplands
- Danube River
- Dinaric Alps
- Iberian Peninsula
- Mediterranean Sea
- North Sea
- Northern Lowlands
- Pyrenees
- Rhine River
- Scottish Highlands
- Strait of Gibraltar
- Ural Mountains
- Western Highlands

2.2. PATTERNS OF HISTORICAL DEVELOPMENT

Europe did not attain its status as a global economic hub with elevated living standards by mere chance. Historical occurrences in global development have benefited this domain due to its geographical and cultural attributes. In Southern Europe, the Greeks contributed concepts, philosophy, and organizational structures. Greek philosophers advocated for the principle of democracy. The Romans elevated the notion of empire to unprecedented heights. From approximately 150 BCE to 475 CE, the Romans amalgamated numerous concepts and dominated a significant expanse of Europe and North Africa. The Roman Empire established a unified infrastructure across Europe. The Romans unified their realm through the construction of roads, bridges, aqueducts, and port infrastructure. They comprehended the principles of governing an empire. By capitalizing on the optimal prospects inside each region they governed, they motivated the most proficient artisans to concentrate on their areas of expertise. This resulted in the specialization of goods and the emergence of a market economy. Individuals were no longer required to produce things independently. They could market their production and acquire superior goods manufactured by others, which would exceed the quality of their home-produced items. Regions that specialized in specific items due to local resources or specialized skills could convey those goods over vast distances to markets. The Roman Empire linked southern Europe with North Africa.

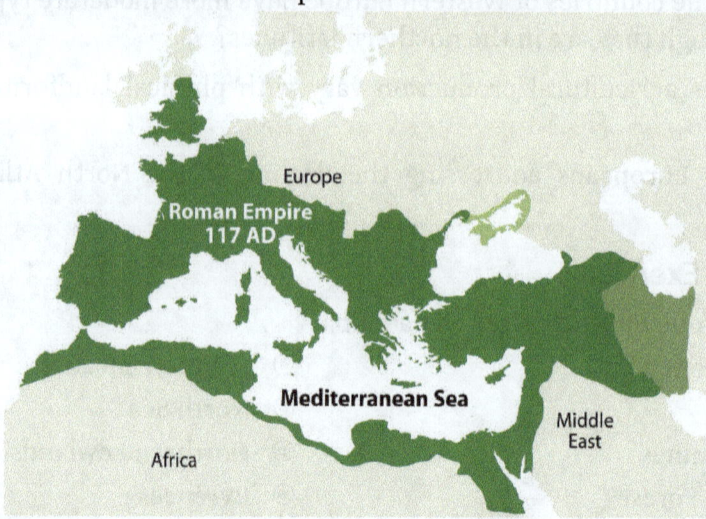

Figure 2.6: Extent of the Roman Empire, 117 CE. Wikimedia Commons – public domain.

The Vikings of Scandinavia (Norway, Sweden, and Denmark; 900–1200 CE) are often inaccurately referred to as rogue bands of armed warriors who pillaged and plundered northern Europe. Though they were fierce warriors in battle, they were actually farmers, skilled craftsmen, and active traders. They developed trade routes throughout the north. Using their seafaring knowledge and skills, the Vikings used Europe's waterways for transportation. They were the early developers of the northern world from Russia to Iceland and even to North America. They developed colonies in Iceland, Greenland, and what is present-day Canada. Their longships were renowned for versatility and provided an advantage on the sea. The Vikings made advances deep into Europe—all the way to **Constantinople**. History indicates that the Byzantine Empire employed Scandinavian Vikings as mercenaries.

In a general sense, the regions to the west of Scandinavia—such as Iceland, Scotland, Ireland, and Greenland—were targeted by the Norwegian Vikings. Southern regions such as England and France were more often destinations of the Danish Vikings. Russia and areas of Eastern Europe were standard trading grounds for the Swedish Vikings, though the different Viking groups could occasionally be found in the same destinations. The present-day Scandinavian countries were established after the main Viking period. The Vikings connected northern Europe with trade during the Dark Ages.

Colonialism

It wasn't until after the Dark Ages of Europe ended that a rebirth of ideas, technology, and progress took hold. The Renaissance of the late fifteenth century prompted activity in Europe that changed the world. In 1492, Columbus and his three ships crossed the Atlantic to land on the shores of the Americas. This event symbolized the beginning of the era of European colonialism, which only diminished after World War II.

Colonialism's effects remain in the colonies or protectorates that European countries still possess.

Colonialism was fueled by the economic concept of mercantilism that included the drive of governments to control trade, promoting the acquisition of wealth by the quick gain of gold or silver from their colonies.

Colonialism included the development of colonies outside the home country, usually for the expansion of imperial power and the exploitation of material gain. The building of larger ships and an understanding of sea travel allowed an exchange of new goods and ideas between continents. North and South America were opened up to the European explorers for colonial expansion. European colonialism brought newfound wealth from the colonies back to Europe. All the regions of the world outside Europe were targeted for colonialism. Africa was divided up, "Latin" America was created, and Asia became a target for resources and trade. The few powerful countries along the Atlantic coast of Europe began the drive to dominate their world. If you live in the Western Hemisphere, consider the language you speak and the borders of your country: both were most likely products of

European colonialism. Most of the current political geographic boundaries were drawn up or shaped through colonial conflict or agreement.

The Agrarian Revolution

The post-Renaissance era introduced a number of agricultural changes that impacted European food production. Before this time, most agricultural methods were primitive and labor intensive, but new technologies were introduced that greatly enhanced agricultural production. Plows, seeders, and harvesting technologies were introduced, and land reform and land ownership transitioned to adapt to the changing times. These innovations supported the expanding port cities that created urban markets for agricultural surpluses. Colonial ships returned from the colonies with new crops such as the potato that revolutionized crop production. This era's progress in agricultural advancements is often referred to as the agrarian revolution. The agrarian revolution led to the industrial developments such as the steam tractor and steel implements that further advanced agricultural production worldwide.

The Industrial Revolution

Great Britain, being an island country, developed the world's largest navy and took control of the seas. Their colonial reach extended from what is now Canada to Australia. The Industrial Revolution, initiated in northern England in the late 1700s, introduced an industrial period that changed how humans produced products. The shift to coal for energy, the use of the steam engine for power, the smelting of iron, and the concept of mass production changed how goods were produced.

Figure 2.7: Mass production during the Industrial Revolution prompted a rural-to-urban shift in the population. Coal for energy and steam for power energized industrial activities.

The development of the steam-powered engine provided a mobile power source. Waterwheels powered by steep-flowing rivers or streams were an early source of power. With coal for fuel and steam for power, the engines of industry were mobile and moved full speed ahead. Power looms converted textiles such as cotton and wool into cloth. Powered by a steam engine, a power loom could operate twenty-four hours a day and could be located anywhere. Industrialization with cheap labor and adequate raw materials brought enormous wealth to the industrial leaders and their home countries.

With the mass production of goods and advancements in technology, there was a major shift in human labor. Fewer people were needed on the farms, and more workers were needed in factories. There was a large rural-to-urban shift in the human population. Europe experienced the development of the major cities of its realm during this period. In Britain, for example, in 1800 only 9 percent of the population lived in urban areas. By 1900, some 62 percent were urban dwellers. As of 2010, it is more than 90 percent. Europe, as a whole, is about 75 percent urban. As a comparison, the US population is about 80 percent urban.

As discussed in earlier sections, the rural-to-urban shift that began with the Industrial Revolution in Europe continues today in developing countries. The Industrial Revolution, which started in northern England's Pennine mountain chain, rippled through Europe and across the Atlantic to the United States. The majority of countries in Europe are currently in stage 5 of the index of economic development. The five stages of the index of economic development illustrate a pattern of development and population dynamics for a country or region. The model outlines how rural societies with an agrarian economy in stage 1 can make the transition to stage 5: the stage that indicates an urban society with a consumer economy. As a general trend, when a country's levels of industrial activity and urban growth rise, the outcome is usually a higher standard of living and smaller family sizes. Additionally, rural-to-urban shift takes place, driven by the pull of opportunities and advantages in the industrializing and urbanizing areas. Countries in stage 5 of the index have small families with a fertility rate below the replacement level. Their incomes, based on a consumer economy, are generally at high levels.

As Europe industrialized and progressed through the stages of the index of economic development, certain core regions reached the postindustrial stages earlier than others. Western Europe established a core industrial region with an extended periphery. The postindustrial activity in this core area continues today in four main centers of innovation: (1) Stuttgart in southern Germany, (2) Lyon in southeastern France, (3) Milan in northern Italy, and (4) Barcelona in northeastern Spain. These four industrial centers have been referred to by some as the **Four Motors of Europe** because they promote business and industry for the European community. The European core region extends as far as Stockholm in the north to Barcelona in the south.

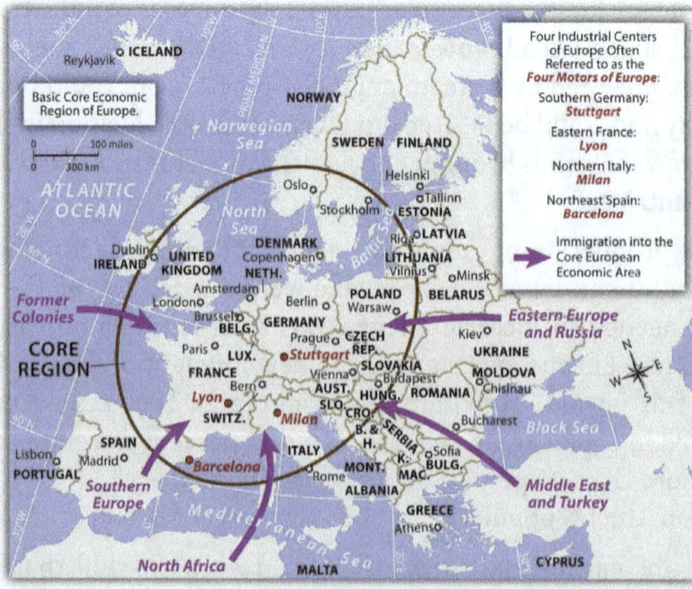

Figure 2.8: The Core Region of Europe and the Four Centers of Industrial Activity

Attractive "pull" forces draw immigrants to the core economic area of Europe, seeking opportunity and advantage. "Push" forces cause people to leave an area due to negative cultural or environmental forces, and/or the lack of opportunity and advantage.

Central Business Districts and Primate Cities

European urban development centered on port cities that had industrial activity. Ships could import raw materials, the factories could manufacture the goods, and the ships could export the products. A central business district (CBD) developed around these activities. Since walking was the main transportation mode, all business activities had to be located in the same vicinity. Banks, retail shops, food markets, and residential dwellings had to be close to the factories and port facilities. Modern cities emerged from this industrialization process, and Europe is one of the more urban realms on the planet.

Many European countries have one main primary city that is more than twice as large as their second city. The term primate city indicates a city that is more than twice as large as their second city and exceptionally expressive of national feeling and heritage. One may instinctively think that the primate city is the same as the capital city of a country; this, however, is not always the case. For example, Washington, DC, is the US capital, but New York City is the country's primate city. Primate cities represent a country's persona and often are symbolic of the country's heritage and character. Though common, not all countries of the world have a primate city. Financial and business centers in cities such as London, Rome, or Paris support the industrial activity that led to their development as primate cities. Most primate cities are ports or are located on a major river.

Rural-to-Urban Shift and Population Growth

Consider the trend that the Industrial Revolution brought to Europe. Do you remember the population growth principle? This principle states that as countries industrialize and urbanize, family size naturally goes down and incomes traditionally go up. Integrate this with the rural-to-urban shift that occurs when countries progress through the five stages of the index of economic development. By understanding these basic trends, one can determine the average family size in Europe and why it is declining.

Because Europe is an urbanized realm, one can understand why family size in Europe is small. As a matter of fact, various countries in Europe have negative population growth rates. Family size in Spain and Italy is around 1.2, with the average family size in all of Europe at 1.4 children. The replacement rate to maintain an even population-growth pattern would be a fertility rate of about 2.1 children. Small families do not provide enough young people to cover the available entry-level service jobs. Europe, the United States, and Japan are all in stage 5 of the index of economic development and are facing low or negative population growth and a deficit in their cheap labor supply. As a core economic global power, Europe has experienced an increase in immigration. With a lower fertility rate and an increase in postindustrial activity, Europe is a magnet for people from poorer peripheral countries and even peripheral regions within Europe who are looking for opportunities and advantages.

With the planet's human population increasing overall, one might think that a smaller family size is a positive trend. It may be, but there are problems along the way. If there are fewer young people in a community—fewer children and fewer people of employment age—consider how this affects the economic situation. With a declining European population, who will apply for the entry-level jobs? Economic core areas attract immigrants seeking opportunities and advantages. Europe follows this pattern. There has been increasing tension between the long-standing European cultural groups and immigrants from developing countries who often speak non-European languages or follow religions other than Christianity. The main religion of immigrants from North Africa or the Middle East is Islam, which is the fastest-growing religion in Europe.

Declining Family Size in Europe

There are six main reasons for declining family size in Europe:

1. The high cost of living makes supporting a large family expensive.

2. Confined living spaces in urban areas are expensive for larger families.
3. Young people are naturally putting off marriage and family until they are older.
4. Family planning and health care for women is available.
5. More people are earning an education and putting off having children.
6. More women choose professional careers and have fewer children.

The following problems arise with the decline in family size in Europe:

- Fewer children results in a shrinking workforce for entry-level jobs.
- The workforce is getting older and more professionals are retiring.
- Pension costs are increasing and there are fewer younger workers to pay for them.
- The tax burden is shifting to fewer younger professionals in the workplace.
- More revenues are needed to pay for the health care of a larger population of senior citizens.
- Immigrants seeking employment may not share the same cultural heritage or values.

Nation-States and Devolution

The agrarian revolution and the Industrial Revolution were powerful movements that altered human activity in many ways. New innovations in food production and the manufacturing of products transformed Europe, which in turn impacted the rest of the world. Even before the agrarian revolution was under way, other transitions in European political currents were undermining the established empire mentality fueled by warfare and territorial disputes. The political revolution that transformed Europe was a result of diverse actions that focused on ending continual warfare for the control of territory and introducing peaceful agreements that recognized sovereignty of territory ruled by representative government structures. Various treaties and revolutions continued to shift the power from dictators and monarchs to the general populace. The Treaty of Westphalia in 1648 and those that followed helped establish a sense of peace and stability for Central Europe, which had been dominated by the Holy Roman Empire and competing powers. The Holy Roman Empire, which was centered on the German states of Cental Europe from 962–1806, should not be confused with the Roman Empire, which was based in Rome and ended centuries earlier. The French Revolution (1789–95) was an example of the political transformation taking place across Europe to establish democratic processes for governance.

The political revolution laid the groundwork for a sense of nationalism that transformed Europe into nation-states. The term *nation* refers to a homogeneous group of people with a common heritage, language, religion, or political ambition. The term *state* refers to the government; for example, the United States has a State Department with a secretary of state. When nations and states come together, there is a true nation-state, wherein most citizens share a common heritage and a united government.

European countries have progressed to the point where the concept of forming or remaining a nation-state is a driving force in many political sectors. To state it plainly, most Europeans, and to an extent every human, want to be a member of a nation-state, where everyone is alike and shares the same culture, heritage, and government. The result of the drive for nation-states in Europe is an Italy for Italians, a united Germany for Germans, and a France for the French, for example. The truth is that this ideal goal is difficult to come by. Though the political borders of many European countries resemble nation-states, there is too much diversity within the nations to consider the ideal of creating a nation-state a true reality.

Various ethnic populations in Europe desire their own nation-states within their countries of residence. They want to devolve or separate from the larger state. The term devolution refers to the process whereby regions or people within a state demand independence and autonomy at the central government's expense. There are now a number of cases where devolution is occurring in Europe. For example, Scotland and Wales seek to devolve from the United Kingdom. The Basque region between Spain and France would like to have its own nation-state. Former Yugoslavia broke up into seven smaller nation-states. Various other minority groups in Europe seek similar arrangements. Thus both cohesive cultural forces and divisive cultural forces are active in the European community.

Table 2.9

Centrifugal Forces Divide a State	Centripetal Forces Hold a State Together
• Ethnic Conflict	• Ethnic Unity and Tolerance
• Social Injustice	• Social/Economic Equity
• Poverty	• Just and Fair Legal System
• Dictatorial Leadership	• Charismatic Leadership
• Religious Intolerance	• Religious Acceptance
• Nationalism	• Nationalism
• Loss of Rights	• Common Heritage
• Any Other Divisive Force	• Common Language
	• Any Other Unifying Force

Centrifugal and Centripetal Forces

Cultural forces continually apply pressure on a country. Some of these cultural forces pull the nation together (centripetal forces) and others pull it apart (centrifugal forces). Primary sources of these cultural forces include religion, language, ethnicity, politics, and economic conditions.

Centrifugal forces divide a state and centripetal forces hold a state together. When there is division, conflict, or confrontation, the centrifugal forces are at play. When unification, agreement, or nationalism are being exercised, centripetal forces are evident. The sources that tie a country together can also be the sources that divide a country. Ethnic unity can be a positive force, while ethnic division and conflict can be a divisive force. If centrifugal forces become strong, the result may be outright civil war, as was seen in the United States in the 1860s. Unity can also be evident through national struggles, such as the nationalism displayed immediately after the 9/11 terrorist attacks in the United States. After the attacks, an outpouring of goodwill and agreement strengthened the bonds within the United States.

To understand our world, it is helpful to understand the cultural forces that are active in any one location. Disagreement, inequity, or injustice related to the cultural factors of ethnicity, religion, language, and economics of a region or country is the cause of most conflicts. The strong personalities that provide leadership can similarly prove to be powerful forces that either divide or unite a nation. An example of cultural forces in conflict in Europe can be found in Northern Ireland, where political forces use religious differences as a means of social division.

Religion and Language in Europe

Europe has historically been considered a Christian realm. The three main branches of Christianity in Europe are Roman Catholic, Protestant, and Eastern Orthodox. Rome has been the geographical base for the Roman Catholic Church since the Roman Empire. Operating on the Romance language, Latin, the Catholic Church has provided southern Europe with a common religion for over 1,500 years.

The Roman Catholic Church split when Constantinople, now called Istanbul, gained preeminence. The Eastern Orthodox Church launched itself as the primary organization in the Slavic lands of Eastern Europe and Russia. The reformation of the fourteenth century, led by people such as Martin Luther, brought about the Protestant Reformation and a break with the Roman Catholic Church. Protestant churches have dominated northern Europe to this day.

Three main Indo-European language groups dominate Europe. Though there are additional language groups, the dominant three coincide with the three main religious divisions. In the east, where the Eastern Orthodox Church is dominant, the Slavic language group prevails. In the north, along with Protestant Christianity, one finds the Germanic language group. In southern Europe, where Roman Catholicism is dominant, the Romance languages are more commonly spoken.

Table 2.1 Three Main Language Groups

Germanic Language Group (*Found Mainly in Northern Europe)*	Romance Language Group *(Latin Based; Found Mainly in Southern Europe)*	Slavic Language Group *(Found Mainly in Eastern Europe)*
English, Dutch, German	French, Italian, Spanish	Polish, Russian, Czech
Danish, Norwegian, Swedish	Portuguese, Romanian, Walloon	Ukrainian, Slovak, Slovene
Icelandic, Faroese, Frysian	Romancsh, Catalan, Provencal	Belarussian, Serbo-Croatian, Lusatian
		Macedonian, Bulgarian

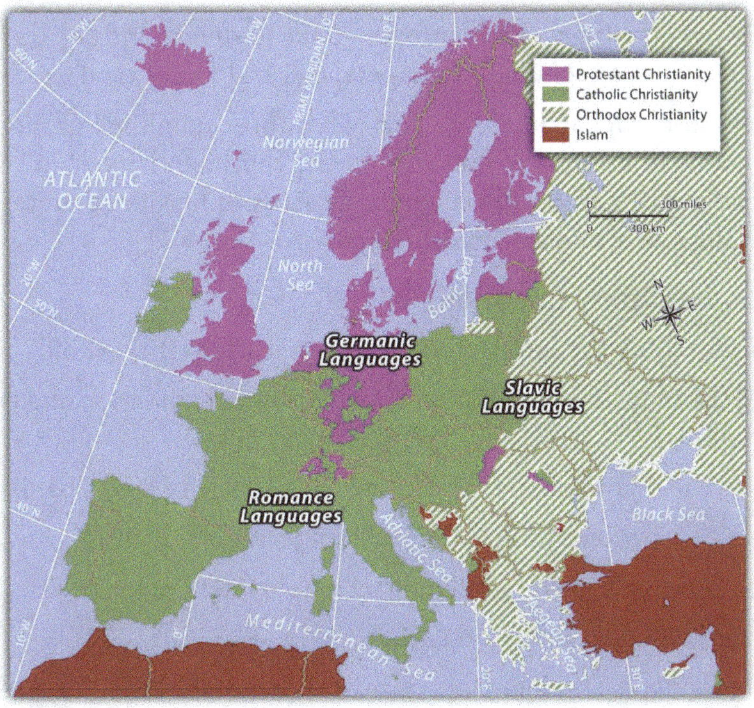

Figure 2.10 Three Dominant Language Groups and Three Dominant Christian Denominations of Europe

Key Takeaways

- The Roman Empire connected southern Europe and created an infrastructure to help promote trade and intercultural connections. The Vikings connected northern Europe through trade and exploitation.
- Technological advancements helped European colonialism dominate other countries and exploit their labor and resources. Coastal European countries created colonies and external sources of wealth.

- The Industrial Revolution was promoted by the development of steam power with coal as a fuel source. The mass production of goods gave the European countries an advantage in the world marketplace.
- The Industrial Revolution prompted a shift in population from the rural agricultural regions to the urban centers. More people were needed in the factories and fewer workers were required on the farms because of improved agricultural methods. This shift resulted in smaller families and more women entering the workplace.
- The early empires of Europe gave way to the concept of a similar people (the nation) unifying under a common government (the state) to create nation-states.
- Divisive centrifugal cultural forces tend to divide and separate people in a state, whereas cohesive centripetal cultural forces tend to unify a state.
- The Indo-European language family has three dominant groups in Europe: Germanic in the north, Romance in the south, and Slavic in the east. The Christian religion has three main divisions in Europe: a Protestant north, Catholic south, and an Orthodox east.

Discussion and Study Questions

1. Why do you think the Industrial Revolution started where it did?
2. How did the Industrial Revolution affect Europe's population and employment opportunities?
3. What was the relationship between industrial activity and urbanization?
4. What is a CBD? Where would you find it in your closest city?
5. How does Europe's fertility rate impact employment opportunities?
6. Why is family size declining in Europe?
7. What are the three dominant language branches of the Indo-European language family in Europe?
8. How are the three main branches of Christianity distributed in Europe?
9. What are some examples of centrifugal and centripetal cultural force where you live?
10. Why do people gravitate toward a nation-state structure as a political unit?

Geography Exercise

Identify the following key places on a map:

- Barcelona, Spain
- Constantinople
- Denmark
- Lyon, France
- Milan, Italy
- Norway
- Stuttgart, Germany
- Sweden

2.3 REGIONS OF WESTERN EUROPE

Europe has historically been categorized into areas based on geographical location according to the four cardinal directions: Eastern Europe, Southern Europe, Western Europe, and Northern Europe. The British Isles are frequently regarded as a distinct region, however they may also be classified as part of Western Europe. These regions are solely geographical areas that may exhibit analogous physical geography or cultural characteristics influenced by physical geographical requirements. Europe is historically categorized into two regions: Western Europe and Eastern Europe. The Western Europe region encompasses Northern Europe, Southern Europe, Central Europe, and the British Isles. Eastern Europe encompasses all territories situated east of Germany, Austria, and Italy. These geographic labels serve mostly for identification rather than for regional analysis. They function as conventional indicators of geographic position.

Figure 2.16: Traditional Regions of Europe

The historical significance of Europe's traditional regions has diminished with the establishment of the European Union (EU). Economic and political linkages are now more interconnected than in previous eras when nation-states and empires held greater significance. Economic realities have frequently overshadowed cultural factors and have heightened the necessity for more integration. Cultural influences have historically bolstered nationalistic movements aimed at safeguarding the culture, legacy, and customs of a populace. Regional cultural distinctions constitute the social fabric of local communities, facilitating the preservation of their identity. Contemporary transportation and communication technologies have consolidated the diverse European identities into a singular realm of global acknowledgment.

Northern Europe

Europe has many different cultural identities within its continent. Northern Europe has traditionally included Iceland, Finland, and the three Scandinavian countries of Norway, Sweden, and Denmark. These countries are often referred to as the Nordic countries. All these countries were influenced by Viking heritage and expansion. Their capital cities are also major ports, and the largest cities of each country are their primate cities. The languages of the three Scandinavian countries are from the Germanic language group and are mutually intelligible. Finnish is not an Indo-European language but is instead from the Uralic language family. Most of Iceland's inhabitants are descendents of Scandinavian Vikings. Protestant Christianity has prevailed in northern Europe since about 1000 CE. The Lutheran Church has traditionally been the state church until recent years. These countries were kingdoms, and their royal families remain highly regarded members of society. The colder northern climate has helped shape the cultural activities and the winter sports that are part of the region's heritage. Peripheral isolation from the rest of Europe because of their northern location and dividing bodies of water have allowed the northern culture to be preserved for centuries and shape the societies that now exist in northern Europe.

Figure 2.17: Port in Stockholm, Sweden. Wikimedia Commons – public domain.

Human rights, education, and social concerns are high priorities of the governments of northern Europe, and the quality of these elements rank highly by global comparisons. Standards of living are among the highest in Europe. Isolation in northern Europe does create an element of economic cost, and transportation technology has been leveraged to address this. A modern bridge has been constructed across the **Baltic Sea** from Denmark to Sweden to increase the flow of people, goods, and materials between the Scandinavian Peninsula and mainland Europe. Iceland is the most remote of the Nordic countries. Its small population—less than a half million people—is connected to Europe by sea and air transportation and communication technologies. Almost all elementary school children in the Nordic countries are taught English as a second language. Fish, meat, and potatoes are traditional dietary staples; fish in particular has been an important food source. The cuisine of the region is not noted for being spicy. Northern Europe has worked diligently to integrate itself with the global community and yet maintain its cultural identity.

As a standard practice, the northern European countries have exploited the opportunities and advantages of their natural resources to expand their economies. Sweden, northern Europe's largest country, has used its natural iron ore supply to develop its manufacturing sector. Sweden was the production base of Saab and Volvo vehicles as well as other high-tech products; however, GM purchased the Saab auto division in 2000 and some of its automobiles were manufactured in Mexico. In 2010, Saab was sold back to European investors and production resumed in Sweden. Volvo Car Corporation was purchased by Ford Motor Corporation in 1999 and then acquired by a Chinese automaker in 2010.

Finland has vast timber resources and is one of Europe's major sources of processed lumber. It was the original manufacturer of Nokia cell phones, an example of its technological advancements. Nokia is the largest manufacturer of mobile phones in the world and has production facilities in eight different countries.

Norway has been benefiting from the enormous oil and natural gas reserves under the North Sea. Fishing and modest agricultural activities had been Norway's traditional means of gaining wealth, but now it is the export of the much-in-demand energy resources. Because of its economic and energy independence, Norway has opted not to join the EU.

Vikings were masters of the seas and colonized **Greenland**, which is located next to Canada and is considered to be the world's largest island. Danish colonization in the eighteenth century included Greenland and the **Faeroe Islands**, which are located between Scotland and Iceland. Both are now under the government of Denmark but retain a high level of self-rule and autonomy, which has aided them in holding on to their cultural identity. Greenland has also opted not to become a part of the

EU even though Denmark is a member. Greenland only had a population of about 57,670 in 2011, and 80 percent of its surface is covered with ice. Fish is Greenland's main export, but minerals, diamonds, and gold are also present in viable amounts.

Denmark has a consumer economy with a high standard of living. This Scandinavian country is often ranked as the least corrupt country in the world and has the happiest people. The country has supported a positive environment and implemented strong measures to protect its natural areas. Denmark's main exports are food products and energy. The country has sizeable oil resources in the North Sea but also receives over 15 percent of its electricity from wind turbines.

The Baltic states of Lithuania, Latvia, and Estonia have often been included in the northern European designation because of their northern location. Estonia has the strongest similarities in religion, traditions, and culture, and geographic literature often has included it as a part of northern Europe. The Baltic states have been associated with Eastern Europe through the Soviet era but, like their neighbors to the north, are becoming more economically integrated with Western Europe.

Northern Europe is a peripheral region. Southern Sweden has an advanced industrial base and resembles a core area. Sweden's northern portion and the main parts of the other Nordic countries act as sources of raw materials for Europe's urban core industrial region. In the core-peripheral spatial relationship, northern Europe most resembles a semiperipheral region that has attributes of both the urban core and the rural periphery.

Norway, Sweden, and Finland are quite large in area but are not densely populated compared with other European nations. Sweden ranks as the fourth-largest European country in physical area. Sweden is larger than the US state of California, but in the 2010 census, it had less than ten million people. In 2010, Finland, Norway, and Denmark all had less than six million people each.

The cultures and societies of northern Europe have progressed along similar paths; that is, they have advanced from once Viking-dominated lands into modern democratic and socially mature nations. Northern Europe is known for its concern for the social welfare of its citizens. Their strong egalitarian ideals have contributed to extensive advancements in free medical care, free education, and free social services for all, regardless of nationality or minority status. Civil rights for minorities, women, and other groups is assured and protected. Denmark doesn't have a legal age for consumption of alcoholic beverages, though tradition sets the age at about fourteen. Culture and the arts are well developed; examples include everything from the Nobel Peace Prize to Hans Christian Anderson to the 1970s chart-topping pop group ABBA. Sweden has become a major exporter of music worldwide. Rock, hip-hop, and pop music are common genres. With English as a widely spoken language in Scandinavia, music and cultural trends have a larger export market in places such as the United States and Britain.

Southern Europe

Southern Europe includes three large peninsulas that extend into the Mediterranean Sea and the Atlantic Ocean. The Iberian Peninsula consists of Spain and Portugal. The Pyrenees mountain range separates the Iberian Peninsula from France. Greece, the most southern country on the **Balkan Peninsula**, includes hundreds of surrounding islands and the large island of Crete. The Italian Peninsula is the shape of a boot with the Apennine Mountains running down its center. Italy also includes the islands of Sicily and Sardinia. Technically, the island country of Cyprus is also included in southern Europe. There are five ministates in this region. The small island of Malta is located to the south of Sicily and is an independent country. Monaco, San Marino, Andorra, and Vatican City are also independent states located within the region. Southern Europe's type C climate, moderated by the water that surrounds it, is often referred to as a Mediterranean climate, which has mild, wet winters and hot, dry summers.

Rural-to-urban shift in southern Europe has not been as strong as that of Western Europe. Only about 50 percent of the people in Portugal are urban; in Spain and Greece, about 60 percent are urban.

Italy is more representative of Europe, with about 68 percent of the population urbanized. Italy is also divided, with northern Italy being more industrialized than southern Italy. The southern regions of Italy, including the island of Sicily, are more rural with fewer industries. Northern Italy has the metropolitan city of Milan as an anchor for its global industrial and financial sector in the Lombardy region, which includes the city of Turin and the port of Genoa. This northern region of Italy has the economic muscle to be one of Europe's leading manufacturing centers. The so-called Ancona Line can be drawn across the middle of Italy from Ancona on the east coast to Rome on the west coast to separate the industrial north with the more agrarian south. The north also has the noted cultural cities of Venice, Florence, and Pisa.

A similar situation exists in Spain. The urbanized Catalonia region around Barcelona in the northeast has high-tech industries and a high standard of living. Southern Spain has large rural areas with economies heavily based on agricultural production. Portugal and Greece are not as industrialized and do not have the same economic opportunities. Historically, southern Europe, Portugal, and Greece in particular each have had a much lower gross domestic product (GDP) per capita than northern or Central Europe. Their economies have been much more aligned with the economic periphery than with the industrial core region of Europe. Greece has had serious economic difficulty in the past few years.

Southern European countries have much larger populations than their northern European counterparts. Italy has about sixty million people in an area smaller than Norway, which has less than five million. Spain has about forty million; both Portugal and Greece have more than ten million. Cultural factors are also different here than in northern Europe. The culture of southern Europe has been built around agriculture. Traditional cuisine is based on locally grown fresh food and wine. Olive oil and wine have been major agricultural exports. The main languages of Iberia and Italy are based on the Romance language group, and Greek is an independent branch of the Indo-European language family. The most dominant religious affiliation in the south is Roman Catholicism, except in Greece, where the Eastern (Greek) Orthodox Church is prominent.

Figure 2.18: Spain's Autonomous Communities

Because of its diversity, Spain is not categorized as a nation-state.

Spain is the most diverse nation in southern Europe with a number of distinct ethnic groups. The Basques in the north along the French border would like to separate and create their own nation-state. The region of Galicia in northwest Spain is an autonomous region and was once a kingdom unto itself. There are many other autonomous communities in Spain, each with its own distinct heritage and culture. Farther east in the Mediterranean is the island state of Cyprus, which is divided by Greek and Turkish ethnic groups. The southern part of the island is dominated by Greek heritage and culture, and the northern part of the island is dominated by Turkish culture and traditions. Islam is the main religion of the Turkish north. The people of southern Europe are diverse and hold to many different traditions but are tied together by the sea and the land, which create similar lifestyles and economic activities.

Special Section: The Cities of Italy has a population of about sixty million people, with about 68 percent living in cities. Italian culture balances the food and lifestyle of the rural countryside with the history and heritage of the cities. Cultural regions such as Tuscany, along the northwest coast, or Sicily, the island to the south, evoke thoughts of Italian heritage and culture. The large cities of Italy each have their own draw and persona that emulates the nation's heritage and history. Italy has four cities with population estimates of over one million people each, and a number of others are major cultural centers. The metropolitan areas around these cities can be extensive.

Table 2.2 Major Italian Cities

Rank by Population	Population Estimate
1. Rome (capital)	3.357 million
2. Milan	2.962 million
3. Naples	2.27 million
4. Turin (Torino)	1.662 million
5. Palermo	872,000
8. Florence	381,762
13. Venice	297,743
Population is for city only and not for metropolitan area.	

Sources: CIA Factbook: https://www.cia.gov/library/publications/the-world-factbook/geos/it.html

Rome, Italy's capital and its largest city, was at the center of the Roman Empire and has a significant historical connection to the culture and heritage of Italy. Rome is also home to Vatican City, which holds a separate political identity. The Vatican City is home to the pope, who is the head of the Roman Catholic Church. Rome provides a historical and political center for the nation of Italy and is located along the Ancona Line, which distinguishes northern Italy from southern Italy.

The northern city of Milan is a core industrial center for the nation and for Europe. With a large metropolitan area, the city of Milan has a long history as a center of industry and power. The city continues to hold its powerful status and has expanded its international reach into the world of high-end fashion and the arts.

Naples is located along the west coast to the south of Rome. Founded by the Greeks as a colony, Naples was transferred to the Roman Empire. This historic city is home to many artistic, musical, and cultural activities that emulate the heritage of the Italian people from the time of the Renaissance.

Figure 2.19: The central part of the city of Naples, Italy, is a UNESCO World Heritage Site. Italy is densely populated with many historical cities emulating the culture and traditions of each part of the country.

Wikimedia Commons – public domain.

The industrial city of Turin (Torino) is located in northwest Italy in the Alpine region on the banks of the Po River. Turin is second only to Milan in its industrial capacity within Italy. Automobile manufacturing as well as prestigious universities, art galleries, and cultural centers can be found here. The hosting of the 2006 Winter Olympics gave witness to Turin's capacity and success.

Palermo is located on the island of Sicily. Founded by the Phoenicians, this city became the capital of the kingdom of Sicily. Palermo is the center of culture and cuisine for Sicily and has been a destination for tourists from around the world. Italy's mild Mediterranean climate has enhanced tourism opportunities for many Italian cities such as Palermo.

Florence and Venice each have their own unique characteristics. Traditionally, Florence has been a city of commerce, trade, and the arts. Located in the Tuscany region, it is often referred to as the birthplace of the Renaissance. Florence is a UNESCO World Heritage Site and is noted for its art and architecture. Venice is located on the northeast coast of Italy at the edge of the Adriatic Sea. The city was a powerful trade and commerce center in the Middle Ages and during the Renaissance. Often referred to with romantic overtones, the city is built on 117 small islands using canals as streets. The uniqueness of the landscape, arts, food, and architecture make this city one of the major tourist destinations in Italy.

Central Europe

In the regional sense of location, when discussing the political geography of the European continent, the specific states located in the western part of the European mainland are often referred to as Central Europe. The term *Western Europe* is also applicable to this region when discussing regional variations within the realm outside of Eastern Europe. Germany and France are the two dominant states, with Belgium, the Netherlands, and Luxembourg making up the Benelux countries. Switzerland and Austria border the Alpine region. The ministate of Liechtenstein is located on the border between Switzerland and Austria. France is the only country with coastlines on both the Atlantic Ocean and the Mediterranean. These countries are located in the core economic region of Europe and have stable democratic governments and a relatively high standard of living by world comparisons.

Figure 2.20: Barges transport goods along the beautiful Rhine River in Germany, where historic towns share the landscape.

DaiLuo - Rail to Heaven - CC BY 2.0.

Central Europe is a powerhouse of global economics. The Rhine River is a pathway for industrial activity from southern Germany to Europe's busiest port of Rotterdam in the Netherlands. Western France has the political capital of the EU along the Rhine at Strasbourg. To the south is France's second-largest city, Lyon, which is a major industrial center for modern technology. Germany had the historical Ruhr industrial complex along the Rhine that supported the high-tech industries in southern Germany in the cities of Stuttgart, Mannheim, and Munich. Germany is the most populous country in Europe, with over eighty-two million people in 2010. Germany is also Europe's largest economy and has the largest GDP overall as a country. Belgium has major business centers in Brussels and Antwerp. Switzerland is noted for its banking and financial markets. Luxembourg has one of the highest GDP per capita in all of Europe. Austria is noted for its high level of cultural activities in Vienna and Salzburg. All these countries complement each other in creating one of the dominant economic core areas in the world.

In the first half of the twentieth century, the political geography of Central Europe was not conducive to the high level of economic cooperation that now exists. In World War Iand World War II, Germany and France were on opposite sides, and the Benelux countries were caught in middle. The cultural differences between the Germans and the French start with the differences in language and religious affiliation. Germany was divided after World War II into East Germany and West Germany, separated by the so-called Iron Curtain. East Germany was under a Communist government, and West Germany was a capitalist democracy. They were reunited in 1990 when the Iron Curtain and the Berlin Wall came down. The two countries merged under one government. Europe is gradually being united economically, but each country or region still retains its cultural uniqueness.

The Benelux Countries

The Benelux countries have a great deal in common historically. Before the economic union that created the term *Benelux*, these countries were collectively referred to as the Low Countries, so called because of their relative position to sea level. The Benelux countries are some of the most densely populated countries. They have managed to work together toward a common economic objective in spite of their cultural differences.

The capital and largest city in Belgium is Brussels, with the other urban areas being the ports of Antwerp and Ghent. Belgium is split into three large geographic areas. The dominant language in the northern region of Flanders is Dutch (Flemish), and the people are known as Flemings. In the southern region of Wallonia, most people speak French and are known as Walloons. German is the third official language and is spoken along the eastern border.

When the Industrial Revolution diffused across the English Channel and arrived in Europe, Belgium was one of the early countries to adapt to the technological developments. Belgium remains heavily industrialized and is a major exporter of products, including finished diamonds, food products, nonferrous metals, technology, petroleum products, and plastics. In general, Belgium imports the raw materials to manufacture these goods for export. Belgium also has a significant services sector. The services sector, including real estate, hotels, restaurants, and entertainment, thrives in part because Brussels is the headquarters of the North Atlantic Treaty Organization (NATO) and components of the EU. Many countries and organizations maintain offices in Brussels to have easy access to these headquarters; therefore, Brussels is the temporary home to many diplomats and foreign business people.

Figure 2.21: Brussels is connected to other European cities through high-speed rail networks.

Wikimedia Commons – public domain.

The European country of the Netherlands also includes the colonies of the Netherlands Antilles in the Caribbean. The Netherlands, sometimes called *Holland*, is actually the name of two provinces (North Holland and South Holland) in the northwest part of the country. The largest city is Amsterdam. The Hague is the seat of government and is home to the United Nations International Court of Justice. Rotterdam is located at the mouth of the Rhine River and is one of the busiest ports on the continent. The country is famous for its Zuider Zee, which is the large inland region below sea level that has been drained of water and surrounded with an extensive dike protecting it from the North Sea. Reclaiming land from the sea in areas called polders has provided this densely populated country with more land area for its people to expand their activities.

For a small country with few natural resources, the Netherlands has an impressive GDP. The Dutch have made good use of their location on the North Sea and of the location of several large navigable rivers. This has facilitated voluminous exports to the inland parts of Europe. The major industries include food processing, chemicals, petroleum refining, and electrical machinery. The Netherlands is a top exporter of agricultural products, which contribute substantially to its economy. Dutch agricultural exports consist of fresh-cut plants, flowers, and bulbs as well as tomatoes, peppers, and cucumbers.

The main languages spoken in the small landlocked country of Luxemburg are French, German, and Luxembourgish. Luxembourg's one major city is Luxembourg City. Luxemburg has an enviable economic situation with a stable and prosperous economy, low unemployment, and low inflation. Thanks to rich iron-ore deposits, this country was able to develop a very robust steel industry, which was the cornerstone of the nation's prosperity until the 1970s. As steel declined, Luxemburg remade itself as an important world financial center. Luxembourg leads Europe as the center for private banking and insurance industries and is second only to the United States in terms of being an investment fund center.

France

France covers 211,209 square miles and is the second-largest European country; Ukraine is slightly larger in physical area. The physical landscapes of France vary widely from the northern low-lying coastal plains to the Alpine ranges of the east. Mont Blanc, the highest mountain in the Alpine range at 15,782 feet, is located in France near the Italian border. In the far south, the Pyrenees run along the border with Spain. The south-central region of the country is home to the Massif Central, which is a plateau and highland region made up of a large stretch of extinct volcanoes.

During the colonial era, France was a major naval power and held colonies around the world. The French Empire was the second largest at the time. The French language is still used for diplomacy in many countries. Though the French Empire no longer exists, France has progressed into a postindustrial country with one of the most developed economies in the world. It is a major player in European affairs, the EU, and the United Nations (UN). France is a democratic republic that boasts a high-quality public education system and long life expectancies.

Figure 2.22

The Eiffel Tower, a symbol of the primate city of Paris, was built in 1889. The city of Paris started as a citadel on an island in the middle of the Seine River.

U.S. Library of Congress – public domain.

In 2010, France's population was about sixty-five million, with about ten million living in France's primate city of Paris. The city of Paris is on an excellent site, it is favorably situated with regard to its surrounding area. It is a major core area of France serving a large peripheral region of the country. The next largest city of Lyon, which is a major high-tech industrial center for Europe's economy, can only boast a population of about 1.4 million. Even with a large population, the country is able to produce enough food for its domestic needs and for export profits. France leads Europe in agricultural production.

France enjoys a robust economy and is one of the world's leading industrial producers. Its industrial pursuits are diverse, including the manufacture of planes, trains, and automobiles, as well as textiles, telecommunications, food products, pharmaceuticals, construction and civil engineering, chemicals, and mechanical equipment and machine tools. Additionally, defense-related industries make up a significant sector of the economy. France's production of military weapons is recognized worldwide. The country has been a leader in the use of nuclear energy to produce electricity. Nuclear energy supplies about 80 percent of the country's electricity, which reduces the need for fossil fuels and imported oil.

Agriculture is an important sector of the French economy, as it has been for centuries, and is tied to industry through food processing. Food processing industries employ more people than any other part of the French manufacturing sector. If you think of cheese and wine when you think of France, you have identified two of its largest food processing endeavors, along with sugar beets, meats, and confectionaries. World-renowned wines are produced in abundance, sometimes in areas that bear their names, such as in Burgundy, around the city of Bordeaux, and in Champagne in the Loire Valley. French cuisine and fashion have long been held in highest esteem worldwide and are a source of national pride. Food and wine are important elements of the French way of life, and each region of France boasts a suite of famous dishes.

Thanks to the climate and favorable soil conditions, agriculture is highly productive and lucrative for France. France is second only to the United States in terms of agricultural exports. Exports mainly go to other EU countries, to the United States, and to some countries in Africa. The plains of northern France are excellent for wheat, which is grown on vast agricultural operations. Dairy products are a specialty in the western regions of France, which also produce pork, poultry, and apples. Beef cattle are raised in the central portion, where a cooler, wetter climate provides ample tracts of grasslands for grazing. Fruit, including wine grapes, is grown in the central and southern regions, as are vegetables. The region around the Mediterranean is blessed with hot, dry weather ideal for growing grapes and other fruits and vegetables.

Special Section: France's Home Front French governments have been challenged by the country's high unemployment rates. In response to decades of high unemployment, the French government has, over time, introduced legislation to try to distribute available work more evenly among the population. Long periods of unemployment and underemployment put France at risk of developing a permanent underclass, along with the social ills, including crime and political unrest, that come with it. The unemployment rate continues to be higher for women, and women tend to have lower-paying jobs. Additionally, women are underrepresented in government. France has a significant immigrant population, which means that, as in many countries with newcomers, the settlement and integration of new arrivals is an issue. Some have blamed the high youth crime rate not just on unemployment but on governmental failure to fully integrate immigrants into French society.

Another demographic matter is a strong rural-to-urban shift. As people move to the cities, rural areas are significantly depopulated. Though more than seventy languages are spoken in France due to its large immigrant population, French has been recognized as the official language by law. Eighty percent of French people identify themselves as Roman Catholics, though the percentage of practicing Catholics is much lower and declining. Approximately one-third of Europe's Jewish population resides in France. About 5 percent of the population self-identifies with Islam, which is a growing population because of immigrants from North Africa, some of whom are Muslim and tend to have larger families. Most of these immigrants target the urban areas in search of employment and opportunity.

In 1995, France experienced three terrorist attacks, which killed eight people and wounded more than a hundred. The attacks were, of course, troubling, but the real issue before France is how to ensure that it does not become a breeding ground for Islamic extremists bent on terrorist activity. The situation of unemployed, disenfranchised youth can provide the perfect setting from which to attract new terrorist recruits. After decades of dealing with this type of unrest, France passed laws in response to the 1995 Islamic terrorist bombings that some civil libertarians found unsettling. Many forms of surveillance are legal and routine, including forms of surveillance that violate private communications. France also passed a law prohibiting "association with wrong-doers involved in a terrorist enterprise." This gives French authorities the right to arrest, detain, try, and convict anyone suspected of any crime that might be even remotely related to terrorist activity. Balancing civil liberties with the quashing of terrorist plans has become a constant challenge for many countries, including France.

Germany

Germany's location in Central Europe has meant that throughout history many peoples—all with their own cultures, ideas, languages and traditions—have traversed Germany at one time or another. Thus Germany's culture has received many influences over the centuries.

Germany's present geopolitical configuration is quite young, as it reunified the eastern and western portions into a single entity in 1990. Germany was formed in 1871 during the leadership of Otto von Bismarck in an attempt to create a Germanic power base. World War I was fought during the last years of the German Empire. Germany, as part of the Central powers (Germany, Austria-Hungary, and Bulgaria), was defeated by the Allies with much loss of life. The German Republic was created in 1918 when, having been defeated in World War I, Germany was forced to sign the Treaty of Versailles.

In 1933, with an environment of poverty, disenfranchisement of the people, and great instability in the government, Germany gave way to the appointment of Adolf Hitler as chancellor of Germany. Within a mouth of taking office, Hitler suspended normal rights and freedoms and assumed absolute power. A centralized totalitarian state quickly resulted. In a move to expand Germany, Hitler started to expand its borders. Germany's invasion of Poland in 1939 kicked off what would become World War II. In 1941, Germany invaded the Soviet Union and declared war on the United States. After Germany's defeat, the country was divided into East Germany, controlled by the Soviet Union, and West Germany, controlled by the Allied powers. The Iron Curtain divided the two Germanys, with the Berlin Wall dividing the city of Berlin. The Iron Curtain and the Berlin Wall were major symbols of the

Cold War. In 1989, the Berlin Wall came down, and the two Germanys were reunited in 1990. Today, Germany is a vibrant country and an active EU member.

Figure 2.23: President Ronald Reagan Speaking at the Brandenburg Gate, Germany, June 12, 1987 From this speech comes his famous quote, "Mr. Gorbachev, tear down this wall." Wikimedia Commons – public domain.

Germany is Europe's largest economy, with strong exports of manufactured goods. To gain national income, Germany has promoted manufacturing as a major component of its economy. Most exports are in automobiles, machinery, metals, and chemical goods. Germany has positioned itself strategically to take economic advantage of the growing global awareness of environmental issues and problems by focusing on improvements and manufacturing of wind turbines and solar power technology. The service sector also contributes heavily to the economy. Deutsche Bank holds the enviable position of being one of the most profitable companies on the *Fortune* 500 list. Germany is also a major tourist destination. The Black Forest, Bavaria, the Alpine south, a variety of medieval castles, national parks, and a vibrant assortment of festivals such as Oktoberfest attract millions of tourists to Germany every year.

German reunification in 1990 posed some challenges for the economy. Notably, the East German infrastructure was far behind that of the West. In an effort to update transportation systems, telecommunications, and other areas needed to support industry and commerce, enormous amounts of money had to be poured into the system. While there has been considerable success in the renewal of infrastructure, unemployment in former East Germany is still significantly higher, and the necessary retraining of the workforce is ongoing and expensive.

Language, religion, and education have been strong cultural forces in Germany. German is the official language of the country and the one spoken by most of its people. More than 60 percent of Germans self-identify as Christian, and another 30 percent self-identify as agnostic or atheistic. During the Cold War, East Germany was under a Communist government that promoted a nonreligious ideology, resulting in a high percentage of people with agnostic or atheistic beliefs in that part of Germany. Historically, there has been a strong connection between the religious reformation and education. The early leaders of the Christian Reformation were generally highly educated themselves and were strong advocates of education, which they viewed as a path to positive moral and social reform. The German tradition of excellence in education continues. Education is provided at no cost (other than taxation) to students at all levels, including the university level, though some universities are now starting to charge very modest amounts for tuition.

The Alpine Center

Landlocked in the center of Europe are the two main states of Switzerland and Austria. Sandwiched on the border of these two states is the ministate of Lichtenstein. This region is dominated by the Alpine ranges. Switzerland, officially known as the Swiss Confederation, is divided into twenty-six cantons (states). Because of its location and close ties with neighboring countries, four languages are spoken in Switzerland: German, French, Italian, and Romansh. Typically, one language predominates in any given canton. Berne is the country's capital, and Geneva, Zurich, and Basel are the other major cities. As of 2010, Switzerland's population was about 7.8 million. Its land area is just slightly larger than the US state of Maryland.

Internationally, Switzerland is well known for its political neutrality. The UN European offices are located there. The Red Cross and the main offices of many international organizations are located in Switzerland. Switzerland joined the UN in 2002 and has applied for EU membership. Swiss culture is thought to have benefited from Switzerland's neutrality. During times of war and political turmoil, creative people found refuge within the Swiss borders. Swiss banking practices and policies are known throughout the world, and Swiss banks have benefited greatly from the country's politically neutral status. Banking is one of the country's top employers and sources of income. The Swiss people enjoy a high standard of living.

Sports are popular in Switzerland and football, or soccer, is popular as it is in much of Europe. Switzerland has also produced excellence in hockey, skiing, and of late, tennis. There is an emphasis on science in Swiss culture with good historical reason: the field of modern chemistry originated in Switzerland. The Bernoulli family, famed for their significant contributions to mathematics over many years, hails from Basel. The Federal Institute of Technology in Zürich, one of the top-ranked universities in the world, has produced an unusually high number of Nobel Prize winners. Albert Einstein, though born in Germany, relocated to Switzerland and later became a Swiss citizen.

Figure 2.24: Einspaenner Coffee A Viennese specialty, it is a strong black coffee served in a glass and topped with whipped cream; it comes with powdered sugar, which is served separately. Lachlan Hardy – Vienna Coffee – CC BY 2.0.

Austria is larger than its neighbor Switzerland and is similar in area to the US state of South Carolina. In 2010, the population was estimated at 8.4 million. Austria has various Alpine ranges, with the highest peak at 12,457 feet in elevation. Only about a fourth of the land area is considered low lying for habitation. The Danube River flows through the country, including the capital city of Vienna. Austria has a well-developed social market economy and a high standard of living.

For more than six hundred years, beginning just before the dawn of the fourteenth century, Austria was tightly associated with its ruling dynasty, the Habsburgs. The Hapsburgs came to power in a new way with the formation of the Austrian Empire in 1804. In Germany's rise to power before World War II, Hitler annexed his native Austria to Germany. Austria regained its independence a decade after the war ended and joined the EU in 1995. Austria is a German-speaking country, and nearly the entire population self-identifies as ethnic Austrian.

Austria is predominantly Roman Catholic and was home to many monasteries in the Middle Ages, influencing a strong Austrian literary tradition. Austria's best-known cities are its capital of Vienna and Salzburg and Innsbruck. Vienna was the center of the Habsburg and Austrian Empires and earned a place as one of the world's great cities. It is famed for its baroque architecture; its music, particularly waltzes; and theater. The Habsburgs were great patrons of the arts, and hence music, drama, and art thrived for centuries in Austria. The country has been home to many famous composers and musicians and is the setting for the story behind the movie *The Sound of Music*. Austria has a worldwide reputation in music and the arts.

The British Isles

The British Isles are an archipelago (group of islands) separated from the European mainland by the English Channel. The British Isles are often included in the region of Western Europe when discussing political geography; however, the fact that they are separated from the mainland of Europe by water provides them with a separate identity. The British Isles consists of two separate, independent countries: the Republic of Ireland and Great Britain.

Figure 2.3.252.3.25: London on the Thames River. Photo by R. Berglee – CC BY-NC-SA.

The United Kingdom (UK) of Great Britain consists of the regions of England, Scotland, Wales, and Northern Ireland. All four regions are now under the UK government. The Republic of Ireland is independent of the United Kingdom and does not include Northern Ireland. The primate city and UK capital is London, which is a financial center for Europe. The capital city of the Republic of Ireland is Dublin.

Influenced by the Gulf Stream, the climate of the British Isles is moderate, in spite of its northern latitude location. The UK and Ireland are located above the fiftieth degree of latitude, which is farther north than the US-Canadian border. The northern latitude would normally place this region into the type D climates, with harsher winters and more extreme seasonal temperatures. However, the surrounding water moderates temperature, creating the moderate type C climate that covers most of the British Isles. The Gulf Stream pulls warm water from the tropics and circulates it north, off the coast of Europe, to moderate the temperature of Western Europe.

The Western Highlands and the Northern Lowlands dominate the islands. Scotland, Wales, and parts of England have highland regions with short mountains and rugged terrain. The lowlands of southern England, Ireland, and central Scotland offer agricultural opportunities. The Pennines mountain chain runs through northern England and was the source of the coal, ores, and waterpower that fueled the Industrial Revolution. To the east of Britain is the North Sea, which provided an abundance of petroleum resources (oil) for energy and wealth.

Though the heritage of the British Isles is unique to this region, the geographic dynamics are similar to Central Europe—that is, smaller families, urbanization, industrialization, high incomes, and involvement with economic globalization. The EU has had an enormous influence on the British Isles. Ireland has embraced EU's economic connections, but the British people have been hesitant to relinquish full autonomy to the EU. This reluctance can be noted in the fact that the United Kingdom kept the British pound sterling as their currency standard after the euro currency was implemented. However, the Republic of Ireland converted to the euro currency.

The regions of the British Isles follow similar dynamics to those of other countries in Western Europe. Though some regions are not as wealthy as others, they all demonstrate a high level of industrialization, urbanization, and technology. These urban societies have smaller families and higher incomes and are heavy consumers of energy, goods, and services. Just as the Industrial Revolution attracted cheap labor, the aging workforce has enticed people from former British colonies to migrate to the United Kingdom in search of increased employment opportunities. The mix of immigrants with the local heritage creates a diverse community. London has diverse communities with many ethnic businesses and business owners.

Devolutionary forces are active in the United Kingdom. Scotland and Wales are already governing with their own local parliaments. Devolutionary cultural differences can be noted by studying the different heritages found in each region. Just as the Welsh language is lingering in Wales, Gaelic continues in Ireland and Scotland. Each region has made efforts to retain local heritage and rally support for its own nation-state. However, this is all done with the overall perspective of remaining under the umbrella of the EU.

Northern England

Figure 2.26: The first iron bridge ever built (1879) was located in central England, home of the Industrial Revolution.

Wikimedia Commons – CC BY 2.0.

Home to the Industrial Revolution, major industrial cities such as Manchester and Birmingham brought together the labor, raw materials, and industry connections necessary to manufacture products. The port city of Liverpool gave access to the world markets established by Britain's extensive colonial empire, and the colonies provided raw materials, new ideas, and cheap labor for the new industrial factories. Great Britain was an avid colonizer and controlled colonies on all inhabited continents. Raw materials such as cotton, which did not grow well in Britain, became a major import that fueled the textile mills of the Industrial Revolution.

Industrialization caused a rural-to-urban shift in Britain. In 1800, only 9 percent of the population lived in urban areas, but by 1900 some 62 percent resided in cities and towns. As of 2010, over 90 percent of Britain's population is composed of urban dwellers. The British colonial empire also caused a migration pattern whereby people moved from colonies to the home country, with the core area located in the Midlands of northern England. Cheap immigrant labor and resources from the colonies provided the manufacturing enterprises of wealthy British industrialists with good fortunes. Raw cotton was brought in from the colonies of India and Egypt. Cheap labor, brought in from the Caribbean and South Asia, resulted in a more diverse population in the industrialized northern England.

As the information age developed, the industrial centers of northern England gave way to a postindustrialized southern England. The north's heavy industries experienced a decline in demand. Factories closed, production became automated, and unemployment increased. The transition from the Industrial Revolution to a postindustrial society turned northern England into the Rust Belt of the British Isles. The port of Liverpool has been updated with modern and automated systems that do not require the high level of manual labor that was necessary during earlier industrial times. Rail service connects Liverpool with London, which is connected to Paris through a tunnel under the English Channel called the Chunnel.

The postindustrial economic activities have shifted the focus of employment away from manual labor to the service sector of information. Many places have looked to tourism to boost their economic situation. Northern England has many attractive physical environments that have been developed into major tourist attractions. The Lake District of northern England is a noted vacation destination, and short mountains and scenic landscapes attracted a number of England's writers. The city of Blackpool on the Irish Sea, just north of Liverpool, is a major vacation destination for the English. The dales and moors of northern England, complemented by the short Pennines, provide a sharp contrast to the urban metropolitan landscapes of London and southern England.

Southern England

Anchored by the primate city of London, southern England also is home to the universities of Oxford and Cambridge. This is Great Britain's most affluent region and is a center of postindustrial activity. Located on the Thames River, London is a central player in the world economic markets. Southern England also houses about one-third of the UK population. With immigration from the former colonies, this region is also becoming more diverse. This is an urbanized region, where the cost of living, transportation, and housing is high. For example, the price of gasoline in the United Kingdom might be two or three times that in the United States. Many urban dwellers do not own automobiles but instead use public transportation.

England is the most populous region of the United Kingdom with a density of about one thousand people per square mile. About half the population increase is because of immigration. The UK population is aging as a result of smaller family sizes and a growing number of senior citizens. This trend is common in countries in stage 5 of the index of economic development. Immigrants from many parts of the world have targeted the England as their new home. Many are from former British colonies in Asia and Africa. England has a diverse population. Islam is the fastest growing religion, even though Muslims make up less than 10 percent of the total population. London is even home to a growing Sikh community.

Figure 2.27: London and the Seat of British Parliament The clock tower of Big Ben is a local landmark. Photo by R. Berglee – CC BY-NC-SA.

London is not only the primate city and capital of the United Kingdom; it is the largest administrative municipality in the entire EU. Only Paris has a larger overall metropolitan area. London is a global center that holds prominence in world markets and the globalization process. About one-fifth of Europe's largest corporations have their headquarters in London. The city's cultural influence is felt around the world in the fashion industry, media, entertainment, and the arts. London is an international city that has a powerful draw for tourists. The core economic region of Europe is anchored by London's international recognition.

Southern England is also home to Great Britain's seat of government The UK government is considered a constitutional monarchy with a king or queen as head of state. The parliament is the official legislative body with a prime minister as head of government. The parliament has two bodies: the House of Commons, whose members are elected, and the House of Lords, whose members are appointed for life. As head of state, the monarch is mainly a figurehead with little actual political power. A number of dependencies remain under the British Crown as far as sovereignty is concerned. Small islands such as Guernsey, Jersey, the Isle of Man, and various islands in the Caribbean and Atlantic and Pacific Oceans remain under the British government for administrative, economic, and defense purposes.

Wales

A highland region to the west of England, Wales holds a Celtic heritage in which the Welsh language and stories of coal mining can still be heard. However, the English language has become more dominant and tourism has replaced coal mining as the main economic activity. The decline in the use of coal depressed the economy but did not depress the culture and heritage of the Welsh people. The largest city and capital of Wales is Cardiff. In the early twentieth century, the port of Cardiff handled the largest amount of coal in the world and handled more tons of cargo than Liverpool or London. The decline in coal mining has reduced the shipping activity in the port of Cardiff in the twenty-first century.

Devolution is alive and well in Wales. Welsh nationalism prompted the declaration of a separate parliament in the capital of Cardiff. The break with London provided local autonomy, but Wales is still reliant on the United Kingdom in national and foreign affairs. Many of the young people in Wales emigrate to find work because of the depressed economy. Emigration has caused a leveling off of population growth, and the number of people who speak Welsh has diminished. Wales and England share a common Protestant Christian religion. Wales is turning to tourism as a means of economic income; the scenic and picturesque landscape of the highland region, with its many castles, provides a pleasant experience for tourists.

Scotland

United with England in 1707, Scotland has been integrated into the United Kingdom while keeping its separate heritage and culture. Scotland has strong centripetal forces uniting the Scottish people, including victories over the British by Scottish clans lead by William Wallace in 1297 and Robert the Bruce in 1314; these forces within Scotland have created a devolutionary split with the UK parliament. In the quest to become a nation-state unto itself, in 1997 Scotland received permission to create its own parliament to govern local affairs. Though Scotland would like to separate into an independent state, they do not wish to isolate themselves from the EU or greater Europe. This contradictory trend, in which local states want autonomy and self-rule yet also want to remain within the larger regional community for economic and national security, is common. The phrase "separate within the Union" is surfacing in Europe with stronger voices. For now, Scotland remains under the British Crown and shares a Protestant Christian heritage with its UK associates.

The Scottish Highlands provide for livestock production, and the central Scottish Lowlands are favorable for agriculture. The North Sea has extensive oil resources. With resources such as these, Scotland is in a position to gain wealth and support its small population of about five million people. As an early export product, scotch whisky has profited many whisky marketers and has become the largest export product of Scotland. Scotland benefited and gained wealth during the Industrial Revolution. As a part of an island, early shipbuilding produced ships that brought about trade and development that coincided with European colonialism.

Postindustrial activities have become a focus of the current economy. High-tech computer industries have concentrated in **Silicon Glen**, an information-age industrial sector that lies between Scotland's largest city, Glasgow, and its capital of Edinburgh. With natural resources and postindustrial opportunities, Scotland is in a good position to compete in the global economic community. Scottish banking firms reach around the globe through their investment holdings. Scotland attracts a healthy tourism market with its Highlands and many castles. Kilts and bagpipes are a part of Scottish history and often distinguish themselves as a part of the region's heritage. The game of golf originated in Scotland and is still popular today.

Northern Ireland

Northern Ireland is a place in which strong cultural forces often erupt into violence. Officially part of the United Kingdom, this small region with about 1.7 million people has developed a unique set of cultural problems. Most people consider the religious differences to be the main problem in Northern Ireland. People of Irish heritage are predominately Roman Catholic, and those of Scottish and English heritages are usually Protestant Christians. But it is not so simple. If the troubles (as they are called in the United Kingdom) in Northern Ireland were based solely on religion, then we should expect to see similar violence in other regions of the United States or of Europe.

Figure 2.3.282.3.28: Protestant Group in a Belfast Parade Celebrating King William of Orange. IrishFireside – Belfast Parades – CC BY 2.0.

The core of the cultural problems in Northern Ireland is actually based on political affiliations. Most of the population in the region is not Irish. About 55 percent of the population is of Scottish or English descent, with only about 45 percent of Irish descent. The real problem centers on the governing of Ireland. The Irish would like to see Northern Ireland join with the Republic of Ireland, which received its independence from the British in 1921. The Irish do not want to be under the UK parliament with the Queen of England as head of state. Irish countrymen want total independence from the United Kingdom. The non-Irish population does not want to be a part of the Republic of Ireland and its Irish parliament and would rather remain under the British Crown. The people of Scottish descent would prefer total independence from all outside forces. English people definitely want to remain with the UK parliament.

Though the real problem is political, religion has become the scapegoat. Because cultural differences can be clearly witnessed between Catholics and Protestants, religion has become the identifier of the two sides. They are not necessarily fighting over religious beliefs but rather over political power and control.

The troubles of Northern Ireland have diminished its economic and employment opportunities. Underground paramilitary groups such as the **Irish Republican Army (IRA)** have heightened tensions between the two groups for decades, with car bombings and other terrorist acts. Hope might lie with a generational change in the population. As the younger generation seeks more opportunities and advantages, the issues that separate the two sides could diminish. The hatred that has been built up over the years can be eased with each new generation if centripetal forces work to bridge the differences and unite the social fabric. If the killing and hatred are passed down to successive generations, it will only take longer to recover.

In 1998, after a series of terrorist acts that were condemned on all sides, a movement took place to create the Northern Ireland Assembly with members of both sides of the division. Obstacles continue to surface to disrupt this calming process, but there is hope that in the future solutions such as this assembly can work toward producing a lasting peace. The devolutionary forces active in Wales and Scotland may act to create a more separate Northern Ireland. The Northern Ireland Assembly was only one step. Progress in Northern Ireland is an uphill battle with high unemployment, a poor resource base, and few economic opportunities. Only by working together will Northern Ireland become a stable, peaceful part of the British Isles.

The Republic of Ireland

The whole island of Ireland was under the control of the British Crown for centuries. In 1921 independence was gained from the British for all but Northern Ireland. This bitterly fought conflict has become well entrenched in Irish culture and literature. As an independent country separate from the United Kingdom, the Republic of Ireland has ascended the economic ladder to become a part of the global economic community. Because the climate is type C, there is adequate rainfall for crops and vegetation, and the green landscape gives it the title, "The Emerald Isle." The island has few other natural resources. There are trees but no large forest reserves for commercial exploitation. Peat, which is an early version of coal, is cut from the bogs and burned as fuel.

Ireland is not a large country. It is just a bit larger than the US state of West Virginia, with a population of about four million. There are no tall mountain ranges in

Ireland. The soils are traditionally rocky with few nutrients. Before colonial times, the traditional food crops included such turnips and rutabagas. When the potato was imported from the Americas, it was well received in Ireland. The potato plant grew well and replaced traditional root crops as the main food source.

In the early 1800s the population of the whole of Ireland (including what is now Northern Ireland) was as high as eight million. Starting in the 1840s, blight and rot destroyed much of the potato crop year after year, causing a serious famine in Ireland. More than a million people died, and another two million people left the island. The potato famine caused losses reminiscent of the Black Death, which had ravaged Europe centuries earlier. The history of the Black Death may have led to the term *Black Irish*, which referred to people who fled Ireland during the potato famine and immigrated to the United States. They were often regarded as lower-class citizens and were discriminated against. Well-established Irish families having immigrated to the United States before the 1840s were not considered Black Irish and assimilated into mainstream American society more easily.

The lack of natural resources and the lack of opportunities and advantages held back the Republic of Ireland from developing a strong economy. It wasn't until the 1990s that conditions improved. With the creation of the EU and advancements in communication, Ireland became an ideal location for expanding North American corporations. The attractive elements included a mild climate, a similar English language, an educated workforce, and a low cost of living. Many high-tech computer firms, communication companies, and automated industries established their base of operations in Ireland—Dublin in particular. Because Ireland was a member of the EU, it was a convenient intermediary location between the United States and the European mainland. Business boomed in the 1990s, and incomes and the cost of living rose. Tourism also has become a growing sector of the economy. The economic growth in Ireland earned it the title **Celtic Tiger** to indicate its growing economic power.

However, the rapidly expanding economic conditions of the 1990s have not extended into the twenty-first century. Since 2007, the global depression has taken its toll on the Irish economy. Starting in 2008, the country witnessed a sharp increase in unemployment that coincided with serious banking scandals. Various banks have been targeted for bailout funds from the government, and the economy witnessed a sharp decline. Property values have plummeted, and protesters have staged demonstrations in the streets demanding the government address the economic situation. Tourism has taken up some of the slack and has expanded by highlighting the Emerald Isle's mild climate and green countryside. Cities such as Dublin are working with the growing pains of recovery in the economic recession. Unemployment remains a concern, as well as declining economic opportunities.

Overall, Ireland is working to reposition itself for future economic growth. The country wants to maintain itself as an important link in the EU's relationship with the United States. Ireland is bracing itself for a sluggish economic road ahead.

Key Takeaways

- Western Europe can be divided into a number of smaller geographic regions, including northern Europe, southern Europe, Central Europe, and the British Isles.
- The differences in climate, terrain, and natural resources provide for a diversity of economic activities that influence cultural development within the European community.
- The countries of Western Europe have high standards of living compared with world standards. The southern countries of Portugal, Greece, and southern Italy are more agrarian and have been struggling economically since the global recession began in 2007.
- The British Isles comprise the two independent countries of Great Britain and the Republic of Ireland. Great Britain is made up of five geographical regions, with Wales and Scotland having more autonomy to govern local affairs.
- Since World War II, the many states of Western Europe have been evolving into a more integrated realm. Each nation has worked to develop its economy to take advantage of its physical geography and its cultural history and heritage. Devolutionary forces remain strong in many areas to counterbalance the forces of supranationalism.

Discussion and Study Questions

1. Outline the main economic activities for each of the regions of Western Europe.
2. How have strong egalitarian ideals shaped northern European countries?
3. How are the differences between northern and southern Italy a reflection of southern Europe as whole?
4. Outline where devolutionary forces are active in Western Europe.
5. Which of the states of Western Europe most resembles a nation-state?
6. Why is there an increased level of cultural diversity in France and other parts of Western Europe?
7. How do the demographic dynamics differ between northern Europe and southern Europe?
8. How does agricultural production vary with physical geography in Western Europe?

9. What are the key factors that make Western Europe an economic core area of the world?
10. How are the troubles in Northern Ireland similar to or different from the Basque situation?

Geography Exercise

Identify the following key places on a map:

- Baltic Sea
- Greenland
- Faeroe Islands
- Balkan Peninsula

2.4. EASTERN EUROPE

Eastern Europe, a region rich in history and cultural diversity, has undergone significant transformation since the end of the Cold War. Once characterized by communist governance and economic isolation, countries in this area have made remarkable strides toward democratization and market-oriented reforms. This transition has brought both opportunities and challenges, as nations navigate the complexities of integration into the European Union and the global economy. From the vibrant cities of Poland and Hungary to the scenic landscapes of the Balkans, Eastern Europe is a tapestry of traditions and modern aspirations, making it a fascinating subject for exploration in the context of contemporary geopolitics, economic development, and cultural identity.

Figure 2.29: Eastern Europe

Transitions of Eastern Europe after the Cold War

After World War II ended in 1945, Europe was divided into Western Europe and Eastern Europe by the Iron Curtain. Eastern Europe fell under the influence of the Soviet Union, and the region was separated from the West. When the Soviet Union collapsed in 1991, all the Soviet Republics bordering Eastern Europe declared independence from Russia and united with the rest of Europe. The transition Eastern Europe has experienced in the last few decades has not been easy; however, most of the countries are now looking to Western Europe for trade and economic development. Cooperation continues between Eastern and Western Europe, and the European Union (EU) has emerged as the primary economic and political entity of Europe.

The collapse of Communism and the Soviet Union led to upheaval and transition in the region of Eastern Europe in the 1990s. Each country in the region was under Communist rule. The countries bordering Russia were once part of the Soviet Union, and those countries not part of the Soviet Union were heavily influenced by its dominant position in the region. When the Soviet Union collapsed in 1991, the bordering countries declared independence and began the process of integration into the European community. Moldavia changed its name to Moldova. The countries of Czechoslovakia and Yugoslavia each broke into multiple countries and, because of the diverse ethnic populations, organized around the concept of nation-states. (See Section 2.2.6 "Nation-States and Devolution" for a discussion of nation-states.) Czechoslovakia peacefully agreed to separate into two states: the Czech Republic and the Republic of Slovakia. Yugoslavia was not so fortunate.

Most Eastern European political borders resemble ethnic boundaries. Each of the regions once resembled nation-states. In principle, Romania is set apart for Romanians, Hungary for Hungarians, and so on. Few are true nation-states because of ethnic minorities located within their borders, but the countries held on to their common heritage throughout the Communist era. In most Eastern European countries, cultural forces have brought people together to publicly support the move to unite and hold onto a heritage that is as old as Europe itself.

Governments that had been controlled by Communist dictators or authoritarian leaderships before 1991 were opened up to democratic processes with public elections. With the fall of Communism came economic reforms that shifted countries from central planning to open markets. Under central planning, the governments dictated which products were produced and how many of each were to be produced. The open markets invited private capitalism and western corporate businesses.

Reforms in Eastern Europe in the 1990s

1. Communism → Democracy
2. Central Planning → Open Markets
3. Government Control → Private Citizen
4. Soviet Union → European Union

Figure 2.30: Reforms in Eastern Europe in the 1990s

The power of the state was transferred from the Communist elite to the private citizen. People could vote for their public officials and could choose businesses and work individually. With the EU looming over the realm, the now-independent countries of Eastern Europe shifted their economic direction away from Moscow and the collapsing Communist state and toward the core industrial countries of Western Europe and the EU.

The stage for the transition from communism to capitalism was further set by Soviet leader Mikhail Gorbachev, who began to liberalize the Soviet Union by allowing multiparty elections and encouraging economic reform and societal openness (concepts known throughout the world by the Russian words **perestroika** and **glasnost**, respectively). At the same time, the world watched as the Chinese government violently cracked down on student protests in the Tiananmen Square protests of 1989, which strengthened the revolutionary sentiments and intentions of many people in the Eastern bloc, the former Communist countries of Eastern Europe.

Countries of Eastern Europe Hungary

During the Communist period, Hungary enjoyed a higher standard of living than its neighbors. It was likely the marked contrast in daily life between the Communist period and the post-Communist period that influenced the outcome of the 1994 elections, which gave the Hungarian Socialist Party, led by former Communists, an absolute majority in parliament. Many were concerned that Hungary was signaling a wish to return to the past, but instead, all three main political parties joined together to work toward the common goals of continued liberalization of the economy, closer ties with the West, and full EU membership.

Economic reforms have not been easy for Hungary and have resulted in a lower standard of living for most people. The removal of government subsidies, a pillar of Communist rule, led to recession. To avoid massive inflation and attract investment, government fiscal policies had to be conservative to the point of austerity. Hungary's large-scale manufacturing and stable government have aided it through the latest global economic downturn.

An important characteristic of post-Communist Hungary is its desire to become an active member of the global community by joining international organizations. In 1990, Hungary was the first Eastern bloc nation to join the Council of Europe. In 2004, Hungary joined the EU. In addition, Hungary joined the other international organizations, including the Organisation for Economic Co-operation and Development (OECD), the International Monetary Fund (IMF), and the World Trade Organization (WTO). Another endeavor that would not have been possible under the Communist rule was the development of tourism as an important economic sector. Budapest's reputation as a city of great elegance has helped it become one of the major tourist attractions in Eastern Europe. Major movie studios have also traveled to the city to film.

Figure 2.4.312.4.31: Parliament Building in Budapest, Hungary, Next to the Danube River. Wikimedia Commons – CC BY 2.0.

The Czech Republic and the Slovak Republic (Slovakia)

Each country in Eastern Europe has a different story to tell regarding its path from communism to independence. Countries such as the Czech Republic and Hungary were more prepared and better equipped to meet the challenges and were early qualifiers for entry into the EU. Historically, the region occupied by the modern Czech Republic was known as Bohemia and Moravia. The Czech Republic and Slovakia were, until recently, part of the same country: Czechoslovakia, which was created in 1918 from part of the Austro-Hungarian Empire after World War II ended. On January 1, 1993, they became two new independent nations. Slovakia is known officially as the Slovak Republic.

Czechs and Slovaks alike descended from Slavic peoples. The national language of the Czech Republic is Czech, while the official language of Slovakia is Slovakian. Both languages are in the Slavic linguistic family. About 94 percent of the population of the Czech Republic identify themselves as ethnically Czech. The others are

Slovenian, Polish, German, Hungarian, and Roma (Gypsy). About 86 percent of the inhabitants of Slovakia self-identify as ethnically Slovakian. Hungarians make up the largest minority population, and about 2 percent of the population identify themselves as Roma.

Under Communist rule, the standard of living was very high in Czechoslovakia. When market reforms began in the Czech Republic and Slovakia, the standard of living declined to some extent. In general, the Czech Republic has more rapidly and effectively transferred state control of industry to private ownership than has Slovakia. Also, Slovakia was hit harder by the move away from defense industries, which had employed many people during the Communist period. Unemployment has consistently been higher in Slovakia. Slovakia is not as industrialized as the Czech Republic but has made strides since independence to provide economic opportunities for its people. Both countries expanded their economic opportunities when they were admitted into the EU in 2004.

The Baltic Republics

The small Baltic states of **Latvia, Estonia** and **Lithuania** transitioned away from their old Soviet connections. After independence, they were quick to look toward Western Europe for trade and development. Most Eastern European countries followed this pattern. Latvia, Estonia, and Lithuania received their independence from the Soviet Union in 1991. Russia withdrew its troops from the region in 1994. The North Atlantic Treaty Organization (NATO) and the EU welcomed them as members in 2004. They have transitioned to market economies with democratic governments. The people of Latvia, Estonia, and Lithuania rapidly expanded their economic conditions after independence but have been hard hit by the downturn in the global marketplace in 2008.

Poland

Poland has a long-standing history of working to oppose the Communist domination of its country. The Solidarity movement in Poland started out as an independent trade union in the 1980s but became a lightning rod for political change in Poland. The Communist Party dominated politics and suppressed any movement to organize labor or the people against the government. The strength of Solidarity became evident by the 1990 election, when Solidarity candidate Lech Wałęsa won the Poland's presidential election.

Solidarity's victory signified the collapse of the power of the Soviet Union and Communism in Eastern Europe. The country has emerged with democratic government and a thriving capitalist economy.

Since the fall of Communism, Poland left behind its old state-directed economy and transitioned to a market economy, in which businesses are privately owned and run. When the Communists controlled Poland, there was a strong emphasis on heavy industry, and that focus remains alive today. Poland produces cars, buses, helicopters, trains, and heavy military equipment, including tanks and ships. Before the Communists industrialized the Polish economy, it was largely agrarian. Though Poland continues to be one of Europe's leading agricultural producers, with a wide variety of crops and dairy and meat production, it is unable to meet the food demands of its large population. Poland's economy is still considered to be under development. Reforms, including privatization, must continue before Poland can adopt the euro, the common EU currency.

Figure 2.4.322.4.32: Royal Castle Square in Warsaw, Poland. Wikimedia Commons – public domain.

Unemployment has at times presented very significant problems for the Polish economy and society. Unemployment skyrocketed to nearly 20 percent in the early 1990s. The situation improved, but unemployment was still at about 14 percent in 2006. By 2010 the situation improved further, and the unemployment rate was around 7 percent. When Poland joined the EU, many people left Poland in search of work. Some of those workers who left have recently returned, as the employment rate and wages are both increasing substantially.

Moldova

A variety of cultural and social forces provided different levels of civility in the transitions of Eastern Europe. The poor, agrarian economy of the small, landlocked country of Moldova provides few opportunities or advantages to grow its economy and provide a stronger future for its people. As a result, young people earning an education or technical skills immigrate to other countries for opportunities or employment.

Belarus

In 1991, independence came to the former Soviet Republic of **Belarus**, but Belarus took a different path from most of the other Eastern bloc countries in that Belarus did not distance itself from its Russian connection. Belarus has consequently experienced authoritarian governments well into the twenty-first century. In this case, Russia and Belarus created a stronger relationship by signing agreements to increase economic integration. Difficulties have hindered implementation of many of these policies. For one thing, the government of Belarus has been slow to move toward democratic reforms. In fact, the president has taken on greater authoritarian powers. Some of the personal freedoms that have been granted in other European countries—freedom of the press, free speech, and the right to peacefully assemble—are still restricted in Belarus. As of 2010, Belarus, Ukraine, and Moldova have not been admitted into the EU. Belarus has indicated that it might join with the Russian Republic and has not even applied for entry into the EU.

Ukraine

With the largest physical area in Europe, **Ukraine** is slightly larger than France. Its population in 2010 was approximately forty-six million, and 77 percent of the population is ethnically Ukrainian and 17 percent is Russian. Most of the population live in the industrial regions of the southeast or eastern parts of the country. Though the official language is Ukrainian, many Ukrainians still speak Russian. As is the case with neighboring states, the Eastern Orthodox Church dominates the religious and cultural reflections of the arts, literature, and architecture. Christian themes are often reflected in the paintings, books and performances.

The demographic trend in Ukraine follows a pattern of industrialization with smaller family sizes but continues to have a higher death rate than most European countries. The country is now losing about 150,000 people per year. The low birth rates are similar to those of Russia and southern Europe. Poor health and childhood poverty in Ukraine are two of the main issues confirmed by the United Nations (UN). In 2010, Ukraine had a negative population growth rate of -0.62 percent. Fertility rates have been in decline throughout Europe over the past few decades. The average fertility rate in Ukraine is 1.1, one of the world's lowest; in Europe as a whole, the average is about 1.3. The causes for some of these trends can be attributed to alcoholism, poor diets, smoking, and the lack of medical care. The average life expectancy for an adult male in Ukraine is about sixty-two years. In comparison, the average life expectancy for men in France is about seventy-eight years.

Of the former Soviet Union republics, excluding Russia, Ukraine was far and away the most important economic component, producing about four times the output of the next-ranking Soviet republic. Its fertile black soil generated more than one-fourth of the overall Soviet agricultural output, and its farms provided substantial quantities of meat, milk, grain, and vegetables to other republics. Likewise, its diversified heavy industry supplied steel pipes and raw materials to Russia's oil and mining industry. Ukraine's political relationship with Russia has been complicated since the dissolution of the Soviet Union, but Russia still supplies enormous amounts of natural gas and oil to fuel the economy, and Russia's markets are still highly integrated with those of Ukraine.

Though Ukraine has transitioned from a Soviet republic to a fully independent country, divisive centrifugal forces have made Ukraine's path to free elections and democratically elected leaders difficult. The country has held political elections, but they have been challenged or tainted with corruption and accusations of fraud. Today, the country has a democratically elected government and is working toward improving its economy and creating stability for its people.

Romania

A number of countries of Eastern Europe have fully transitioned from Communist dictatorships to modern, integrated economies. **Romania** is one of those countries. Romania encompasses an area equivalent to the US state of Minnesota. Its population of 21.5 million people includes two million who live in Bucharest, the capital and largest city of the country. The Carpathian Mountains circle Romania, with the Transylvanian Alps to the south. The Danube River runs across the region and creates a natural border with Bulgaria and Serbia before flowing into the Black Sea. The Romanian forests are some of the largest in Europe, with about half (13 percent of the country) set aside from logging and placed in watershed conservation programs. The integrity of the ecosystems in the Romanian forests provide diverse habitats for plants and animals. Romania claims to have the most European brown bears and about 40 percent of all European wolves living within its borders.

Figure 2.33 Bran Castle Near Brasov in Transylvania

Tourism is growing in Romania. Almost 5 percent of Romania is placed in protected areas, including thirteen national parks and three biosphere reserves, all of which are attractive to tourists. Tourist attractions also include medieval castles as well as historic Transylvanian cities. Rural tourism focuses on folklore and traditions including such sites as Bran Castle, referred to locally as the castle of Dracula, a mythical person patterned after the stories and legends of Vlad III the Impaler.

Wikimedia Commons – public domain.

Romania's Communist dictator, Nicolae Ceauşescu, ruled from 1965 to 1989. The end of his domination came when the government was overthrown by a revolution. Ceauşescu and his wife were arrested and shot to death. Even after his death, the Communist Party maintained strong ties to the government until the mid-1990s.

The transition to integrate Romania's economy with that of the greater European economy was delayed because of Romania's obsolete industrial infrastructure, established during the Communist era. Since joining the EU in 2007, Romania has developed a stronger export market with Western European countries. Investments and consumer confidence have fueled the growth of the domestic economy. Romania has worked through a number of difficult issues in its attempts to provide a stable government and a growing market economy.

Albania

The southern Adriatic is home to the small country of **Albania**. The rugged mountainous country of Albania has a Muslim majority. Poverty, unemployment, and a lack of opportunities to gain wealth have plagued the country. Albania has even received Communist support from China. As a result of the war in Kosovo, Albania suffered a major setback in its progress toward an improved standard of living and integration with the rest of Europe. When stability is established, Albania can progress toward becoming more integrated with the European economy and raise its standard of living for its people. A parliamentary democracy has been installed since the Communist era, and foreign investments have aided in developing updated transportation and power grids.

Bulgaria

Located in the crossroads of the continents, **Bulgaria** has a major trans-European corridor running through its territory that connects all the way to Asia. The country is home to diverse landscapes, which include the sunny Black Sea coast and the higher elevations of the Balkan Mountains, which reach an elevation of 9,596 feet. The Danube River flows across the border with Bulgaria on its way to the Black Sea. About one-third of the country consists of plains, which provide for extensive agricultural activity. Ore and minerals can also be found in Albania, which has allowed the country to gain wealth.

Upon declaring independence from Russia, Bulgaria held multiparty elections. Its economy is emerging, but the transition to a capitalist system has not been without the difficulties of unemployment, inflation, and corruption. Bulgaria became a member of NATO in 2004 and was accepted for EU membership in 2007. The transition to a free market economy is still in progress, with mining, industry, and agriculture as the main economic activities. Tourism is an emerging segment of the economy that has been gaining international attention in recent years. The country has a milder climate than the northern states of Eastern Europe and has been marketing itself as a major tourist destination. Main points of interest include historical monasteries, coastal resorts on the Black Sea, and the capital city of Sofia.

Special Section: A Story of "Yes" and "No" in Bulgaria

Here is an example of cultural differences between the United States and Bulgaria. This story is from Elizabeth Kelly, a US Peace Corps volunteer working in Bulgaria (2003–5).

"I'll have coffee," I tell the waitress at a cafe during my first week in Bulgaria. She shakes her head from side to side. "OK, tea," I say, thinking that maybe there's something wrong with the coffee machine. Again, she shakes her head. "Um. cola?" Once more, she shakes her head. By now, she's looking at me like I'm crazy, and I'm totally confused. Then I remember: A shake of the head by a Bulgarian means "yes," and a nod—what the rest of the world does for "yes"—means "no."

Early on, when I communicated with Bulgarians, it seemed like my head was moving in ways my brain hadn't told it to. Sometimes I wanted to grab my ears and use them as controls. Learning a language with a completely different alphabet was challenging enough without trying to figure out whether to nod or shake.

When I began teaching, all this head bobbing made communication in the classroom interesting. Although I had made sure my students knew about this cultural difference on the first day of school, we all frequently forgot what we were doing. My students would answer a question correctly or say something really great, and I'd nod. A second later, they were trying to change their answer, since they thought the nod meant they had been wrong. But the confusion went both ways. Sometimes I'd ask a student a yes-or-no question and he or she would answer with a nod or a shake, without saying anything. Not remembering the difference, we'd have to go through the motions several times before I understood. Frequently I found myself saying, "Da or ne—just tell me one or the other!"

I also had to deal with confused colleagues who couldn't figure out why I kept nodding my head while they talked, as if I were arguing with them. In truth, I was just trying to show that I understood and was following along with the story. And then there was the even greater problem of how to act with Bulgarians who spoke English and were aware of the nodding-shaking problem. Was I supposed to nod or shake for "yes" when I was speaking English with them? And what was I supposed to do when we were speaking Bulgarian? What if we were in a situation where both languages were being spoken? To make matters even more complicated, after going a couple of weeks without any contact with other Americans, we'd finally get together and I'd find myself shaking when I should have been nodding. My head was spinning!

Tuning in to how the people around me communicate has brought me closer to the people and the culture of Bulgaria. And whenever we slip up and forget to control our heads, the laughter that follows brings us together. Luckily, a smile is a smile the world over.

Source: Peace Corps Coverdell World Wise Schools program. Used by permission.

Former Yugoslavia
A Valuable Lesson in Political and Cultural Geography

The name *Yugoslavia*, applied to the region along the Adriatic in 1929, means **Land of the South Slavs**. From 1918 to 1929, the region had been called the Kingdom of Serbs, Croats, and Slovenes. Non-Slavic populations surround Yugoslavia. The region's core is mountainous. The Dinaric Alps, with the highest peak at just below nine thousand feet in elevation, run through the center of the Balkan Peninsula. The rugged mountains separate and isolate groups of Slavic people who, over time, have formed separate identities and consider themselves different from those on the other sides of the mountain ridges. Distinct subethnic divisions developed into the Slovenes, Macedonians, Bosnians, Montenegrins, Croats, and Serbs, with various additional groups. These differences led to conflict, division, and war when the breakup of former Yugoslavia began.

World War I started in the city of Sarajevo, Bosnia, when a Serb advocate assassinated Archduke Ferdinand of the Austria-Hungarian Empire. In the next conflict, World War II, there was also divisiveness within Yugoslavia: Croatia sided with Nazi Germany, but Serbia was an ally with the Communist Soviet Union. The region of Bosnia, with a Muslim majority, faced religious opposition from its mainly Christian neighbors. A group headed by Marshal Tito (a.k.a. Josip Broz) led Yugoslavia after World War II ended in 1945. Tito created a Communist state that attempted to retain its own brand of neutrality between the Warsaw Pact nations led by the Soviet Union and the NATO nations of the West.

Tito was a centripetal force for the region of Yugoslavia. For over forty years, he held the many ethnic Slavic groups together under what he called *Brotherhood and Unity*, which was actually the threat of brute military force. It appears to have been effective. The 1984 Winter Olympics were held in Sarajevo as witness to the progress and unity of Tito's Yugoslavia. At the same time, Yugoslavia started manufacturing a model of automobiles called the **Yugo**. While the Yugo was not in the same league as high-end European luxury cars, the ability to make and purchase automobiles was a testimony to the rising industrial level of the Yugoslav economy. This progressive trend, unfortunately, was not to last.

Figure 2.34: The Balkan Peninsula and Former Yugoslavia

Macedonia is officially called the Former Yugoslav Republic of Macedonia (FYROM) because of a name conflict with Greece. Albania is an independent country and was never a part of Yugoslavia.

Table 2.3 Status of States in Former Yugoslavia

Country	Year of Independence	Religion	Population	Size (Sq. Mi.)	World War II Ally
Slovenia	1991	Catholic, 57%; Muslim, 2.4%;Orthodox, 2.3%	2 million	7,827 (about the size of Vermont)	—
Croatia	1991	Catholic, 88%; Orthodox, 4.4%; Muslim, 1.3%	4.5 million	21,831 (about the size of West Virginia)	Nazi Germany
Bosnia	1991	Muslim, 40%; Orthodox, 31%; Catholic, 15%	4 million	19,741 (about the size of West Virginia)	—
Serbia	2006	Orthodox, 85%; Muslim, 4%; Other, 10%	10 million	29,915 (about the size of South Carolina)	Soviet Union
Kosovo*	2008	Muslim, 90%; Orthodox, 9%	1.8 million	4,203 (about the size of Connecticut)	—

Country	Year of Independence	Religion	Population	Size (Sq. Mi.)	World War II Ally
Macedonia	1991	Orthodox, 67%; Muslim, 30%	2.1 million	9,781 (about the size of Vermont)	—
Montenegro	2006	Orthodox, 74%; Muslim, 18%; Catholic, 3.5%	678,177	5,423 (about the size of Connecticut)	—
*Note: As of 2008, Kosovo was not accepted by the UN. Its status is pending.					

The Breakup of Former Yugoslavia

Tito died in 1980. The unity that had helped hold the country together began to break down in the early 1990s with the Soviet Union's collapse. With the dual loss of the Soviet Union and Tito's strong policies as centripetal forces, the power struggle for dominance among the various ethnic groups began.

In 1991, Slobodan Miloševik began pushing for the nationalistic goal of uniting all the ethnic Serbs that lived in the various parts of Yugoslavia into a **Greater Serbia.** The efforts were not approved or supported by the UN, which rejected Yugoslavia for membership in 1992. At that time, Kosovo and Montenegro were part of the Serb state. Miloševik first sent the Yugoslav military to Kosovo to take control from the majority Albanian population and secure the region for Greater Serbia. Fearing war, Slovenia, Croatia, and Macedonia declared independence in 1991. Yugoslavia was breaking up.

Miloševik then moved into Croatia to secure Serb areas for his Greater Serbia. After a brief but bitter war between Serbia and Croatia, the UN stepped in to halt the conflict. The first signs of ethnic cleansing were reported during this conflict. Serb military units would roll into a town or village and claim it as a Serb-only location. The Croats and any other people living there were forced to leave. Reports of assaults on women and the systematic killing of men of fighting age were documented. Sadly, ethnic cleansing began to be reported on all sides of the war. The bitter hatred between Croats and Serbs, which had been handed down from the World War II era, surfaced for a new generation. After the UN stopped Miloševik in Croatia, he turned to the Serb areas of Bosnia to expand his Greater Serbia. Bosnia immediately declared independence and was approved by the UN in 1992.

The bitter battle for Bosnia extended from 1992 to 1995, and Europe wondered if the Bosnian war would develop into World War III. The region of Bosnia included people of Serb, Croat, and Bosnian ethnic backgrounds. Bosnia broke up along ethnic lines.

The Serb group supported Greater Serbia; the Bosnian group wanted independence. In March 1994, the Croat group signed an agreement joining with the Bosnian group against the Serbs. In 1995, with strong pressure from Europe and the United States, the warring groups signed a peace agreement known as the Dayton Accord, named after a meeting in Dayton, Ohio. The Dayton Accord accepted Bosnia's borders and supported the creation of a unity government that was democratic and included the multiethnic groups. The country of Bosnia was divided into three parts: Serb, Croat, and Bosnian.

Figure 2.4.352.4.35: Street Checkpoint in Kosovo in 2004. U.S. Military – public domain.

Military forces continue to monitor and secure the regions of Bosnia. Portions of Bosnia under Serb control have declared themselves the Republic of Srpska and have attempted to create their own nation-state. The remaining territory of Bosnia (51 percent) consists of a joint Bosnian/Croat federation. Both regions have sublevel governments within the formal country of Bosnia and Herzegovina. Herzegovina is a small region extending from the city of Mostar to the southern border with Montenegro. In June of 2006, the region of Montenegro declared itself independent of Serbia. Montenegro uses the euro as its currency and has applied for WTO membership. Montenegro has privatized its main industries and is soliciting a tourism industry.

The War for Kosovo

Yugoslavia, under the leadership of President Miloševik, sought to ensure that Kosovo would never leave the umbrella of Greater Serbia. The 1.8 million Albanian Muslims who lived in Kosovo constituted 90 percent of the population by 1989. They did not want to live under Serb control. Serbia claimed that Kosovo was the heart of the Serb Orthodox Church and the cradle of the medieval Serbian Empire. The historic battles against the Islamic Turkish Ottoman Empire in Kosovo have been memorialized in Serb tradition and history.

The main opposition to the Serb power structure in Kosovo in the early 1990s was the Kosovo Liberation Army (KLA). Many in the KLA wanted an independent Kosovo and an alliance with Albania. In 1998, Miloševik sent troops into areas controlled by the KLA. The civil war in Kosovo between Serbs and Albanian Muslims was devastating the region and creating thousands of refugees. The tension, hatred, and massacres by both forces further complicated the peace process. In March of 1999, NATO implemented a unified mission, called **Operation Allied Force**, to force Miloševik to discontinue the ethnic cleansing campaign and end the violence in Kosovo.

In June of 1999, Miloševik yielded to a peace agreement that brought Kosovo under the auspices of the UN and NATO forces, removed the Serb military, and allowed for the safe return of over eight hundred thousand refugees from Albania, Montenegro, and Macedonia. Miloševik lost his bid for the presidency in the elections of 2000, was arrested for crimes against humanity in 2001, and was transferred to the International Criminal Tribunal in The Hague, the Netherlands. The charges brought against him included ethnic cleansing and torture. He died in 2006 while in custody before the trial ended.

The wars in former Yugoslavia were fueled by strong cultural forces.

NATO forces continue to be stationed in Kosovo to keep the peace and work toward restoring order. After much negotiation between the UN and regional entities, the Kosovo Assembly declared its independence from Serbia in 2008. Serb officials overwhelmingly opposed Kosovo's independence, as they had in 1991. Kosovo's status is now pending in the UN.

Conclusion of Former Yugoslavia

Kosovo and the independent republics of former Yugoslavia provide an excellent study in the dynamics of ethnicity, culture, and political geography. The lessons learned from this region could be applied to many other areas of the world suffering similar conflicts, such as Chechnya, Iraq, Afghanistan, Rwanda, Congo/Zaire, or East Timor. Former Yugoslavia represents an example of how divisive centrifugal and devolutionary forces can lead to nationalism and eventually to war. The drive toward a nation-state has fueled nationalism and conflict in the Balkans. The civil wars within former Yugoslavia have cost thousands of lives and destroyed an infrastructure that had taken decades to build. Geographers have called Eastern Europe a shatterbelt because of the conflicts and divisions that have occurred there.

In spite of the problems with the transition in Eastern Europe, the region has nonetheless seen enormous economic gains. Even Slovenia, once part of former Yugoslavia, has rebounded with strong economic growth. Many of the progressive Eastern European countries have been accepted into the EU. The map of the EU includes many of the developing Eastern bloc countries. Eastern Europe has experienced many transitions throughout its history. The transition from communism to capitalism is only one part of the geography and history of Eastern Europe.

Key Takeaways

- After World War II ended in 1945, Europe was divided into Western Europe and Eastern Europe by the Iron Curtain. Western Europe promoted capitalist democracies, and Eastern Europe came under the Communist influence of the Soviet Union. After the collapse of the Soviet Union in 1991, Eastern Europe began to transition toward Western European ideals.
- Eastern Europe has been shifting toward democratic governments, open market economies, private ownership, and the EU rather than the old Soviet Union.
- Countries with stable governments and industrial potential have been accepted into the EU and have expanding economies. Other countries that have not reached that level of economic development or political reforms have not been admitted into the EU.
- The transition of former Soviet republics to capitalist systems has not been without difficulties, including unemployment, inflation, corruption and crime, and poverty. Political infighting has been evident in the transitioning countries that do not yet have a stable democratic government or economy.
- The breakup of former Yugoslavia was an example of how strong devolutionary forces can promote nationalism resulting in open war. The once stable country of Yugoslavia split into seven separate countries patterned after the nation-state concept. Bosnia remains the most diverse state, with a majority Muslim population.

Discussion and Study Questions

1. What were four of the main reforms that occurred in Eastern Europe with the collapse of the Soviet Union?
2. Which Eastern European countries were once members of the Soviet Union?
3. What factor has most heavily influenced most of the political boundaries of Eastern Europe?
4. How would the change from central planning to open markets influence the use of natural resources?

5. Which countries of Eastern Europe are members of the EU?
6. How did the larger Slavic population of former Yugoslavia five hundred years ago become the existing subgroups of the Slovenes, Croats, Bosnians, Montenegrins, and Serbs?
7. Why was Slobodan Miloševik so determined to keep the region of Kosovo under Serb control?
8. What was the purpose of ethnic cleansing in former Yugoslavia?
9. What role did religion play in the wars in Bosnia and Kosovo?
10. Why would the United States be so concerned about the wars in Bosnia and Kosovo—so much so that the United States spent billions to help end the war and then billions more to keep the peace and rebuild the region?

Geography Exercise

Identify the following key places on a map:

- Adriatic Sea
- Albania
- Baltic Sea
- Belarus
- Black Sea
- Bosnia
- Croatia
- Czech Republic
- Estonia
- Hungary
- Kosovo
- Latvia
- Lithuania
- Macedonia
- Moldova
- Montenegro
- Poland
- Romania
- Serbia
- Slovakia
- Slovenia
- Ukraine

2.5: CONCISE SUMMARY

The European continent stretches from the North Atlantic Ocean to the Ural Mountains in Russia. The European portion of Russia is typically analyzed in conjunction with the entirety of the nation. Europe is delineated by the Arctic Ocean to the north, the Atlantic Ocean to the west, and the Mediterranean Sea to the south.The Gulf Stream contributes to the establishment of a type C climate across a significant portion of Western Europe. Type D climates predominate in the northern and eastern regions of Eastern Europe. Europe comprises four primary physical landforms that offer a variety of resources for human endeavors.

European colonialism generated heightened wealth and commercial activity in Western Europe. The Industrial Revolution originated here as well. Europe evolved into an industrialized domain with robust economic dynamics that persist in propelling the postindustrial engine of globalization. The transition from rural to urban areas and urbanization resulted from industrialization. The outcome for Europe has been reduced family sizes and increased wages.

The Roman Empire established first infrastructure networks in southern Europe, but the Vikings linked northern Europe through commerce and conflict. The most recent endeavor to consolidate European nations is via supranationalism in the establishment of the European Union (EU). Countries with robust economies and stable governments are permitted to join the EU. Western Europe can be categorized into distinct geographic regions according to cardinal directions. The distinctions frequently pertain to the nature of economic activities individuals partake in or are founded on cultural characteristics, such as the diverse forms of Christianity or the branches of the Indo-European language family.

Since the dissolution of the Soviet Union in 1991, Eastern Europe has been shifting from Communist regimes to democratic governance accompanied by capitalist economy. Certain countries have facilitated this shift more effectively than others. A number of more progressive nations have been admitted to the EU. Other nations persist in their efforts to develop stable democratic governance and a burgeoning economy.The disintegration of former Yugoslavia exemplifies the influence of cultural factors on geographic regions. The Slavic territory of the former Yugoslavia was divided by nationalism and potent devolutionary forces. The largest state of Serbia led the dissolution with the advocacy for a Greater Serbia exclusively for Serbs. The outcome was the establishment of seven smaller states based on a nation-state model.

3 Chapter RUSSIA

3.1: INTRODUCTION: AN OVERVIEW

The vast territory of Russia encompasses diverse physical habitats, including tundras, steppes, mountains, and birch woods. Type D (continental) climates predominate across the majority of the nation and define extensive landmasses like Eurasia and North America. Inland areas situated at the heart of a vast continent, distant from the tempering influence of seas, typically experience swift heating during summer and rapid cooling in winter. These regions are characterized by sweltering summers and frigid, severe winters. Northern Russia adjoins the Arctic Ocean, with chilly air masses from the Arctic descending southward throughout Russia each winter. Furthermore, Russia's northern latitude results in a brief growing season, preventing it from becoming an agricultural giant; consequently, the nation typically relies on food imports to sustain its population. Mountain ranges to the south obstruct summer precipitation and warm air masses that would otherwise originate from South and Central Asia, hence resulting in deserts and steppes in southern Russia.

Figure 3.1 Map of Russia

The majority of Russia's population resides in the European region, namely on the Eastern European Plain, often referred to as the Western Russian Plain or the Russian Plain, which is the most agriculturally fertile area in the country. The eastern boundary of the plain is delineated by the Ural Mountains, a modest mountain range (about 6,000 feet) that traverses Russia from the Arctic Ocean to Kazakhstan. The mountains house quantities of coal, iron ore, and both valuable and semiprecious stones, serving as the demarcation between Europe and Asia. South of the Russian Plain lies the Caucasus Mountains, which connect the Caspian and Black Seas. East of the Urals lie the West Siberian Plain, the Central Siberian Plateau, the Yakutsk Basin, the Eastern Highlands, and the Central Asian Ranges. Russia possesses abundant natural resources, including petroleum, natural gas, and timber products.

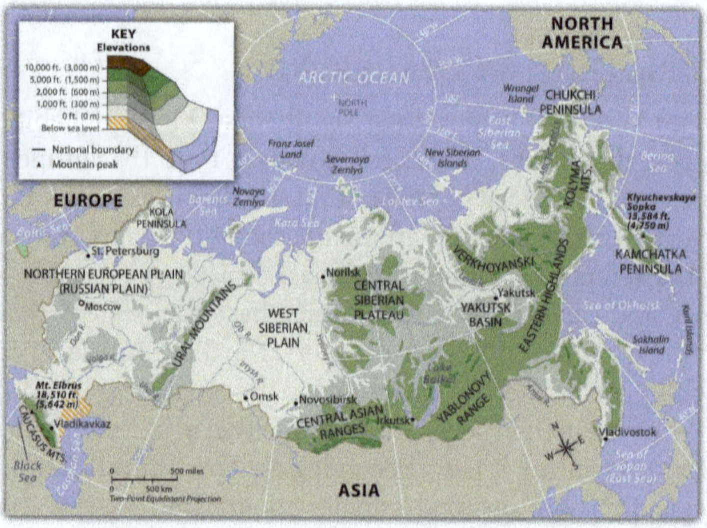

Figure 3.2 Physical Regions of Russia

Expansion of the Empire

The territory that makes up the Russian Federation was gradually conquered by the Russian Empire as the country expanded from its political core around Moscow /St. Petersburg during the sixteenth through the nineteenth centuries. By the end of the eighteenth century, Czarina Catherine the Great had expanded Russia to include the area that is now Ukraine (the north side of the Black Sea), the northern Caucasus Mountains, and Alaska (which Russia later sold to the United States). During the next century, the Russian Empire expanded eastward into Central Asia (what is now Kazakhstan, Uzbekistan, and the other Central Asian republics), southward into the rest of the Caucasus region, and westward into Poland and Finland. In the twentieth century, when the Russian Empire disintegrated and was replaced by the Soviet Union (the Union of Soviet Socialist Republics or USSR), the central government continued to expand and strengthen its control of the vast area from Eastern Europe to the Pacific Ocean.

Both the Russian Empire and the Soviet Union were imperial powers. In other words, these governments ruled a large variety of ethnic groups in distant places:

people who spoke many languages, people who worshiped different gods in different ways, people who had various skin and hair colors, and people who did not consider themselves to be Russian. Although the British and some other European powers had an arguably more difficult task of ruling empires that were widely scattered around the world, Russia had the largest empire in terms of territory. Ruling this diverse, immense empire was an incredible challenge.

The czars ruled this empire with Russification and the sword. Russification refers to the attempt to minimize cultural differences and turn all Russian subjects into Russians, or at least to make them as Russian as possible. As they were all subject to the Russian czars, people were taught the Russian language and were encouraged to convert to Russian Orthodoxy. Russification was not very successful, and the farther people were from Moscow the less likely they were to be Russified. When the Soviets took over the Russian Empire, millions of Muslims still lived in Central Asia, on the Crimean peninsula of southern Ukraine, in the Caucasus Mountains, and elsewhere.

The Soviets took a different tack when it came to taming the diversity of the empire. Instead of emphasizing unity under the Russian czar, the Russian language, and the Russian Orthodox religion, the Soviets decided to organize—and thus try to control—the diversity of ethnic groups found in the Soviet Union.

Figure 3.3: Former Soviet Union

They chose some of the major groups (Uzbek, Kazakh, and so forth) and established Soviet Socialist Republics that corresponded to these major groups. Thus they created the Uzbek Soviet Socialist Republic (Uzbek SSR), the Kazakh Soviet Socialist Republic (Kazakh SSR), the Ukrainian Soviet Socialist Republic (Ukrainian SSR), and a different republic for each of fourteen ethnic groups, plus the Russian Soviet Federative Socialist Republic. About eighty-five other ethnic groups were not given their own republics, although some of them were allocated regions within the Russian Republic. In theory, each of the Soviet Socialist Republics was an independent state choosing to ally with the Soviet Union. In practice, of course, these republics were part of a totalitarian, centrally ruled state with far fewer autonomous rights than states in the United States.

The creation of these republics strengthened certain ethnic/national identities and weakened others. There had not been a fully developed Uzbek national identity before the formation of the Uzbek SSR. The same was true for the Kazakh SSR, the Turkmen SSR, and others. Although people in a certain area might have spoken the same language, they did not think of themselves as belonging to a nation of fellow Uzbeks, Kazakhs, or Turkmen until they were put into one by the Soviet rulers. In 1991, when the Soviet Union collapsed, these Soviet Socialist Republics were able to declare their independence from Russia, and the national identities fostered during the Soviet era came to fruition.

At the same time that the Soviets were organizing minority ethnic groups into republics, they were also sending ethnic Russians to live in non-Russian parts of the Soviet Union. Some were sent by force—such as Russians who were sent to prison camps in Siberia and stayed in the area after they were eventually freed. Other Russians were sent around the empire to work in factories, power plants, and other industries, or they were sent to help administer the government. By sending Russians to the far reaches of the Soviet Union, the Russian government hoped to consolidate its control over the various republics and to dilute the strength of the minority ethnicities. This policy also had unintended consequences: when the Soviet Union collapsed after 1991 and the various republics became independent countries, they each had to deal with sizable Russian minorities. For example, at the time of its independence, nearly as many ethnic Russians lived in Kazakhstan (38 percent) as ethnic Kazakhs (40 percent). In the twenty years since then, many Russians moved to Russia from the former Soviet republics. In 2010, Kazakhstan's population was only about 24 percent ethnic Russian[1].

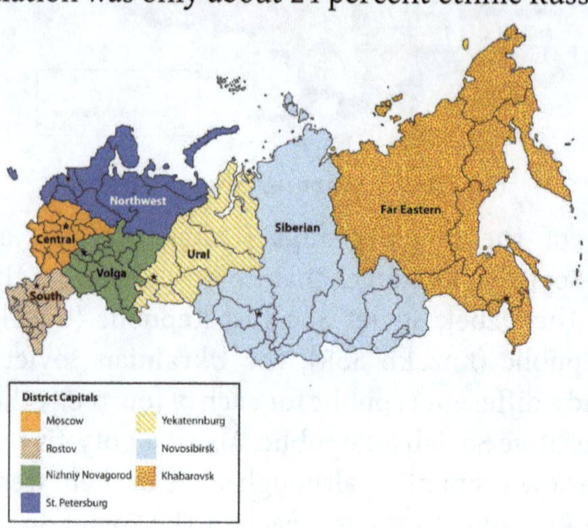

Figure 3.4: New Russian Federal Districts. Wikimedia Commons – CC BY 2.5.

The Russian federation was created with eighty-three federal subjects: two autonomous federal cities; forty-six provinces (**oblasts**) and nine territories (**krais**) that function in the same way and are the most common type of federal unit; twenty-one republics; four autonomous districts (**okrugs**); and one autonomous oblast. Moscow and St. Petersburg are the two federal cities that function as their own units. The oblasts and krais each have a governor appointed by the central government and a locally elected legislature. The governorship was an elected position in the 1990s, but President Vladimir Putin changed the structure to strengthen the power of the central state. The republics, designed to be home to certain ethnic minorities, are allowed to have their own constitutions and governments and to select an official language that will be used besides Russian, but they are not considered independent countries with the right to secession. The autonomous districts were also formed for ethnic minorities and are administered either by the central state or by the province or territory in which they are located. The only autonomous oblast was created in the 1930s to be a home for Jews in the Russian Far East, but only about 1 percent of the population remains Jewish today.

Regional Environmental Problems

Each region of the Russian republic has its own environmental issues. The core region surrounding Moscow, with all its industrial activity and large urban expanses, introduces sewage and chemicals to the country's waterways, contributing to serious water pollution. The same water pollution is found east of the Ural Mountains—and in the waterways in that region—because of the industrial cities found there. Moscow and the ring of industrial cities surrounding it have seen a dramatic increase in automobile use since 1991, contributing to air pollution. Russia is blessed with abundant natural resources, but significant environmental damage has been the price of exploiting and extracting those resources. Massive oil spills have occurred in the taiga and tundra areas, where the lack of safety management has increased environmental damage during oil exploration and development. The taiga is the large expanse of evergreen or boreal forests in the north just south of the tundra in North America, Europe, and Asia. The taiga is most common in type D climates and is one of the largest biomes on earth. The taiga is the largest biome in Russia. Mining and smelting processes in Siberian cities have added to the region's air and water pollution. These ecosystems are rather fragile and will take years to recover from such damage.

Water pollution from the rivers extends into the Black Sea, the Caspian Sea, and other bodies of water. **Lake Baikal**, described as the largest freshwater lake in the world, was at one time pristine, but pollutants have entered its waters from nearby industrial activity. Increased pollution in the Black and Caspian Seas, as well as overfishing, the lack of fishing regulations, and the lack of law enforcement, has resulted in the devastation of fish populations such as the caviar-producing sturgeon. The Arctic waters of the **Barents Sea** off the northern coast of Russia have been a dumping ground for nuclear waste products and expended nuclear reactors from naval vessels. The consequences of this nuclear pollution are not widely known or studied. Many additional aging nuclear reactors from the Soviet era dot the landscape, and they will need to be decommissioned at some point, adding to the nuclear waste issues. Various regions in Siberia were used for nuclear testing and are also contaminated with nuclear radiation.

Key Takeaways

- Russia's climate is characterized by long, cold winters and short summers, and because of a short growing season, the country has such poor agriculture that usually it needs to import grain.
- The Russian Empire gradually expanded its territory to the east, west, and south of Moscow and by the end of the nineteenth century had accumulated a vast area of land and a great variety of people of many ethnicities, languages, and religions.
- The Soviets sent ethnic Russians across the Union of Soviet Socialist Republics (USSR) to better control and govern the territory and its people.
- When the Soviet Union collapsed, its internal nationality-oriented republics declared independence, and the Russians who had been sent to live around the USSR were now living in non-Russian countries. While most of the ethnic Russians stayed in their current countries, many others returned to Russia after 1991.

Discussion and Study Questions

1. What physical feature provides a divide between European and Asian Russia?
2. What are Russia's main physical regions?
3. In what ways was Russia a colonial empire?
4. What strategies did the Russian Empire use to govern a multiethnic state?
5. What were Soviet Socialist Republics, and why were they created?
6. What was a long-term consequence of the formation of Soviet Socialist Republics for the current geopolitical organization of this world region?
7. What are some of the major environmental problems in Russia?
8. Why did so many of the Soviet republics separate and become independent in 1991?
9. Why wasn't the Russian Orthodox Church more prominent or powerful during the Soviet era?

10. Why do more people live on the smaller European side of Russia than the larger Asian side?

Geography Exercise

Identify the following key places on a map:

- Barents Sea
- Black Sea
- Caspian Sea
- Caucasus Mountains
- Central Asian Ranges
- Central Siberian Plateau
- Eastern Highlands
- European Plain
- Lake Baikal
- Ural Mountains
- West Siberian Plain
- Yakutsk Basin

3.2 THE USSR AND THE RUSSIAN FEDERATION

Section 3.1, "Introducing the Realm," elucidates that the Russian Empire was established by the czars over several centuries. Nonetheless, the economic and political frameworks of the Russian Empire were not viable in the contemporary age. The overwhelming majority of the populace was impoverished, predominantly consisting of landless peasant farmers; in a region characterized by brief growing seasons, agriculture was not a viable route to wealth. Political decisions were determined by a minuscule elite cohort. At the onset of the twentieth century, a century after the Industrial Revolution transformed Great Britain and Western Europe, Russia continued to be an agrarian nation and had not initiated extensive industrialization. Beyond the nobility, minimal support existed for the status quo in Russia, accompanied by a pervasive aspiration for a novel political structure and governance.

Nevertheless, consensus on the structure of a new administration remained elusive. Following the First World War, a civil war ensued in Russia. In these tumultuous times, the final czar, Nicholas II, was ousted from power, and he along with his family was executed. The most formidable faction vying for dominance in Russia was the Communist organization known as the Bolsheviks, which translates to the "larger group." Other factions, notably the lesser faction, the Mensheviks, were defeated in the civil war. Vladimir Lenin, the Bolshevik leader, and his supporters initiated a campaign in 1917 to transform Russia into a Communist state.

The capital was relocated to Moscow from St. Petersburg, where it had resided since the era of Czar Peter the Great in the eighteenth century. St. Petersburg was renamed Petrograd and subsequently Leningrad to honor Vladimir Lenin, as the atheistic Soviets sought to eliminate allusions to Christian saints. The complete expanse of the Russian Empire was transformed into the Union of Soviet Socialist

Republics (USSR). The Russian populace exchanged a monarchy for a Marxist totalitarian regime (see to the elucidation of Marxism in Section 3.2.1 "Marxist-Leninist Central Planning").

Figure 3.5 Vladimir Lenin and Josef Stalin Wikimedia Commons – public domain

The Soviet Union lasted from 1922 to 1991. **Josef Stalin**, the Soviet dictator who took over after Lenin, was incapacitated in 1922 (and died in 1924). He was a ruthless leader who murdered his way to power and killed or exiled anyone who got in his way. Stalin is famous for initiating economic plans that helped move Russia from a poor, agrarian state to a large, industrial superpower. He pushed for rapid industrialization, the eradication of family farms in lieu of large communal farms, the end of personal ownership of land or businesses, and the dramatic weakening of organized religion. All these changes came at a great price. During his reign of terror, an estimated thirty million people lost their lives. The forced collectivization of agriculture brought about a devastating famine in 1932–33, in which between six and eight million people starved to death or were killed outright, many of them in Ukraine. Stalin led periodic purges of his perceived political enemies. The largest of these is known as the Great Purge. At that time (1936–38), about one million so-called enemies of the state were executed. More people lost their lives under Stalin than in all the concentration camps of Adolf Hitler's Nazi regime. The full extent of Stalin's purges of his people may never be fully known. Stalin's rule ended in 1953, when he reportedly died of natural causes. However, some historians believe he was poisoned by his close associates.

Marxist-Leninist Central Planning

The Soviet Union espoused the philosophies of **Karl Marx**, a nineteenth-century German theorist. Marx wrote that all political and economic life can be understood as a struggle between the various classes in society. People who adhere to Marx's philosophy are called Marxists, and the Soviet version of Marxism is called Marxism-Leninism. In Marxist thought, capitalism is an oppressive economic system in which the working class (the proletariat) is oppressed by the bourgeoisie (the wealthy middle class). Marxists believe that the proletariat should revolt, rise up against the bourgeoisie, take the property away from the rich, and give it to the government to control it for the benefit of the common people. Ultimately, a pure Communist system would result, with no social or economic classes, no private property, no rich people, and no poor people. In real life, governments that adopt these ideas practice socialism and are said to be socialist.

Table 3.1 Soviet Leaders and Russian Presidents

Soviet Union Leaders	Time as Leader	Life Span
Vladimir Ilich Ulyanov (*Lenin*)	1917–24d	1870–1924
Josef Vissarionovich Djugashvili (*Stalin*)	1924–53d	1879?–1953
Nikita Sergeyevich Khrushchev	1953–64	1894–1971
Leonid Ilyich Brezhnev	1964–82d	1906–1982
Yuri Vladimirovich Andropov	1982–84d	1914–84
Konstantin Ustinovich Chernenko	1984–85d	1911–85
Mikhail Sergeyevich Gorbachev	1985–91	1931–
Russian Federation presidents (1991+)		
Boris Nikolayevich Yeltsin	1991–99	1931–2007
Vladimir Vladimirovich Putin	2000–2008	1952–
Dmitry Anatolyevich Medvedev	2008–	1965–
d = died while in office		

As a socialist state, the Soviet Union did not include open markets. The Soviet Union was a command economy, in which economic decisions were made by the state and not left to the market to decide. During the Soviet era, for example, industrial production was planned by the central government. The government would decide what would be produced, where it would be produced, the quantity produced, the number of workers who would produce it, where the raw materials would come from, and how the final product would be distributed. By mobilizing the entire country to work toward common goals, the USSR was able to achieve the rapid industrialization that it so desired. However, the Soviets underestimated the power and efficiency of free-enterprise capitalism, and their socialist system was undermined by waste, fraud, and corruption.

Another main economic feature of the Soviet Union was collectivized agriculture. The Soviet leaders did not want individual, capitalist farmers to become rich and threaten their economic system. Nor did they want thousands of small, inefficient farms when the country was perpetually unable to feed itself. Instead, they decided to streamline agricultural production into large farm factories. All the farmland in each area was consolidated into a government-owned collective operation. Some collective farms were run by the state, while others were run by private cooperatives. During the transition period to collective farming, individual farmers were forced to give up their land, animals, farm equipment, and farm buildings and donate them to the collective farm in their area. The state also demanded a high percentage of the crops produced. At times, the government collected the entire harvest, not even allowing seed crops to be held for the following season. This brought about widespread famine in 1932–33. Collectivized agriculture remained the norm in the Soviet Union until the country's dissolution in 1991 and even afterward in some areas.

The Cold War

From the end of World War II in 1945 until the collapse of the USSR in 1991, the Soviet Union and the United States competed in the global community for the control of labor, resources, and world power. Each side attracted allies, and most countries were on the side of either the United States or the Soviet Union; very few remained neutral. This era, known as the Cold War, did not involve direct military armed conflict between the United States and the Soviet Union, but it transformed the world into a political chessboard, with each side wanting to block the other side from gaining ground. Whenever the Soviets would enter into an alliance with a certain country, the United States was right there to try to counter the move. Wars, armed conflicts, sabotage, spying, and covert activities were the methods of the Cold War. Both sides stockpiled as much deadly weaponry as possible, including nuclear warheads and missiles. They also competed in the race to put people in outer space.

Figure 3.6: October Revolution Celebration 1983, Moscow, during the Cold War. Wikimedia Commons – public domain.

The Cold War led to wars fought in Vietnam, Korea, Grenada, Afghanistan, Angola, and the Middle East, with the Soviet Union funding one side and the United States supplying the other. Covert wars or guerilla wars with secret agents and political assassinations were fought in Cuba, Nicaragua, Chile, Guatemala, Mozambique, Laos, Cambodia, and a host of other third-world countries. The Cold War divided the world into two main camps, each with a high number of nuclear weapons. Eastern Europe was sectioned off by the **Iron Curtain**, and the Berlin Wall divided the city of Berlin, Germany. These physical barriers divided the communist countries of Eastern Europe with the capitalist democracies of Western Europe. Germany itself was divided into two separate countries, as explained in Chapter 2 "Europe".

Various Soviet dictators came to power and died in office before the end of the Cold War. The last Soviet leader was **Mikhail Gorbachev**, who assumed power in 1985. The US president at the time was Ronald Reagan. During the 1980s the United States was outspending the Soviets militarily, and its economy was growing at a much faster rate than that of the USSR. At the same time, the Soviets were engaged in a costly war in Afghanistan, and their economy was faltering and in danger of collapse. Gorbachev realized that reforms had to be implemented to modernize the Soviet system: political life needed to be more open so that people would feel ownership of the country, and the economy needed to be restructured. Gorbachev implemented perestroika (restructuring of the economy with market-like reforms) and glasnost (openness and transparency of all government activities). The restructuring exposed fundamental problems in the economy, and by 1990 the Soviet economy was in worse shape than ever before.

Figure 3.7: The first McDonald's restaurant in the former Soviet Union was in Moscow, Russia.

Fran Urbano – First McDonalds in Russia – CC BY 2.0.

The end came in 1991: the Soviet Union collapsed when fourteen of the Soviet republics broke away and declared their independence. At this point, the Soviet state was too weak to prevent it. All the republics, including Russia itself (now called the Russian Federation), became independent countries. The only territories that did not achieve independence were the smaller republics and autonomous regions that existed within the Russian Federation's boundaries. The Iron Curtain melted away seemingly overnight, and people were free to travel to and from the former Communist countries. The old Russian flag flew over the Kremlin—the seat of the Russian government—for the first time since Czar Nicholas II had been in power seventy-three years earlier. The Communist era of the Soviet Union and the Cold War were over.

Figure 3.8: Walls of the Kremlin in Moscow *Kremlin* is a Russian word for a walled city. Inside Moscow's Kremlin is the famous Red Square. Larry Koester – Kremlin – CC BY 2.0.

The post-Soviet transition was filled with political, economic, and social turmoil. **Boris Yeltsin**, the first president of the new Russian Federation, ushered in a series of economic reforms that privatized state-owned enterprises. Russian leaders tried to reverse socialism rapidly through what they called "shock therapy," which they knew would be painful but hoped would be brief. These reforms created a new class of capitalist entrepreneurs.

Wealth, once controlled by the political elite, was now being shifted to the business elite, a pattern found in most capitalist countries. Many ordinary workers faced unemployment for the first time as the new owners of various companies trimmed unnecessary staff. Private ownership forced housing costs to skyrocket, and while families were generally allowed to keep the homes they already lived in, purchasing houses or condominiums became out of reach for many people. The value of the **ruble**, Russia's currency, declined rapidly, and older people watched as their life savings evaporated overnight. Western goods were more easily accessible, and all kinds of consumer products became much more widely available than they were during the Soviet era. Bread lines and empty store shelves became distant memories as the former Communist state shifted over to a type of capitalist democracy. Yeltsin became increasingly less popular as citizens became dissatisfied with corruption and the high social costs of the post-Soviet transition. He resigned in 1999 and was replaced by **Vladimir Putin.**

Outline of Russia's Historical Geography

1. **The Region's Early Heritage**
   - Vikings created fortified trading towns called gorods
   - Genghis Khan's Mongol Empire invaded (1240 CE)
   - Feudal states arose around dominant trading centers
2. **Czarist Russia, 1547–1917**
   - Czars unified empire by internal colonialism
   - Forward capital of St. Petersburg created
   - Pioneers pushed eastward to Siberia and North America
3. **Bolshevik Revolution, 1917–22**
   - Czar Nicholas II and his family executed
   - Russian Civil War fought
   - Vladimir Lenin created Communist imperial state
   - Capital moved to Moscow; republics broke away
4. **The Soviet Union (USSR), 1922–91**
   - Czarist Empire became the Soviet Union
   - Central planning, collectivization, and the Cold War began
   - Republics kept together by military force
   - External interaction of glasnost initiated (1980s)
   - Economic restructuring and reforms of perestroika introduced (1980s)
5. **The Russian Republic, 1991–Present**
   - Independent republics lost with internal unrest
   - Economics privatized
   - Democracy introduced (1990s)
   - Central state strengthened (twenty-first century)

Figure 3.9: Moscow and the Moskva River Behind the Borodinsky Bridge on the right are the government buildings of the Russian Federation. Kirill Vinokurov – Borodinsky Bridge – CC BY 2.0.

Twenty-First-Century Russia

A Russian style of capitalism replaced the social, political, and economic system of the Communist era with a growing market economy. The export of Russia's vast quantities of natural resources, such as oil, natural gas, and timber, to Europe and the rest of the world helped the country rebound from the economic collapse of the 1990s. Russia has benefited from the recent increase in energy prices, and oil, natural gas, metals, and timber account for more than 80 percent of exports and 30 percent of government revenues. However, Russia still needs to modernize its dilapidated manufacturing base if it is to economically compete against the European Union, North America, or eastern Asia. During Vladimir Putin's presidency (2000–2008), Russia witnessed substantial economic growth that inspired foreign investors to pump money into the Russian economy and catapulted Moscow into an investment haven and one of the richest cities in the world. In 2008, Moscow claimed to have more billionaires than any other city in the world. Russia has reestablished itself as a major player in the global economy, although much of its population still suffers from poverty and social problems.

Russia's population grew steadily during the Soviet era, except during periods of famine or warfare, and the country underwent a rural-to-urban shift as farm workers moved to cities to labor in factories. However, when the USSR collapsed, Russia's population began a steep decline, falling from a peak of 149 million in 1991 to about 143 million in 2005. The trend of low birth rates actually began during the Soviet period. The population decline occurred because birth rates always decline during periods of economic and social crisis as people delay or decide against having children, and the country experienced particularly high death rates because of alcoholism, heart disease, and the collapse of the social safety net. Any time death rates surpass birth rates, a country's population will decline unless the difference is offset by immigration. Russia's birth rate was also impacted by very high rates of abortion: in 1992, for example, there were 221 abortions in Russia for every 100 live births (Johnston's Archive). Although Russia's population has seemingly bottomed out, the only reason it is not continuing to decline is immigration from the former Soviet republics. Russia still has a negative rate of natural increase.

To put Russia's demographic profile in context, its fertility rate was only about 1.5 in 2010, meaning that the average woman would have 1.5 children in her lifetime. This is below the 2.1 children each woman would need to have for the population to remain stable. The fertility rate in Russia is similar to that of countries in Europe (1.5) but lower than that of the United States (2.1) (Central Intelligence Agency). Life expectancy for Russian men is variously reported as sixty years up to sixty-three years, while women can expect to live seventy-three to seventy-five years. In Western Europe, life expectancies are about eighty years.

About 80 percent of Russia's population is ethnically Russian. The next largest group is Tatar (3.8 percent), a group that traditionally has spoken the Tatar language and practiced Islam. More than 150 ethnic groups are represented in Russia, including indigenous people of the Arctic who herd reindeer for a living. Each of the nationalities of the former Soviet republics has a presence in Russia, and because of its relatively strong economy, Russia (especially Moscow) is an immigration magnet for residents of those countries. Most of these groups have their own language and cultural traditions.

Figure 3.10: The Cathedral of Intercession of the Virgin on the Moat, Also Known as the Cathedral of St. Basil the Blessed, on the Red Square, Moscow. Wikimedia Commons – public domain.

The Russian Orthodox Church is the dominant religious denomination in Russia. For generations, it was the country's official religion, and Russian people were automatically considered to be Orthodox, no matter what their personal beliefs. During the Soviet era, the government did much to weaken the church, including killing tens of thousands of priests, monks, and nuns and closing most churches. The much smaller church that survived was largely controlled by the state. Because of state-sanctioned atheism during the Soviet era, only 15 to 20 percent of Russia's population today actively practices Orthodoxy, although a much greater number claim to be Russian Orthodox Christians. Another 15 percent of the country's population practices Islam, especially in places such as the Caucasus region and the southern Ural Mountains, and about 2 percent practice other forms of Christianity, such as Catholicism and Protestantism (Central Intelligence Agency). Even though Orthodoxy is practiced by a minority of Russia's population, the church has increased its influence since 1991 and often acts as an official church. A small percentage of the population are Buddhists, and various ethnic groups in Siberia and the Far East practice tribal religions and pagan rites.

The Russian Language

Like English, Russian is an Indo-European language. Russian is on the Slavic branch of the language tree, along with languages such as Polish, Ukrainian, and Serbo-Croatian. A relatively new language, it developed from a language called Old East Slavic that was spoken around 1100 CE. Medieval Russian developed in the thirteenth century, and modern literary Russian is usually traced back to the early nineteenth century. Russian uses the Cyrillic alphabet, which was developed from Greek. The Russian language is the official language of the Russian republic. With over one hundred languages used in the country, twenty-seven of them are officially recognized in various regions within the republic.

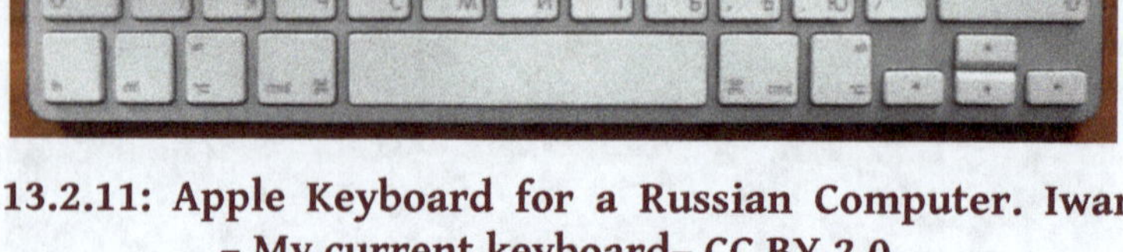

Figure 3.2.113.2.11: Apple Keyboard for a Russian Computer. Iwan Gabovitch – My current keyboard– CC BY 2.0.

Key Takeaways

- The Communist state was created in 1917 as the result of a civil war, which evolved into the creation of the Soviet Union in 1922. During this era, the Russian people traded a monarchy for a Marxist totalitarian state.

- One of the primary goals of the new USSR was rapid industrialization, and this goal was achieved through central planning and the collectivization of agriculture. Inefficiencies in the system persisted throughout the Soviet era.
- The Soviet Union's Communist system was maintained at great cost: millions were killed by purges and in government-produced famines.
- A weakened USSR tried to reform in the 1980s through Mikhail Gorbachev's policies of perestroika and glasnost but instead collapsed in 1991.
- Early post-Soviet years were ones of democratization, rapid privatization, and the unraveling of the social safety net. The shift to capitalism restructured the Russian economy.
- The Russian economy strengthened after 2000, and Russian president Vladimir Putin strengthened the power of the central state.
- Russia has a low fertility rate and a negative rate of natural increase. High rates of abortions and alcoholism have been contributing factors.
- The Russian Orthodox Church is the dominant religious denomination in Russia.

Discussion and Study Questions

1. By what process did the Russian Empire become the Soviet Union?
2. Who were the first two leaders of the USSR, what were some of their accomplishments, and what were some of the human costs?
3. What were some of the main features of Karl Marx's philosophy?
4. What were two main economic policies of the Soviet Union?
5. Who was fighting the Cold War? Where were major conflicts fought during this war?
6. How did the Cold War end? What leaders were in power in the United States and the Soviet Union when it ended?
7. What two policies did the last Soviet leader implement to assist in reforming the USSR?
8. What happened to the many external republics after 1991?
9. How did the economic system change for Russia after 1991?
10. What are some reasons for Russia's population decline?

Geography Exercise

Identify the following key places on a map:

- Moscow
- St. Petersburg

References

Central Intelligence Agency, "The World Factbook," https://www.cia.gov/library/publications/the-world-factbook/geos/xx.html.

Johnston`s Archive, "Historical Abortion Statistics, Russia," http://www.johnstonsarchive.net/policy/abortion/ab-russia.html.

3.3: REGIONS OF RUSSIA

The majority of Russia's populace and principal industry are situated west of the Ural Mountains on the Russian Plain. This area, recognized as Russia's geographic core, encompasses the Moscow region, the Volga region, and the Ural Mountain region. Moscow, the capital of Russia, serves as the core of an industrial region inhabited by over fifty million individuals. Moscow has around ten million inhabitants, with over thirteen million in its metropolitan region, rendering it marginally smaller than the Los Angeles, California, metropolitan area. A ring of industrial cities encircling Moscow include essential production hubs of Russian manufacturing. During the Communist era, Moscow expanded from its nineteenth-century nucleus, despite the city's origins dating back to at least the twelfth century, evolving into an industrial metropolis characterized by planned communities. This premier metropolis possesses a comprehensive subway and road network that is being augmented to accommodate present expansion requirements. Despite the set prices of rents, commodities, and home products during the Communist era, the dissolution of the Soviet Union altered this situation. Moscow currently ranks among the costliest cities globally, with prices determined by supply and demand dynamics. Numerous individuals aspire to reside in Moscow; nonetheless, it remains monetarily unattainable for a significant portion of the Russian populace.

St. Petersburg, Russia's second-largest city, with a population of approximately five million. Situated on the Baltic Sea, it is the principal port city in western Russia. The city was designated as Petrograd (1914–24) and Leningrad (1924–91), but is currently frequently referred to as Petersburg, or simply "Peter" for brevity. In the early eighteenth century, Peter the Great constructed the city with assistance from European architects to compete with other European capitals, designating it as the capital of the Russian Empire. Named for St. Peter from the Bible (distinct from Peter the Great), it serves as a cultural hub for Russia and a prominent tourist attraction. It is also recognized for shipbuilding, oil and gas commerce, industry, and finance. The most significant catastrophe occurred during a twenty-nine-month siege by the German forces in World War II. Approximately one million residents perished from starvation or shelling, while hundreds of thousands evacuated the city, rendering it almost desolate by the siege's conclusion.

Figure 3.12: Russia's Core Region. Source: Map courtesy of the CIA World Factbook.

To the far north of St. Petersburg on the Barents Sea are the cities of **Murmansk** and **Arkhangelsk.** Murmansk is a major military port for Russia's navy and nuclear submarine fleet. Relatively warm water from the North Atlantic drift circles around Norway to keep this northern port city fairly free of ice. Arkhangelsk (which literally means "archangel"), used as a port for lumber exports, has a much shorter ice-free season than Murmansk. Both of these cities are in Russia's far north, with long winters and exceedingly brief summers.

The **Volga River** flows through the core region of Russia, providing transportation, fresh water, and fishing. The Volga is the longest river in Europe at 2,293 miles, and it drains most of Russia's western core region. This river has been a vital link in the transportation system of Russia for centuries and connects major industrial centers from the Moscow region to the south through an extensive network of canals and other waterways. The Volga River flows into the **Caspian Sea**, and a canal links the Volga with the **Black Sea** through a connecting canal via the **Don River.**

At the eastern edge of Russia's European core lie the Ural Mountains, which act as a natural divide between Europe and Asia. These low-lying mountains have an abundance of minerals and fossil fuels, which make the Ural Mountains ideal for industrial development. The natural resources of the Urals and the surrounding area provide raw materials for manufacturing and export. The eastern location kept these resources out of the hands of the Nazis during World War II, and the resources themselves helped in the war effort. Oil and natural gas exploration and development have been extensive across Russia's core region and have greatly increased Russia's export profits.

Figure 3.13: Russia's Eastern Frontier, the Far East, and Siberia. Source: Map courtesy of the CIA World Factbook.

The Eastern Frontier

East of the Urals, in south-central Russia, is Russia's **Eastern Frontier**, a region of planned cities, industrial plants, and raw-material processing centers. The population is centered in two zones here: the **Kuznetsk Basin** (or Kuzbas, for short) and the **Lake Baikal** region. The Kuzbas is a region of coal, iron ore, and bauxite mining; timber processing; and steel and aluminum production industries. Central industrial cities were created across the Eastern Frontier to take advantage of these resource opportunities. The most important of these is **Novosibirsk**, the third-largest city in Russia after Moscow and St. Petersburg and home to about 1.4 million people. The city is not only noted for its industries but it is also the region's center for the arts, music, and theater. It is host to a music conservatory and a philharmonic orchestra, a division of the Russian Academy of Sciences, and three major universities.

Agriculture, timber, and mining are the main economic activities in the eastern Lake Baikal region, which is more sparsely settled than the Kuzbas. Lake Baikal (400 miles long, 50 miles wide) holds more fresh water than all the US Great Lakes together and about 20 percent of all the liquid fresh water on the earth's surface. Its depth has been recently measured at 5,370 feet (more than a mile). Some of the longest river systems in the world flow through the **Eastern Frontier**. The **Irtysh**, **Ob**, **Yenisey**, and **Lena** are the main rivers that flow north through the region into Siberia and on to the Arctic Ocean. To the east, the **Amur River** creates the border between Russia and China until it flows north into the **Sea of Okhotsk.** In addition to waterways, the Trans-Siberian Railway is the major transportation link through the Eastern Frontier, connecting Moscow with the port city of **Vladivostok** in the Far East.

Siberia

Siberia, as a place name, actually refers to all of Asian Russia east of the Ural Mountains, including the Eastern Frontier and the Russian Far East. However, in this and some other geography textbooks, the term *Siberia* more specifically describes only the region north of the Eastern Frontier that extends to the **Kamchatka Peninsula.** The word *Siberia* conjures up visions of a cold and isolated place, which is true. Stretching from the northern Ural Mountains to the **Bering Strait,** Siberia is larger than the entire United States but is home to only about fifteen million people. Its cities are located on strategic rivers with few overland highways connecting them.

Type D (continental) climates dominate the southern portion of this region, and the territory consists mainly of coniferous forests in a biome called the taiga. This is one of the world's largest taiga regions. Type E (polar) climates can be found north of the taiga along the coast of the Arctic Sea, where the tundra is the main physical landscape. No trees grow in the tundra because of the semifrozen ground. Permafrost may thaw near the surface during the short summer season but is permanently frozen beneath the surface. On the eastern edge of the continent, the mountainous Kamchatka Peninsula has twenty active volcanoes and more than one hundred inactive volcanoes. It is one of the most active geological regions on the Pacific Rim.

Figure 3.14: Mount Koryasky, an Active Volcano, and Petropavlovsk-Kamchatsky, a Town on Russia's Kamchatka Peninsula. Wikimedia Commons – public domain.

The vast northern region of Russia is sparsely inhabited but holds enormous quantities of natural resources such as oil, timber, diamonds, natural gas, gold, and silver. There are vast resources in Siberia waiting to be extracted, and this treasure trove will play an important role in Russia's economic future.

The Far East

Figure 3.15: Vladivostok Tram Opposite the Admiral's Club on Its Way to the Railway Station, Vladivostok. Wikimedia Commons – CC BY 3.0.

Across the strait from Japan is Russia's **Far East** region, with the port of **Vladivostok** (population about 578,000) as its primary city. Bordering North Korea and China, this Far East region is linked to Moscow by the Trans-Siberian Railway. Before 1991, Vladivostok was closed to outsiders and was an important army and naval base for the Russian military. Goods and raw materials from Siberia and nearby **Sakhalin Island** were processed here and shipped west by train. Sakhalin Island and its coastal waters have oil and mineral resources. Industrial and business enterprises declined with the collapse of the Soviet Union in 1991. Today, the Far East is finding itself on the periphery of Russia's hierarchy of productivity. However, it has the potential to emerge again as an important link to the Pacific Rim markets.

Southern Russia

In the southern portion of the Russian core lies a land bridge between Europe and Southwest Asia: a region dominated by the Caucasus Mountains. To the west is the Black Sea, and to the east is the landlocked Caspian Sea. The Caucasus Mountains, higher than the European Alps, were formed by the Arabian tectonic plate moving northward into the Eurasian plate. The highest peak is **Mt. Elbrus** at 18,510 feet.

Located on the border between **Georgia** and Russia, Mt. Elbrus is the highest peak on the European continent as well as the highest peak in Russia.

Most of this region was conquered by the Russian Empire during the nineteenth century and held as part of the Soviet Union in the twentieth. However, only a minority of its population is ethnic Russian, and its people consist of a constellation of at least fifty ethnic groups speaking a variety of languages.

Since the collapse of the Soviet Union, the Caucasus region has been the main location of unrest within Russia. Wars between Russia and groups in the Caucasus have claimed thousands of lives. Some of the non-Russian territories of the Caucasus would like to become independent, but Russia fears an unraveling of its country if their secession is allowed to proceed. To understand why the Russians have fought the independence of places such as **Chechnya** but did not fight against the independence of other former Soviet states in the Caucasus such as **Armenia**, it is necessary to study the administrative structure of Russia itself.

Figure 3.16: Republics of Southern Russia, Including Chechnya. Wikimedia Commons – CC BY-SA 3.0.

Of the twenty-one republics, eight are located in southern Russia in the Caucasus region. One of these, the **Chechen Republic** (or Chechnya), has never signed the Federation Treaty to join the Russian Federation; in fact, Chechnya proposed independence after the breakup of the Union of Soviet Socialist Republics (USSR).

Although other territories to the south of Chechnya, such as Georgia, Armenia, and Azerbaijan, also declared their independence from Russia after 1991, they were never administratively part of Russia. During the Soviet era, those countries were classified as Soviet Socialist Republics, so it was easy for them to become independent countries when all the other republics (e.g., Ukraine, Belarus, and Kazakhstan) did so after 1991. However, Chechnya was administratively part of the USSR with no right to secession. After 1991, Russia decided that it would not allow territories that had been administratively governed by Russia to secede and has fought wars to prevent that from happening. It feared the consequences if all twenty-one republics within the Russian Federation were declared independent countries.

Chechnya has fought against Russia for independence twice since the USSR's collapse. The First Chechen War (1994–96) ended in a stalemate, and Russia allowed the Chechens to have de facto independence for several years. But in 1999, Russia resumed military action, and by 2009 the war was essentially over and Chechnya was once more under Russia's control. Between twenty-five thousand and fifty thousand Chechens were killed in the war, and between five thousand and eleven thousand Russian soldiers were also killed (Wikipedia). In 2003, the United Nations called **Grozny**, Chechnya's capital, the most destroyed city on Earth. Reconstruction of Grozny has slowly begun.

Even before the recent wars, Chechnya had a difficult past. Over the course of its history, it has been at the boundary between the Ottoman Empire, the Persian Empire, and the Russian Empire. Most of the people converted to Sunni Islam in the 1700s to curry favor with the Ottomans and seek their protection against Russian encroachment. Nevertheless, Chechnya was annexed by the Russian Empire. During Soviet leader Josef Stalin's reign of terror, more than five hundred thousand Chechens were loaded on train cars and shipped to Kazakhstan, where as many as half died.

Transcaucasia

The independent countries of **Georgia**, **Armenia**, and **Azerbaijan** make up the region of **Transcaucasia**. Although they are independent countries, they are included in this chapter because they have more ties to Russia than to the region of Southwest Asia to their south. They have been inextricably connected to Russia ever since they were annexed by the Russian Empire in the late eighteenth and early nineteenth centuries, and they were all former republics within the Soviet Union. When the Soviet Union collapsed in 1991, these three small republics declared independence and separated from the rest of what became Russia.

Geographically, these three countries are located on the border between the European and Asian continents. The Caucasus Mountain range is considered the dividing line. The region known as Transcaucasia is generally designated as the southern portion of the Caucasus Mountain area.

Figure 3.17: Southern Russia and Transcaucasia. Wikimedia Commons – CC BY-SA 3.0.

The country of Georgia has a long history of ancient kingdoms and a golden age including invasions by the Mongols, Ottomans, Persians, and Russians. For a brief three years—from 1918 to 1921—Georgia was independent. After fighting an unsuccessful war to remain free after the Russian Revolution, Georgia was absorbed into the Soviet Union. Since it declared independence in 1991, the country has struggled to gain a stable footing within the world community. Unrest in the regions of **South Ossetia, Abkhazia**, and **Adjara** (where the populations are generally not ethnic Georgian) has destabilized the country, making it more difficult to engage in the global economy. Russia and Georgia had a military conflict in 2008, when Russian troops entered the South Ossetian region to support its move toward independence from Georgia. Georgia considered South Ossetia to be a part of Georgia and called the Russians an occupying force. Many other countries, including the United States, condemned Russia for their action. Russian troops pulled out of Georgia but supported the independence of South Ossetia and Georgia's westernmost region of Abkhazia. Neither South Ossetia nor Abkhazia are considered independent states by most of the world's countries.

A democratic-style central government has emerged in Georgia, and economic support has been provided by international aid and foreign investments. The country has made the switch from the old Soviet command economy to a free-market economy. Agricultural products and tourism have been Georgia's main economic activities.

In 2010, Armenia, to the south of Georgia, had a population of only about three million in a physical area smaller in size than the US state of Maryland. It is a country with its own distinctive alphabet and language and was the first country in the world to adopt Christianity as a state religion, an event traditionally dated to 301 CE. The Armenian Apostolic Church remains the country's central religious institution, and the Old City of Jerusalem in Israel has an Armenian Quarter, an indication of Armenia's early connection with Christianity.

The small landlocked country has experienced invasions from every empire that controlled the region throughout history. The geographic area of the country decreased when the Ottoman Empire took control of western Armenia, and that region remains a part of Turkey to this day. A bitter conflict between Turks and Armenians during World War I resulted in the systematic deaths of as many as a million Armenians. This genocide continues to be commemorated annually on April 24, the traditional date of the Armenian Martyrs' Day, but Turkey still denies the events were genocide.

Like the other former Soviet republics, Armenia has shifted from a centrally planned economy to a market economy. Before independence in 1991, Armenia's economy had a manufacturing sector that provided other Soviet republics with industrial goods in exchange for raw materials and energy. Since then, its manufacturing sector has declined and Armenia has fallen back on agriculture and financial remittances from the approximately eight million Armenians living abroad to support its economy. These remittances, along with international aid and direct foreign investments, have helped stabilize Armenia's economic situation.

Azerbaijan is an independent country to the east of Armenia bordering the **Caspian Sea.** It is about the same size in area as the US state of Maine. This former Soviet republic has a population of more than eight million in which more than 90 percent follow Islam. Azerbaijan shares a border with the northern province of Iran, which is also called Azerbaijan. Part of Azerbaijan is located on the western side of Armenia and is separated from the rest of the country.

Located on the shores of the Caspian Sea, **Baku** is the capital of Azerbaijan and is the largest city in the region, with a population approaching two million. During the Cold War era, it was one of the top five largest cities in the Soviet Union. The long history of this vibrant city and the infusion of oil revenues have given rise to a metropolitan center of activity that has attracted global business interests.

Wealth has not been evenly distributed in the country, and at least one-fourth of the population still lives below the poverty line.

Azerbaijan is rich with oil reserves. Petroleum was discovered here in the eighth century, and hand-dug oil wells produced oil as early as the fifteenth century. Since the Industrial Revolution, the rising value of petroleum for energy increased the industrial extraction of oil in Azerbaijan. At the end of the nineteenth century, this small country produced half the oil in the world. Oil and natural gas are the country's main export products and have been a central focus of its economy. Large oil reserves are located beneath the Caspian Sea, and offshore wells with pipelines to shore have expanded throughout the Caspian Basin. As much as the export of oil and natural gas has been an economic support for the country, it has not been without costs to the environment. According to US government sources, local scientists consider parts of Azerbaijan to be some of the most devastated environmental areas in the world. Serious air, soil, and water pollution exist due to uncontrolled oil spills and the heavy use of chemicals in the agricultural sector.

Figure 3.18: Oil Fields of Azerbaijan. Wikimedia Commons – CC BY 2.0.

Key Takeaways

- The vast majority of Russia's population live in the western core area of the country, the region around the capital city of Moscow.
- Most of the cities on the Barents Sea and in the Eastern Frontier were established for manufacturing or for the exploitation of raw materials.
- The Volga River and its tributaries have been an important transportation network for centuries. The Volga is the longest river in Europe.
- Very few people live in Siberia, but the region is rich with natural resources.

- The most contentious region in Russia is the Caucasus Mountain region, especially the area of Chechnya. The Caucasus is characterized by ethnic and religious diversity and by a desire for independence from Russia.
- South of Russia in the Caucasus is the region of Transcaucasia. It is ethnically, religiously, and linguistically diverse. Countries there are independent of Russia, although they have a long history of being part of the Russian and Soviet Empires. Some of the countries are rich in petroleum reserves.

Discussion and Study Questions

1. What are Russia's four main regions and what are the main qualities of each?
2. What connects Moscow with Russia's Far East, and why is this important?
3. What are some of the major environmental problems in Russia and Transcaucasia?
4. What is the relationship between the countries of Georgia and Russia? Describe their 2008 conflict.
5. What happened to Chechnya? Why has Russia fought two wars in Chechnya since the Soviet Union's collapse?
6. What are the three independent countries of Transcaucasia, and when did they gain independence from Russia?
7. Which of the three Transcaucasian countries has the least opportunity to gain wealth? What is the largest source of income in this region?
8. What physical feature demarcates the boundary between Europe and Asia in southern Russia?
9. What is the largest city in Transcaucasia? On what sea is it located?

Geography Exercise

Identify the following key places on a map:

- Abkhazia
- Adjara
- Arctic Ocean
- Amur River
- Arkhangelsk
- Armenia
- Azerbaijan
- Baku
- Baltic Sea
- Bering Strait
- Black Sea
- Caspian Sea
- Chechnya
- Don River
- Eastern Frontier
- Georgia

- Grozny
- Irtysh River
- Kamchatka Peninsula
- Kuznetsk Basin
- Lake Baikal
- Lena River
- Murmansk
- Mt. Elbrus
- Novosibirsk
- Ob River
- Sakhalin Island
- Sea of Okhotsk
- Siberia
- South Ossetia
- Transcaucasia
- Vladivostok
- Volga River

References

Wikipedia, "Second Chechen War," en.Wikipedia.org/wiki/Second_Chechen_War.

3.4: CONCISE SUMMARY

Russia is an extensive nation that spans the divide between Europe and Asia. It possesses plentiful natural resources, diverse continental and polar climates, mountainous regions, expansive plains, and extensive river systems.The wide expanse of Russia has rendered governance difficult for both the Russian Empire and the Soviet Union. Each administration addressed the magnitude and cultural heterogeneity in distinct manners.Both the Russian Empire and the Soviet Union were empires, characterized by extensive territories where Russian governmental authority prevailed over other cultures and ethnicities inside their borders. The Union of Soviet Socialist Republics (USSR) was established after a violent insurrection and civil conflict. The Bolshevik Party, a socialist faction led by Vladimir Lenin, governed it. Josef Stalin, the second leader of the USSR, was infamous for the millions he executed while consolidating his power and pursuing economic advancement for his nation.

The USSR operated as a command economy, wherein economic decisions were dictated by the central government. The economic ambitions of the early leaders were swift industrialization and agricultural collectivization. The Soviet economy was fundamentally corrupt and inefficient—two elements that, alongside further issues, contributed to the dissolution of the Soviet Union in 1991. The changes implemented by the last Soviet leader, Mikhail Gorbachev, were insufficient to avert the dissolution of the Soviet Union. The initial post-Soviet years were characterized by swift privatization, significant wealth accumulation for a select few, economic adversity for the majority, and the erosion of the social safety net.

Since 1999, Russian presidents Vladimir Putin and Dmitry Medvedev have fortified Russia's economy and centralized state authority. The majority of Russians reside in the western region of the country, particularly near Moscow, with additional significant population concentrations situated in the European core. While there are few industrial centers in the Eastern Frontier region, the majority of Russia east of the Urals comprises extensive wilderness.

The Caucasus Mountain region in southern Russia has been the primary source of turmoil for Moscow. Non-Russian factions, including the Chechens, want for independence; nonetheless, Russia has resorted to military action to thwart their secession.The nations of Georgia, Armenia, and Azerbaijan were formerly constituents of the USSR but are currently sovereign entities. They are located south of the Russian border, among the southern Caucasus Mountains. These countries have challenges due to the influences of Russia and the Middle East to the south. They host significant populations of Muslims and Christians, together with other ethnic groupings. Armenia is the least affluent of the three nations, although Georgia and Azerbaijan possess wealth derived from petroleum exports.

4 Chapter

NORTH AMERICA

4.1: INTRODUCTION: AN OVERVIEW

Figure 4.1

North America is segmented into several geographic regions characterized by unique landforms. The western region of the continent features north-south mountain ranges, specifically the Rocky Mountains and the Pacific Mountains and Valleys physiographic provinces, separated by the Intermontane Basins and Plateaus.

The climates of the United States and Canada include the frigid type E climate of the tundra of northern Canada and Alaska, the tropical type A climate of southern Florida and Hawaii, the type C climates of the humid eastern United States, the seasonal type D climates of the northern United States and most of Canada, and the arid type B climates of the Southwest and Great Plains. In general, there are two different climate patterns common in North America. The first pattern is that temperatures get warmer as you travel from north to south and get closer to the equator. The second pattern is that there is a decrease in precipitation as you move from east to west across the continent until you reach the Pacific Coast, where rainfall is abundant again.

The second climate pattern is created by the rain shadow effect of the western mountain ranges. As wet air masses move from the Pacific Ocean over the North American continent, they run into the Cascades and the Sierra Nevada. The Cascade ranges of Washington State and Oregon cut off moisture from falling on the leeward side of the mountains; thus eastern Washington State and eastern Oregon are semiarid. The western United States experiences a strong rain shadow effect. As the air rises to pass the mountains, water vapor condenses and is released as rain and snow. This means that west of these mountain ranges there is much more precipitation than to their east, resulting in arid and semiarid lands. The entire Great Plains of the western United States are affected by the rain shadow effect and have a semiarid type B climate.

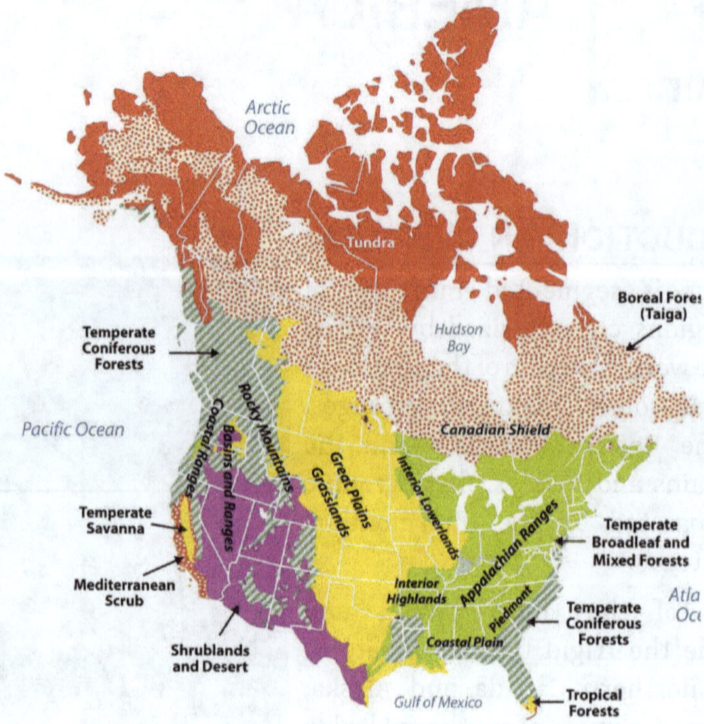

Figure 4.2 Physical Regions of the United States and Canada

Wikimedia Commons - CC BY-SA 3.0.

European Realms in North America

Both the United States and Canada are products of European colonialism. North America was inhabited by many Native American groups before the Europeans arrived. Complex native societies, federations, and traditional local groups faced the European invasion. While the indigenous population of North America was robust at the time of the European encounter, within a few generations, these native peoples were overwhelmed by the diseases, weapons, and sheer numbers of the European arrivals.

The Europeans—mainly the Spanish, French, and British—left a strong imprint on their North American colonies. The oldest colonial city in North America is St. Augustine, Florida (1565), founded by Spain when Florida was a remote portion of the Spanish Americas. Spain also had outposts in what are now California, Arizona, New Mexico, and Texas. The forms of settlement characteristic of those areas were similar to the Spanish colonies of Central America. While Spain governed what is now the southern United States, France ruled Canada and much of the interior of the North American continent. The French first came to Canada in the late 1500s to engage in fishing in the North Atlantic and soon expanded their reach by creating a fur trade in the area surrounding the **Great Lakes** and throughout the **Mississippi River** system.

Although there were fewer settlers from France than from other European countries—especially in what became the United States—this French era left behind place names (Baton Rouge and Detroit), patterns of land use, and a French-speaking population in Canada. Despite the early influence of Spain and France in North America, most North Americans speak English as their native language as a result of Britain's colonial dominance in the United States and Canada. The earliest permanent British colony, Jamestown, was founded in 1607 in what became Virginia. The British built up a successful empire in the New World. Their thirteen American colonies became populous, economically robust, and militarily strong enough to gain independence in 1776. Canada functions as an independent country but remains part of the British Commonwealth.

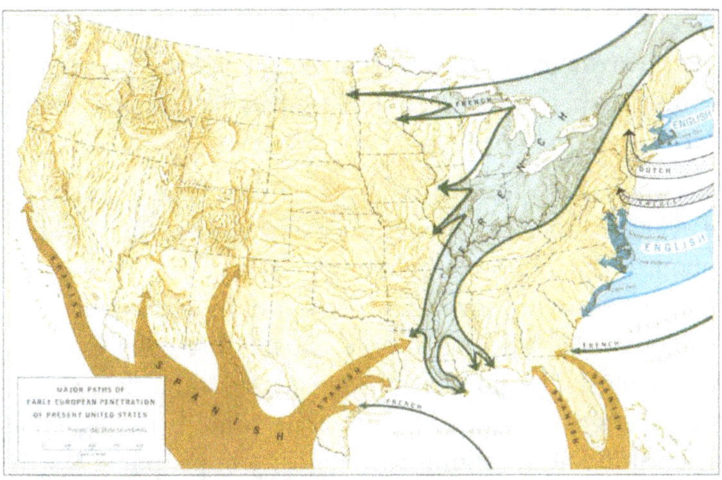

Figure 4.3: European Influence in the Colonial United States. Map courtesy of the National Park Service.

Population Distribution in North America

The US population surpassed the three hundred million mark in 2006. Canada now has over thirty-four million people. The US population is growing by about 2.5 million people each year. A little less than half the growth can be attributed to immigration and the rest to birth rates. The pace of growth is slower than the world average but more rapid than many other industrialized countries such as those in Europe.

The population is not uniformly spread over North America, nor are the population growth rates the same in all locations. Most Canadians live in near proximity to the US border. The North American population tends to be clustered in cities, with about 80 percent of US citizens residing in urban/suburban areas. Additionally, over time, the population has been moving southward and westward. US states experiencing the greatest rates of population growth include those located on the southern portion of the eastern seaboard, as well as Texas, Nevada, Utah, California, Oregon, and Washington. Three states—California, Texas, and Florida—accounted for about a third of the entire US population growth since 1990. Still, the Northeast is the most densely populated area of the country thanks, in large part, to the megalopolis that forms the corridor and encompasses the cities from Washington, DC, north to Boston. The largest concentration of Canadians lives in the most southern-reaching province of Ontario. For this reason, the province of Ontario is often referred to as South Canada.

In general, the population of minorities is growing most rapidly. Some of the fastest-growing populations in the United States are Hispanics. Another interesting factor in population growth is the increase in life expectancy. As more people live longer, the growth of the segment of the population aged sixty-five has doubled in the last fifty years. However, it appears that the growth of this population segment is slowing. Of this group, the greatest increase was seen in people aged eighty-five years and older.

The American population tends to be on the move. The US Census Bureau data show that the average American moves once every seven years; these data further predict that about forty million people move each year (US Census Bureau). Data also indicate Americans will move to a metropolitan area. Urbanization has been a trend since about 1950. Until that time, most Americans lived in small towns or more rural settings. The population density of the cities, and especially the suburban areas, has grown steadily since that time, bringing about a rural-to-urban population shift. Now a significant majority of people in North America live in suburban areas.

Urbanization has brought some challenges. The layout of these areas often makes owning a car a necessity; thus traffic congestion is a major problem in many suburban and urban areas. Other issues that have arisen are overcrowded schools, racial tensions, and a widening economic gap between the wealthy and impoverished. As people move to the cities, housing and other resources might not be able to meet demand, forcing prices upward. The gap between the cost of living in an urban area and the population's ability to pay has contributed to poverty and homelessness. Environmental issues also abound, including how to reduce or eliminate smog, manage waste, and ensure adequate clean water supplies.

Key Takeaways

- The United States and Canada have mountain ranges along their eastern and western portions, with lowlands in the middle.
- In general, temperatures get cooler as you move from south to north, and the climate gets more arid as you move from east to west across the continent.
- The Spanish were the earliest Europeans to establish a permanent settlement in the United States or Canada. They controlled the territory in the southern edge of what is now the United States, and their influence is still felt today through the Mexican American culture in that region.
- The French colonized eastern Canada, the Great Lakes, and the Mississippi River valley. Although the number of settlers was small outside Quebec, French place names and French land-use patterns are still evident.
- The British colonized the eastern coast of what became the United States. The number of English-speaking settlers was so high that the English culture dominated the region and left a strong long-term impact in terms of language, religion, and many other cultural aspects.
- The more than 310 million people who live in the United States and the more than 34 million people who live in Canada are not evenly distributed across North America. The realm continues to urbanize, and minority groups are the fastest-growing segment of the population.

Discussion and Study Questions

1. Who were the three main European colonizers of what became the United States and Canada?
2. Where was each European colonizer most influential?
3. Describe long-term impacts of each European power in North America.
4. What is a main reason the Great Plains and the western part of the United States are so arid?
5. What three states had the highest population growth since 2000?
6. What segment of the population has doubled in the past fifty years?
7. What are the current populations of the United States and Canada?
8. What problems have been caused by urbanization in North America?
9. Why do about forty million people move within the United States each year?
10. Why do most Canadians live near the US border?

Geography Exercise

Identify the following key places on a map:

- Appalachian Highlands
- Canadian Shield
- Great Lakes
- Great Plains
- Gulf-Atlantic Coastal Plain
- Interior Lowlands
- Intermontane Basins and Plateaus
- Mississippi River
- Pacific Mountains and Valleys
- Rocky Mountains

References

US Census Bureau, "Geographical Mobility between 2004 and 2005," www.census.gov/population/www/pop-profile/files/dynamic/Mobility.pdf.

4.2. UNITED STATES- EARLY DEVELOPMENT AND GLOBALIZATION

The early development of the United States was marked by a complex interplay of cultural, economic, and political factors that laid the groundwork for its emergence as a global power. From the colonial era through the Revolutionary War, the nation was shaped by diverse influences, including Indigenous cultures, European settlers, and the transatlantic slave trade. This rich tapestry fostered an entrepreneurial spirit and a quest for independence, culminating in the establishment of a unique democratic identity. As the nation expanded westward and industrialization took hold in the 19th century, the United States increasingly engaged in global trade and diplomacy, setting the stage for its role in the modern world. The dynamics of early American development not only transformed the domestic landscape but also positioned the nation within the broader context of globalization, with lasting impacts that continue to resonate today.

Early Development Patterns

With abundant resources and opportunity, the original thirteen colonies prospered and expanded into what became the fifty US states. The political geography of this nation was a product of various treaties and acquisitions that eventually resulted in the country extending from the Atlantic to the Pacific Ocean. Fueling the expansion was the concept of Manifest Destiny: the belief of some Americans that the new nation was divinely predestined to expand across the continent. The United States negotiated with France for the Louisiana Purchase in 1803, acquiring millions of acres in the central United States (see Figure 4.5 “Geopolitical Map of the Creation of the United States from the Atlantic to the Pacific”). Florida was acquired from Spain in 1819, and Texas was annexed in 1845. The British sold portions of the Pacific Northwest to the United States, and the exact northern boundary between the United States and Canada was settled in 1846. Through conflicts with Mexico, large portions of the West were ceded to the United States in the mid-nineteenth century. Alaska was purchased from the Russians in 1867 for only $7.2 million. Alaska and Hawaii were the last two possessions to enter into statehood, which they did in 1959.

Figure 4.4 The Thirteen Original Colonies

Figure 4.5: Geopolitical Map of the Creation of the United States from the Atlantic to the Pacific. Wikimedia Commons – public domain.

Manifest Destiny

The concept of Manifest Destiny came from the works of John O'Sullivan in 1839. O'Sullivan wrote, The far-reaching, the boundless future will be the era of American greatness. In its magnificent domain of space and time, the nation of many nations is destined to manifest to mankind the excellence of divine principles; to establish on earth the noblest temple ever dedicated to the worship of the Most High—the Sacred and the True. Its floor shall be a hemisphere—its roof the firmament of the star-studded heavens, and its congregation an Union of many Republics, comprising hundreds of happy millions, calling, owning no man master, but governed by God's natural and moral law of equality, the law of brotherhood—of "peace and good will amongst men" (John L. O'Sullivan).

As the United States developed, it acquired external colonial possessions. With victory over Spain in the Spanish-American War of 1898, the US government gained control of the Philippines, Cuba, Puerto Rico, Guam, and various Pacific islands. Cuba and the Philippines later became independent countries, but Puerto Rico and Guam continue to be part of the United States. The US Virgin Islands were purchased from Denmark in 1918 after World War I as a location to provide strategic military support to protect the shipping lanes through the Caribbean and the Panama Canal.

Westward Settlement Patterns and European Immigration

The thirteen original colonies are often grouped into three regions, each with its own economic and cultural patterns. These three areas—New England, the Mid-Atlantic, and the South—are considered culture hearths, or places where culture formed and from which it spread. The three regions were source areas for westward migration, and migrants from these regions carried with them the cultural traditions of their culture hearths. New England was characterized by poor soils, subsistence agriculture, and fishing communities and was the birthplace of North America's Industrial Revolution. Its largest city was Boston. Settlers from New England traveled west across New York State and into the upper Midwest and the Great Lakes region. The Mid-Atlantic region, focused on Philadelphia, Pennsylvania, was known for its fertile soils, prosperous small-scale agriculture, and multinational population. Prosperous farming led to a vibrant economy and a robust network of towns and cities. People who wanted to migrate west from this region traveled down the **Great Valley** into the **Appalachian Mountains** and across the **Cumberland Gap** into Kentucky, or they crossed Pennsylvania and traveled west via the Ohio River valley. The heart of the South was Virginia, a region oriented around plantation agriculture. The South was overwhelmingly rural, and in time the bulk of its agricultural workforce consisted of slaves brought to the United States from Subsaharan Africa.

Westward migration was spurred along by the gold boom in California (1849) and by the completion of the transcontinental railroad (1869). The settlement frontier pushed westward during the course of the nineteenth century and was declared "closed" by the Bureau of the Census in 1890. This did not mean that settlers were spread uniformly across the continent by 1890; indeed, vast areas of the Great Plains and the mountain west remained sparsely populated by Europeans at that time. The Homestead Act of 1862 also encouraged westward migration by offering 160 acres of free land to households willing to move west. The continental United States had been organized into official states by the end of the nineteenth century, except Oklahoma (1907), Arizona (1912), and New Mexico (1912).

Figure 4.6: The Holy Trinity Serb Orthodox Church in Butte, Montana This church gives testimony to the impact that European immigration has had on the North American continent and the freedom of religion that the US Constitution provides. Photo by R. Berglee – CC BY-NC-SA.

Most US residents at its founding in 1776 had roots in Great Britain, with large numbers from other northern and western European countries and many others from Africa (most of whom were slaves in the South). During the nineteenth century, migrants continued to immigrate to the United States as its economy grew, especially after the 1830s. Germans and Irish began arriving in large numbers, joining others from Britain and other countries, predominantly those in western Europe. As the century progressed, others from southern and eastern Europe, from countries such as Italy, Russia, and Austria, became the most significant stream of immigrants to the United States. The new arrivals were different from the early British immigrants: they practiced Roman Catholic or Eastern Orthodox

Christianity (not Protestantism), they primarily moved to urban areas, and they found work in the new manufacturing sector growing rapidly in the Northeast and around the Great Lakes. Very few immigrants came from Latin America or Asia at that time.

Industrial Development and Urbanization

The **Industrial Revolution** that began in Great Britain in the late eighteenth century eventually moved across the Atlantic and took hold in the United States. Rapid industrial growth emerged in the nineteenth century and was focused in the northeastern United States around the Great Lakes in an area called the Manufacturing Belt (Figure 4.7 "Manufacturing Belt Turned Rust Belt"). Mechanized manufacturing began with textiles (New England), moved to steel and other metals (Pennsylvania and Indiana), and later was dominated by the manufacture of automobiles (Michigan). Of course, manufacturing would not have been possible without an abundant supply of power. Coal mining became an important industry in western Pennsylvania and in Appalachia.

Manufacturing took place in the cities and towns of the Manufacturing Belt. Not until the second half of the twentieth century did manufacturing move to rural areas; until then, it was almost entirely an urban activity. As the United States went through its Industrial Revolution, its population shifted from being almost entirely rural to being mostly urban. In 1790, only about 5 percent of the US population lived in urban areas; by 1920, about 50 percent lived in cities. As the rural to urban shift took place, the function and form of US cities also changed.

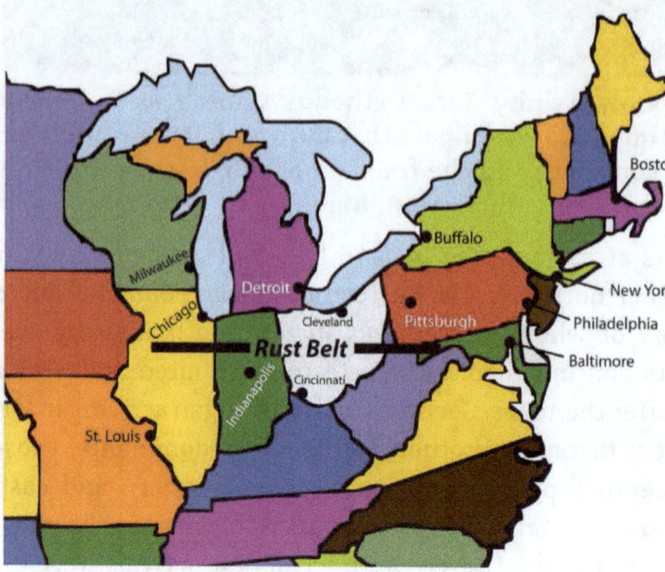

Figure 4.7: Manufacturing Belt Turned Rust Belt

Table 4.1 US Population and Percentage Urban

Year	Percentage Urban	Population in Millions
1790	5.1	3.9
1810	7.3	7.2
1830	8.8	12.9
1850	15.4	23.2
1870	25.7	38.6
1890	35.1	63
1910	45.6	92.2
1930	56.1	123.2
1950	60	151.3
1970	73.6	203.3
1990	75.2	248.7
2010	82.0	308

Source: "The World Factbook," Central Intelligence Agency, www.census.gov/population/censusdata/table-4.pdf.

From the colonial era until the late nineteenth century, US cities were walking cities. Because most Americans lived on farms, cities were small, compact, and centrally oriented: everything was located within walking distance. Only wealthy people had access to transportation by horse, and city dwellers needed to live within a short distance of where they worked, shopped, and carried out all their activities. The invention of the electric streetcar (1888) allowed cities to increase in size. People could live farther from their place of employment as long as they lived within walking distance of a streetcar line. Streetcar suburbs grew up along streetcar lines, and these neighborhoods were often segregated by ethnicity and race. Fewer people lived in downtowns, which became dedicated to retail and manufacturing. Cities remained oriented around a central business district (CBD), which was often located near the railway station. Factories needed to be near modes of transportation for both shipping in parts and shipping out completed products and so that workers could easily get to work.

Large numbers of middle class Americans began acquiring automobiles after about 1920; this eventually led to a complete rethinking of the spatial layout of the city.

Automobile suburbs sprang up outside the traditional city limits as people were able to buy homes far from streetcar lines or railway stations. Cities became increasingly decentralized: people could go shopping in suburban malls instead of downtown department stores, factories could spring up at highway interchanges and not only near rivers and the railroad, and people could live in one suburb and work in another instead of living in the suburbs and working downtown.

Neighborhoods became even more racially and economically segregated than they had in the past as middle-class whites moved into the new automobile suburbs and left the poorer African Americans behind in the cities.

By the late twentieth century, the automobile had led to a new urban form: the edge city. Edge cities are areas of dense urban development outside the boundaries of the traditional city. They often form at the intersection of major interstate highways and contain shopping malls, office complexes, high-rise apartment buildings, industrial parks, restaurants, and hotels. Sometimes edge cities are called **suburban downtowns**. Edge cities have supplanted the CBD as the destination of choice for Americans, whether they are heading to work or to play.

Economic Changes

For the purpose of understanding economic geography, all economic activities can be grouped into one of four categories, each with its respective terms, depending on the nature of what is being produced:

1. Primary economic sector activities include everything that pertains to the collection of raw materials, such as agriculture, forestry, fishing, and mining—in other words, growing and extracting activities.
2. Secondary economic sector activities involve the processing of those raw materials through manufacturing, which has been the mainstay of economic growth for most developed countries.
3. Tertiary economic sector activities are those that produce services, not physical products.
4. Quaternary economic sector activities are those that deal with information collecting and processing, as well as management.

The tertiary and quaternary economic sectors are often thought of together as the **service sector**. In the explanation of how countries gain national income (Section 1.4 "Globalization and Development"), only primary and secondary activities produce actual physical products, and manufacturing traditionally earns the highest value-added profits. Tertiary activities are selective in gaining national wealth. For example, service activities such as tourism can bring in national wealth if the visitors are from outside the country. Tourism within a country can also influence economic conditions by increasing the amount of consumer spending.

During the colonial era and into the nineteenth century, when the majority of Americans lived on farms and worked in agriculture, most economic activity in the United States took place within the primary economic sector. Today, the primary sector is still an important component of the US economy, but far fewer people are employed in it. For example, less than 1 percent of Americans make their living by farming, but agricultural output has continued to grow because of advancements in mechanization and the development of high-tech seeds, fertilizers, and pesticides. The United States has been able to export surplus agricultural output to other parts of the world. Fewer people work in coal mines than in the past, but because of new mining technologies and methods such as mountaintop removal, coal production remains high.

The geographic distribution of primary activities depends both on the location of natural features such as physical geography and climate and on the location of the market for a particular crop or resource. The nineteenth-century German economist Johann von Thünen created a model that predicted land use around a central market. In his theory, land closest to the market would be used to produce crops that were expensive to transport, such as dairy. Land far from the market would be used for the production of crops that were less expensive to transport and less perishable, such as grain. The von Thünen model predicts a series of concentric rings surrounding a central market, with each ring producing a different kind of crop. If the von Thünen model is applied at a much larger scale to the United States as a whole, with the densely populated urban zone from Boston to Washington, DC (called a **megalopolis**), used as the central market, the model does a fairly good job predicting the United States' agricultural land use. Dairy farms are found close to the market, grain farms are farther away, and ranch lands used for livestock production are even farther away.

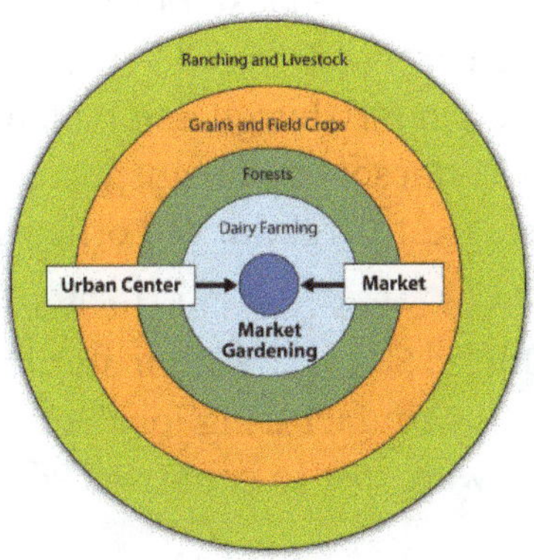

Figure 4.8: The von Thünen Model

The von Thünen model as it relates to the development west of the East Coast megalopolis. For example, New Jersey is called the Garden State, Appalachia has its forests, the Midwest has its agricultural production, and the Great Plains has its large cattle ranching operations. This model was more applicable in past centuries for local communities when modern transportation technology was not available.

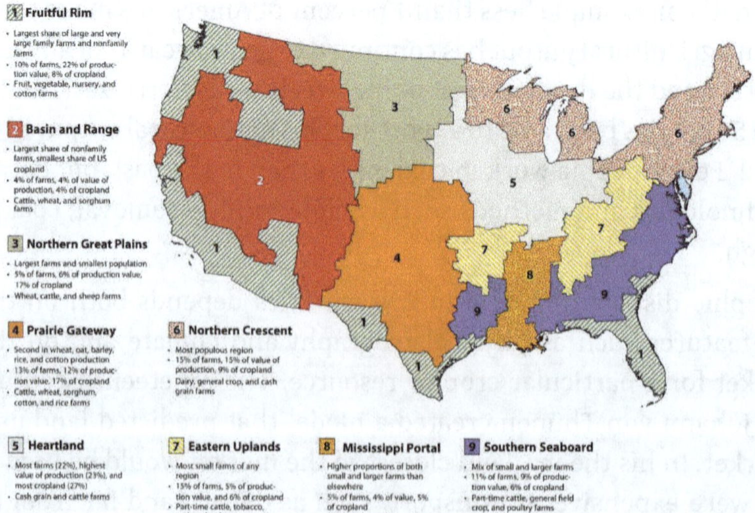

Figure 4.9 Farm Resource Regions Source: Map courtesy of the US Department of Agriculture.

Anything that involves the processing of raw materials—manufacturing—is a secondary activity. As the United States moved into the Industrial Revolution and into the mid-twentieth century, the percentage of the US workforce involved in manufacturing grew from almost nothing until it peaked in the late 1970s. It was the main area of economic growth for decades. Although manufacturing was present in most areas of the country, it was focused in the northeastern United States and along the Great Lakes. Factories were close both to the reserves of labor and to the markets for manufactured products found in the densely populated Northeast. The steel industry was located in Pittsburgh and its environs because of the area's access to iron ore (mined in Minnesota and transported via the Great Lakes) and to coal (mined in Pennsylvania, West Virginia, and other parts of Appalachia).

As manufacturing has grown in other parts of the world, the secondary economic sector has declined in the United States. US labor statistics indicate that the United States lost about five million manufacturing jobs between 2000 and 2010 (Nance-Nash, S.). Many of these jobs were lost to countries with lower labor costs, such as Mexico or China.

The third group of economic activities takes place in the tertiary and quaternary sectors, commonly known as the **service sector**. Tertiary and quaternary activities create services, not physical products. Service jobs include everything from engineering to finance, restaurants to sports, and childcare to medicine. The tertiary sector makes up more than three quarters of the US economy, as measured by its share of the **gross domestic product (GDP)**, which is the total value of all goods and services produced in a country in a given year. The GDP is then divided by the country's population to provide a GDP per capita statistic. The 2010 estimated **GDP** composition by sector for the United States is shown in Table 4.2 "US GDP by Sector (2010 Data)".

Table 4.2 US GDP by Sector (2010 Data)

Economic Sector	GDP (%)
Agriculture	1.2
Industry and mining	22.2
Services	76.7

Source: "The World Factbook," Central Intelligence Agency, https://www.cia.gov/library/publications/the-world-factbook/geos/us.html.

These figures show that the United States has shifted to a postindustrial service economy. The rise of the **information age** in the latter part of the twentieth century shifted the workforce into the information sector. By the start of the twenty-first century, less than 2 percent of the US workforce was employed in agriculture, 15 percent in industry, and the rest in services (18 percent) and information activities (65 percent).

The locations of service-sector jobs are much more flexible than are jobs in the primary or secondary sectors. They are called footloose jobs: an accountant can live in New York or in Denver, whereas it is much more difficult for factories to move from one place to another and it is impossible for farms to relocate. Many of the information-technology jobs are emerging in the southern regions of the United States called the Sun Belt. Southern cities such as Atlanta, Dallas, and Phoenix are centers of innovation and population growth. The warmer climate, combined with a lower cost of living and less congestion, makes the Sun Belt an attractive location for emerging information-based companies. Note that the popularity of the South and West for service-sector jobs only came about after the invention and adoption of air-conditioning. Air-conditioning was not widespread until after the Second World War in the 1950s.

While the population of the Southern states has increased, the population of some Northern states has decreased. The Sun Belt has always been a destination for people escaping the harsh winters of the Northern states. This has usually been only a seasonal transition. However, the new trend is one of permanent growth because of the increase in information technologies and in the service industry. Emerging companies looking to establish their businesses have targeted major cities from the Carolinas to the Southwest.

Migration Patterns

The United States has not only undergone a massive rural-to-urban shift in its population; intermigration within the United States from one region to another has also been prevalent. Each of the US regions has witnessed changes in demographics because of migration patterns.

In the agricultural regions of the United States, such as the **Midwest**, the migration pattern has been caused by changes in farm technology. Portions of the United States were opened up for agriculture because of the Homestead Act of 1862, where each person could receive 160 acres from the government to start a farm. They could keep the acres if they lived on them and farmed them for a period of years. In the 1800s, 160 acres was enough land to support a family if conditions were appropriate. The Industrial Revolution brought about improved farm equipment and technology. Larger and more expensive tractors and improved farming methods pushed the small farmers to sell out. Farms increased in size and fewer people were required to operate them. Since fewer farm workers are needed in rural areas, there has been a major rural-to-urban shift in the population. Central cities are increasing in population, while small towns and rural areas in the Midwest and across the nation are decreasing in population.

Americanism and Globalization

The freedom of personal expression in the United States has supported individual ingenuity and creative ambition to create the largest economy in the world. US citizens have pushed American corporations to become a major force in the world markets. Products and franchises from the United States are being distributed throughout the world. Items such as fast food, computers, news networks, and Hollywood movies have become the products of choice in countries across the globe. The English language dominates the Internet, which has been heavily influenced by US corporations. The power of the American Dream—the idea that through hard work anyone can achieve upward mobility and financial success—as it is portrayed in the US media holds sway in the minds of people both in the United States and abroad.

US news networks, such as CNN, are so dominant that small countries, having no resources to create networks, rely on the US networks to deliver their world news. US fast food franchises, of which McDonald's is the largest, exist in over one hundred countries. Despite humble beginnings in Arkansas, Walmart grew to become the world's largest corporation. It has become the buyer and seller of retail trade that shapes and molds cultural attitudes and fashions internationally.

The size of the US population (more than 310 million strong as of 2010) and the country's vast resource base have allowed it to become a world military superpower. After the fall of the Soviet Union, the United States became the most powerful military force in the world. The United States has also dominated the world's economy and its communications networks. The advancements of multinational corporations have in essence enabled the sale of America to the rest of the world. The selling of American products and the large consumer market in the United States have provided the profits that have fueled global economic markets.

Figure 4.10 US Corporate McDonald's Franchise in Israel

Photo by R. Berglee – CC BY-NC-SA.

The United States has become a worldwide franchise of its own. Corporate colonialism has advanced the American brand to a level that is now synonymous with consumerism, success, and power worldwide. Media advancements have promoted the concept of the American Dream across the seven seas. The reaction of the global community includes both admiration and disdain. Many view Americanism as interchangeable with globalization. Some welcome it; others reject it. The country of Iran is an example of this dichotomy. Young people in Iran wearing blue jeans gather in secret to watch American television programming from a hidden illegal satellite dish, while at the same time the anti-American forces in their government condemn America as decadent, immoral, and imperialistic.

Corporate colonialism has become a dominant force impacting the global cultural fabric. Supporters appreciate access to American goods and services, while opponents claim that the English language and the American corporate franchise system are destroying the culture and heritage of untold millions who see their unique traditional ways of life being overshadowed and destroyed.

Some argue that American television advertisements exemplify a trend that supports conformity and uniformity in American culture. They contend that America's unique cultural diversity, which historically has provided ingenuity and creativity, is being eroded by the franchising of similar retail products, fast food, professional sports, and Hollywood entertainment that stifle the creative will of the American people. Others continue to see opportunities to pursue the American Dream and believe that innovation and ideas continue to emerge in spite of these trends.

Figure 4.11 Walmart is one of the greatest success stories of the American Dream. It started in Arkansas with one store and grew to become the largest corporation in the world. Now it is one of the most aggressive forces of corporate colonialism across the globe.

Photo by R. Berglee – CC BY-NC-SA.

Many people worry about the future of the American Dream. American culture continues to evolve as people face changing economic and social conditions. Over the course of their history, Americans have faced both difficult and prosperous times, and now the future of this vibrant country is in the hands of the current generation. The United States has developed into one of the most powerful countries on the planet. Will the American Dream continue to motivate people in the future? Only time will tell.

Key Takeaways

- The United States' territory expanded gradually through various treaties and land acquisitions and was influenced by the concept of Manifest Destiny.
- The three main colonial regions in the United States—New England, the Mid-Atlantic, and the South—had their own distinct economic foundations, settlement patterns, and social structures. People from these regions moved westward in particular migration patterns.
- In the beginning of the nineteenth century, most immigrants were from western and northern Europe. By the end of the nineteenth century and the beginning of the twentieth century, immigrants were coming in large numbers from southern and Eastern Europe and moving to industrial cities of the Northeast.

- City structure changed from the walking cities of colonial America, to the railroad and streetcar cities of the late nineteenth century, to the automobile cities of the mid- and late-twentieth century.
- The US economy was initially based in the primary economic sector (particularly farming), then was based in the secondary economic sector (manufacturing), and is now oriented around the tertiary and quaternary economic sectors (services and information).
- The diverse immigrants who have created American society have been unified by common aspirations and common ideals that created the concept of the American Dream. The concept indicates that regardless of one's station in life, by working hard, applying oneself, and following the rules, one can obtain upward economic mobility.

Discussion and Study Questions

1. What did "Manifest Destiny" mean for the United States?
2. What European region did most early nineteenth-century immigrants come from? How did that change by the end of the nineteenth century?
3. How did modes of transportation affect the layout of American cities?
4. What is an edge city, and how did edge cities form?
5. How does the von Thünen model relate to the spatial pattern of land use in the United States?
6. How have economic sectors shifted in the United States in the past two hundred years?
7. What is the American Dream, and how has it been expanded globally?
8. What has been the major migration pattern in the Midwest region of the United States? Why is this occurring?
9. What part of the United States has been a target for the location of information age–based businesses? What was the role of air-conditioning technology in making this possible?
10. Outline some examples of how US culture influences the rest of the world.

Geography Exercise
Identify the following key places on a map:

- Cumberland Gap
- Great Valley
- Manufacturing Belt
- Midwest

- Original thirteen colonies
- Sun Belt

References

"John L. O'Sullivan on Manifest Destiny, 1839," http://www.mtholyoke.edu/acad/intrel/osulliva.htm. Nance-Nash, Sheryl "'Made in America' Leaders That Really Make a Difference," Daily Finance, www.dailyfinance.com/2011/07/01/made-in-america-leaders-that-really-make-a-difference/.

4.3. UNITED STATES- POPULATION AND RELIGION

Figure 4.12: The Statue of Liberty has long been a symbol of the American ideals that welcome immigrants to America.

Wikimedia Commons – public domain.

The American "Melting Pot"

Early immigration to America was dominated by people from the British Isles, resulting in an American population for whom speaking English and practicing Protestant Christianity was the norm. There were some regional exceptions to this, such as Catholicism in Maryland and the widespread speaking of German in Pennsylvania, but by and large English and Protestantism were standard in the American colonies. As migrants arrived in the United States from non-English-speaking countries, within a generation they learned English and assimilated into American society, giving rise to the idea of the United States as a cultural melting pot. People were drawn to the United States by the hope of economic opportunities; most immigrants were poor and came to the United States to make a living and improve their financial well-being. They viewed assimilation into mainstream society as a necessity for success. They believed in the American Dream—that through hard work, you could achieve upward mobility and financial success no matter your background. The dream came true for millions of Americans but remains out of reach for many who live in poverty.

As of 2010, the United States was home to approximately 310 million people and was the third-most populated country in the world after China and India. Among developed countries, the US population is one of the fastest growing, at about 1 percent each year. This is thanks to a fertility rate of about 2.1 that is higher than the 1.5 for that of most European countries, as well as a positive net migration rate (more people immigrating to the United States than emigrating from it). In terms of human well-being, life expectancy is more than seventy-eight years for men, and the average woman can expect to live more than eighty years. While this may seem high, especially when compared with a century ago, life expectancy in the United States is lower than in forty-nine other countries.

Although English has remained the dominant language, as a country of immigrants, the United States is home to people from all corners of the world and home to many cultural or ethnic minority groups. According to the 2010 census, the ethnic minority groups in the United States included 16.3 percent Hispanic (who can be of any race); 12.6 percent black or African American; 5.0 percent Asian and Pacific Islander; and 1.0 percent Native American (American Indians, Eskimos, and Aleuts) (US Census Bureau). An interesting trend is that Asians are growing faster than any other ethnic group in California, the nation's largest state. The 2010 census reported that the Asian population in California had increased 31.5 percent since 2000 to a total of five million.

The US Hispanic Population

One of the most striking shifts in immigration patterns of the past few decades has been the dramatic increase in Hispanic/Latino immigrants to the United States. While there have long been Spanish speakers living in the United States (recall that Spain colonized Florida, Texas, and the Southwest before those regions became part of the United States), for most of American history the Hispanic minority had little impact outside of a few areas of the country. In 1970, Hispanics made up less than 5 percent of the US population, but by 2010, forty-eight million Hispanics made up about 16 percent of the population (Figure 4.13 "Hispanic Population in the United States and the US Sun Belt"). For the first time, Hispanics were the largest ethnic minority in the United States, surpassing blacks as the largest minority starting with the 2000 US census (12.5 percent Hispanic compared with 12.3 percent black). The US Hispanic population doubled between the 1990 and 2000 censuses. Between 2000 and 2006, Hispanic population growth accounted for about half the nation's growth and grew about four times faster than the country's population as a whole. The growth of the US Hispanic population is a direct result of increased immigration from Latin America to the United States in the late twentieth and early twenty-first centuries (Hoefer, M., et. al., 2011) and the Hispanic population having higher fertility rates than the non-Hispanic US population (Archibold, R. C., 2011).

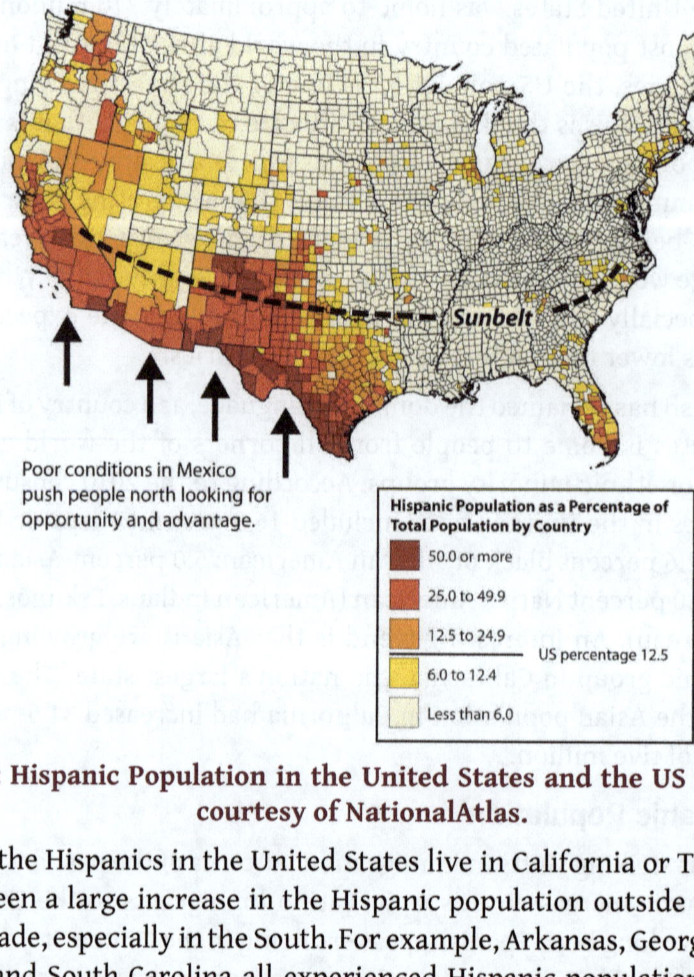

Figure 4.13: Hispanic Population in the United States and the US Sun Belt. Map courtesy of NationalAtlas.

Nearly half the Hispanics in the United States live in California or Texas, although there has been a large increase in the Hispanic population outside those states in the past decade, especially in the South. For example, Arkansas, Georgia, Tennessee, and North and South Carolina all experienced Hispanic population growth rates between 55 and 61 percent from 2000 to 2006. All regions of the country saw double-digit growth rates of their Hispanic populations during that time. In places such as California, the large Hispanic population has an especially significant impact on the economy, politics, and every aspect of social life: more than one-third of Californians are Hispanic (37 percent), while 42 percent are non-Hispanic white, and a much smaller minority are African American (7 percent).

Who are the Hispanics living in the United States? Most were born in the United States (60 percent), while the rest are immigrants. Two thirds are either from Mexico or of Mexican descent, while others hail from the US territory of Puerto

Rico, Cuba, or the Dominican Republic. Note that all Puerto Ricans are US citizens and can move to and from the US mainland without any special documentation requirements. More Hispanics come from Central America than from South America. Hispanics work in all professions but are found in professions such as agriculture, construction, and food service at higher rates than the country's non-Hispanic population.

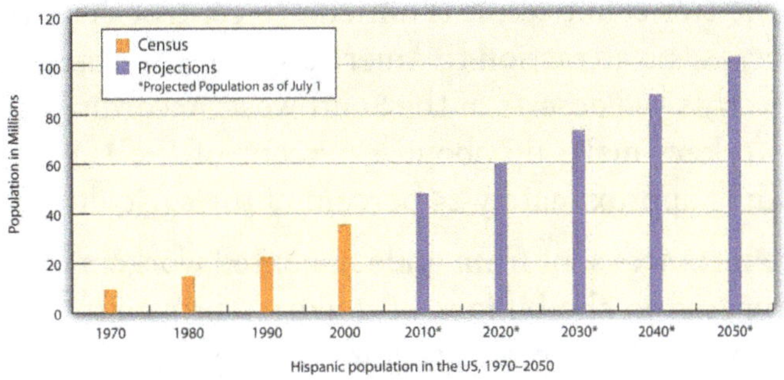

Figure 4.14: Hispanic Population in the United States, 1970–2050. Source: Data from the US Census Bureau, 1970, 1980, 1990, and 2000 Decennial Censuses; Population Projections, July 1, 2010, to July 1, 2050.

Figure 4.15: Two men tried to cross the US border from the Mexican side with a homemade ladder built of rebar. They hooked the ladder to the wall. One man climbed over and was arrested by the US Border Patrol; the other man ran away and was arrested by the Mexican border police.

Omar Barcena - La Frontera - CC BY 2.0.

The draw of opportunities and advantages has always pulled people toward the United States. While many of the Latin American immigrants enter the United

States legally, according to the US Department of Homeland Security's Office of Immigration Statistics, there were about 10.8 million undocumented immigrants residing in the United States as of 2010; 62 percent are from Mexico (US Department of Homeland Security). This number is lower than it was in previous years, possibly because of the economic recession and higher-than-usual rates of unemployment. The US-Mexican border is about 1,970 miles long and runs through an arid and open region between the two countries. It is difficult to control the illegal immigration across this border, as the attraction to American jobs is so compelling that people will risk death to cross the deserts of the Southwest. According to 2010 estimates, undocumented workers make up about 5 percent of the United States' civilian workforce, including approximately 24 percent of the agricultural workforce.

The amount of remittances sent from undocumented workers in the United States to Mexico is estimated in the billions of dollars. The remittances from Mexican nationals living outside Mexico and sending money home to their families are Mexico's second-largest source of foreign income. Without remittances, many Mexican families would have a difficult time making ends meet.

The African American Population

Most African Americans were concentrated in the South before the Civil War, where they worked as slaves in the cotton and tobacco plantations that supported the region. In some counties, blacks made up most of the population, and this did not change when the war was over. Many of the newly freed slaves remained as poor agricultural workers in the South well into the twentieth century. Even as late as 1910, seven out of every eight African Americans lived in the South.

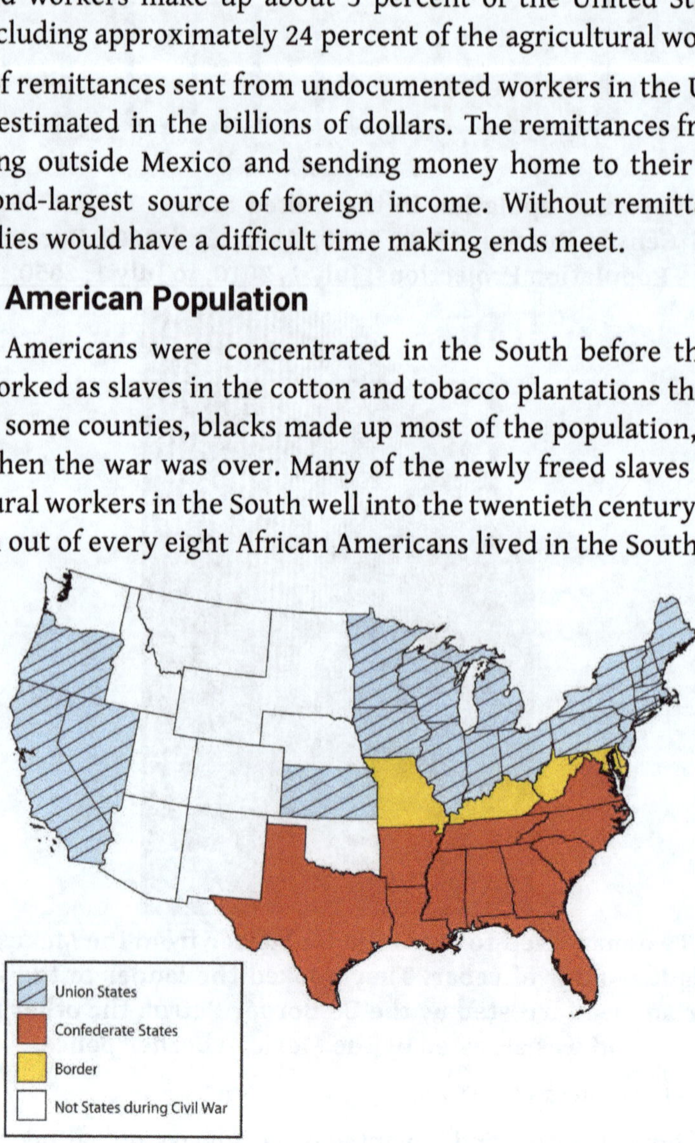

Figure 4.16: Civil War Division in the United States, 1861–1865

Union States are in the north, Confederate States are in the south, and border states are in the middle.

In the late nineteenth and early twentieth centuries, as the industrialization of northern cities was accelerating, the increased need for factory workers was largely met by immigration from Europe, especially from southern and eastern Europe. However, when the First World War began (1914), European immigration began to slow down. European immigration then nearly ground to a halt in the 1920s as Americans set quotas in place to reduce the number of Eastern European immigrants. At that time, the factories in the Manufacturing Belt continued to need workers, but instead of European workers, they recruited African Americans from the South.

This led to a massive migration of blacks from the South to cities of the North and West. This migration was so significant to African Americans in the United States that it is called the Great Migration. Between 1910 and 1925, more than 10 percent of African Americans made the journey north, and even more followed. Examining a map of the distribution of African Americans today shows the legacy of the Great Migration, as blacks live in many parts of the United States, both in the South and in postindustrial cities of the north and the Midwest. Blacks also now live in Sun Belt cities, as people of all races look for jobs related to the new information technologies and service industries.

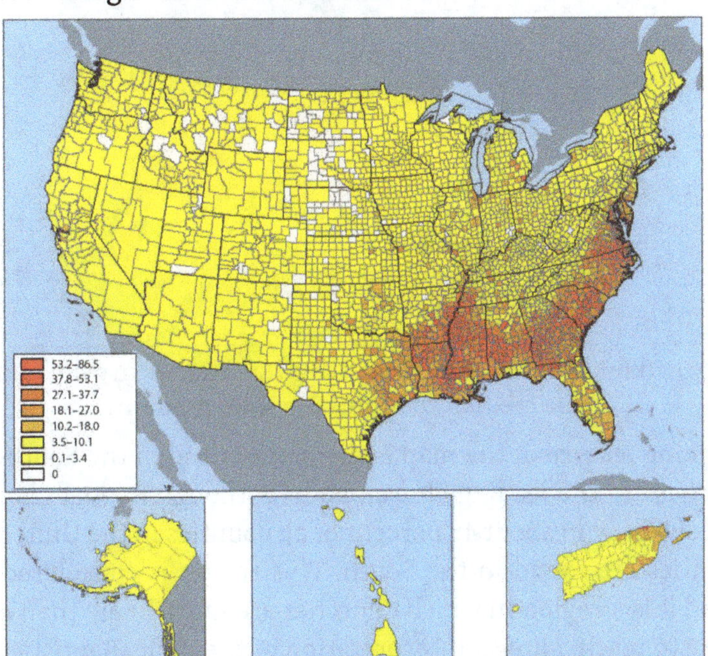

Figure 4.17: Black/African American Population by County, 2000. Map courtesy of NationalAtlas.

Geography of Religion in the United States

Most early settlers to the United States were Protestant Christians: Puritans lived in New England and Anglicans (later called Episcopalians) lived in Virginia. Roman Catholic immigrants settled in Maryland, and members of the Society of Friends (Quakers) founded Pennsylvania. Even within that overall picture, there was a great deal of religious diversity in the United States, and that diversity increased as new arrivals came from different countries with different religious backgrounds. The current pattern of religious affiliation in the United States remains quite complex, and one can find observers of nearly every major religion, and many minor ones, in virtually every area of the country. That being said, there are clear patterns to the geography of religion in the United States that tell stories of immigration and migration history, as well as stories about other aspects of American history. The map of leading church bodies (Figure 4.18 "Distribution of Religious Organizations in the United States") shows regions of religious observance that are worth examining.

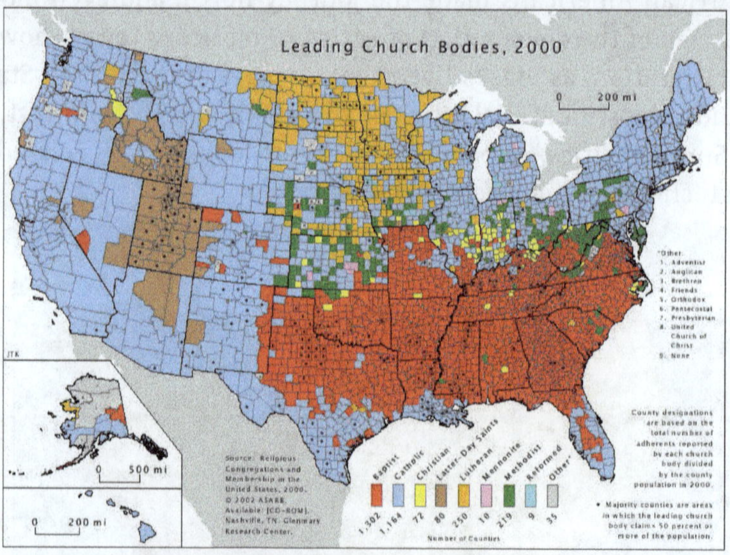

Figure 4.18: Distribution of Religious Organizations in the United States. Wikimedia Commons – public domain.

The most striking feature of the map is the block of red in the Southeast in which Baptist churches are the leading church body. Although Baptist churches are the leading religious body in about 45 percent of all counties in the United States, most of those counties are found in the South. This region is considered the nation's Bible Belt, and it is a region in which churches are more likely than in other parts of the country to teach a literal interpretation of the Bible. Baptist churches grew in popularity in the South after the Civil War as more liturgical denominations such as Methodists went into decline, and Baptist churches are popular among both African American and white residents.

Figure 4.19: Dehart's Bible and Tire is located in the peripheral region of eastern Kentucky in the Bible Belt. Customers get a free Bible with every set of new tires.

Photo by R. Berglee – CC BY-NC-SA.

Another interesting Protestant region is northern Appalachia and the lower Midwest from Ohio to Iowa and Kansas. As seen on the map, some of these counties are Baptist, in some of them the strongest church presence is Methodist, and in others it is Christian churches (Disciples of Christ and historically similar denominations) that prevail. The Methodist and other Christian areas were heavily influenced by the Second Great Awakening of the early nineteenth century, which promoted the theology that every person could be saved through revivals. The movement provided for strong showings of Protestant denominations that arose during that time. The third Protestant region is the northern Midwest and Great Plains: Minnesota, the Dakotas, and surrounding areas. This was the destination of German and Scandinavian Lutheran settlers during the late nineteenth and early twentieth centuries, and the leading denominations today in much of that area remain Lutheran.

The Roman Catholic Church, which is the leading religious body in 40 percent of US counties, is well represented in the Northeast, West, and Southwest. In the Northeast and Midwest, the Catholic dominance points to nineteenth- and early twentieth-century immigration from Roman Catholic countries in Europe such as Italy, Ireland, and Poland. Those earlier Catholics have been joined more recently by large numbers of Hispanic immigrants. The dominance of Roman Catholics in the western United States, the Southwest, and even Florida in the Southeast are a reflection of the strong Hispanic presence in those parts of the country.

In the western United States, the Church of Jesus Christ of Latter-Day Saints (Mormon church or LDS) dominates a region including Utah and surrounding states. Utah was the destination of Mormon members as they migrated westward during the mid-nineteenth century.

Major urban centers are home to people who follow all the major religions of the world. Muslim populations are found in cities as a result of late twentieth century immigrants arriving from countries such as India, Pakistan, and places in the Middle East and moving to urban areas for employment. Detroit and its surrounding counties in southeastern Michigan are one part of the country with a high concentration of Muslim immigrants.

Key Takeaways

- US immigrants learned English and assimilated into American society, giving rise to the idea of the United States as a cultural melting pot.
- Immigrants created the American Dream—the idea that by working hard and following the rules, one could achieve upward mobility and financial success regardless of one's background or heritage.
- The Hispanic population is the largest minority group in the United States, and its population is growing.
- Hispanics have an especially strong presence in California and Texas but are found all over the South as well as in rural and urban areas around the country.
- One-third of Hispanics are immigrants to the United States, and most undocumented workers to the United States are Hispanic (58 percent are from Mexico).
- Mexican nationals are an important component of the US workforce, and they send home billions of dollars each year in the form of remittances.
- African Americans are heavily represented in former plantation agricultural states of the South, as well as in cities of the Rust Belt and far West.
- Patterns of religious observance in the United States reflect immigration patterns.

Discussion and Study Questions

1. Why has the United States been called a melting pot? What are the largest minority groups in the country?
2. What parts of the United States have the highest representation of the four largest minority groups?
3. How did the US Hispanic population change in size and location from 2000 to 2010? Why is the Hispanic population growing faster than other ethnic groups in the United States?
4. What two states have about 50 percent of the Hispanic population?
5. What have been the push-pull factors for Hispanic immigration to the United States?
6. What is a remittance, and how are remittances significant for immigrants to the United States?

7. Describe the spatial pattern of the African American population. Why do so many blacks live in the Southeast?
8. What was the Great Migration, and what were some of its long-term impacts on the United States?
9. List three of the main Protestant religious denominations in the United States, describe their spatial patterns, and explain what caused these patterns.
10. Name the states or provinces that have high percentages of Catholics.

References

Archibold, R. C., “Arizona Enacts Stringent Law on Immigration,” *New York Times*, April 23, 2010, accessed September 20, 2011, http://www.nytimes.com/2010/04/24/us/politics/24immig.html.

Hoefer, M., Nancy Rytina, and Bryan C. Baker, “Estimates of the Unauthorized Immigrant Population Residing in the United States: January 2008,” Department of Homeland Security, accessed September 20, 2011, http://www.dhs.gov/xlibrary/assets/statistics/publications/ois_ill_pe_2008.pdf.
US Census Bureau, “2010 Census Data,” 2010.census.gov/2010census/data/.
US Department of Homeland Security, “Immigration Statistics,”http://www.dhs.gov/files/statistics/immigration.shtm.

4.4. CANADA

Figure 4.20: Map of Canada. Wikimedia Commons – public domain.

Canada's democratic system exhibits a comparable developmental history and economic standing to that of the United States. In 2011, Canada had a population exceeding thirty-four million, which is lower than that of California. Canada surpasses the United States in size, rendering it the second-largest nation globally. Nonetheless, despite the extensive land relative to its little population, almost 90 percent of Canadians reside within 150 miles of the US border. Northern Canada is excluded from Canada's ecumene, or the area designated for permanent human habitation. A limited expanse of land in southern and eastern Canada possesses the climate and topography conducive to agricultural cultivation and extensive habitation. Furthermore, Canada's economy is intricately linked to that of the United States, rendering proximity to the US border advantageous for residents. Canada consists of ten provinces and three territories (see Table 4.3 "Canadian Provinces, Territories, and Capitals"). Ottawa is the nation's capital, and Toronto is its largest city.

Table 4.3 Canadian Provinces, Territories, and Capitals

	Population	Province/Territory	Capital
Maritime Provinces of the east	2.3 million	Newfoundland	St. Johns
		Nova Scotia	Halifax
		New Brunswick	Fredericton
		Prince Edward Island	Charlottetown
French Canada	7.9 million	Quebec	Quebec (City)
South Canada	13.2 million	Ontario	Toronto
Prairie Provinces	6.0 million	Manitoba	Winnipeg
		Saskatchewan	Regina
		Alberta	Edmonton
Pacific Canada	4.5 million	British Columbia	Victoria
Territories	About 110,000	Yukon	Whitehorse
		Northwest Territories	Yellowknife
		Nunavut	Iqaluit

Physical Geography of Canada

As you can see from Table 4.3 "Canadian Provinces, Territories, and Capitals", Ontario has by far the largest population of any of Canada's provinces, with about 39 percent of the total. Quebec, the dominantly French-speaking province, is home to about 23 percent. Almost everyone else lives in British Colombia, west of the Rocky Mountains (13 percent); in the prairies (18 percent); or along the Atlantic coast (7 percent). The wide-open areas of Canada's far north are occupied by only one-third of 1 percent of the population.

Type D (continental) climates dominate most of central Canada, with their characteristically warm summers and cold winters, although the farther north you go, the cooler the summers are. Canada's west coast receives the most rainfall—

between eighty and one hundred inches a year—while coastal areas in the Maritime Provinces can receive up to sixty inches per year. The northern territory of Nunavut barely receives ten inches per year, usually in the form of snow. Of course, far northern Canada has an arctic type E climate, and conditions there are so harsh that only a very few people inhabit it. The cultural influence of the colder climates and the long winters on the people is evident by the sports that are enjoyed by most Canadian citizens. Ice hockey is Canada's most prevalent sport and its most popular spectator sport. Other sports such as curling are also common in Canada.

Figure 4.21: The Canadian Curling Team during the 2006 Winter Olympics in Torino. Wikimedia Commons – CC BY-SA 3.0.

Canada has abundant natural resources for its population. The **Canadian Shield** is an area of rock and forest that covers much of central Canada around the **Hudson Bay**. This region, and the area to the east and west of it, provides timber and minerals for Canadian industries and for export. In the Maritime Provinces of the east, the main economic activities include fishing and agriculture. Some of Canada's best farmland is located along the **St. Lawrence River** and in the southern Prairie Provinces of Manitoba, Saskatchewan, and Alberta. The St. Lawrence River region includes dairy farms and agriculture, which provide food for the larger cities of the region. The province of Ontario has fertile farmland on the north and east sides of the Great Lakes. The farmland in the Prairie Provinces has much larger grain and beef operations.

Centered in the province of Alberta is a large region of fossil fuel exploration. Coal, oil, and natural gas are found there in abundance, and much of it is exported to the United States for profit. Oil is found absorbed in surface soil called tar sands and is being extracted for energy. When tar sands are heated, the oil is separated and refined for fuel. Projections are that there is more oil in the tar sands of Canada than in the underground reserves of Saudi Arabia. Natural resources have even filtered into the cultural arena: Edmonton's professional hockey team is called the Edmonton Oilers.

Figure 4.22: Physical Features of Canada. Wikimedia Commons – CC BY 3.0.

The Rocky Mountains and the coastal ranges located in western Canada provide for mining and lumber mills. Vancouver, on the coast in British Columbia, has become a major port for Canada to export and import goods to and from the Pacific Rim. The Yukon Territory, also located in the mountains, has experienced a gold rush in years past. Canada has adequate natural resources to provide for its people and gain wealth.

British versus French Canada

As explained in Section 4.1 "Introducing the Realm", French fishermen and fur traders initially colonized Canada, the British later took it from the French, and immigrants from various other countries moved there to farm and otherwise make a living. Although none of the provinces retain French names, countless cities—especially in Quebec—have French names, among them Montreal, Trois-Rivières, Charlesbourg, and Beauport.

The names of several provinces indicate the British connection: Nova Scotia means "New Scotland," and it was so named by the British when they took over the island from the French. Prince Edward Island was named for the father of the famous nineteenth-century British queen, Victoria. You can see remnants of British colonialism in the way Canadian government is organized. Canada, like many countries of Europe, including Great Britain, is a parliamentary democracy. The monarch of the United Kingdom is still the top-ranking government official in Canada, but only as a figurehead. The queen (or king) appoints a governor general to be her (or his) representative in the Canadian federal government. Again, this is a symbolic position. There are two chambers, a House of Commons and a Senate. Members of the House of Commons are elected and are called members of Parliament (MPs). Senators are appointed to a lifelong term by the prime minister.

As of 2006, about 58 percent of Canadians spoke English as their primary language, French was the mother tongue of 22 percent, and another one-fifth of the population (20 percent) spoke a mother tongue other than English or French (Province and Territory, 2006). For most of Canada the lingua franca remains English. The French-speaking portion of Canada is a reminder of Canada's history as a French colony. Many of today's French speakers are descendants of those earlier French settlers. The proportion of French speakers in Canada is declining as more and more immigrants (who also have higher fertility rates than Francophones [French speakers]) arrive from other parts of the world and as more Francophones begin using English instead of French as their primary language. The new immigrants (along with native peoples) make up the 20 percent of Canada's population who speak neither English nor French as their native language. About 90 percent of Canadian Francophones live in Quebec, which is a center of French culture in Canada.

The separation between French Canada and British Canada goes back to colonial times. Beginning in the 1530s, the French were the first to develop fur-trading activities in the region and colonize what is present-day Canada, calling it *New France*. The French claimed much of the St. Lawrence River valley and the Great Lakes region, including the region that is now Ontario. When Britain began to dominate the eastern coast of North America in the 1680s, they entered into a series of wars with France. As a result of these wars, New France was eventually turned over to Britain.

Not wanting continued war with France, Britain allowed the French-dominated region to retain its land ownership system, civil laws, and Catholic faith. The Revolutionary War in 1776, which granted the United States independence from Great Britain, also pushed many people of English descent—especially those who had sided with the British during the Revolutionary War—from the United States into Canada. British North America no longer included the United States; Canada became the main British colony in North America.

Figure 4.23 Bilingual Stop Sign in Ottawa, Ontario

Wikimedia Commons – public domain.

In an attempt to keep the peace between French and English settlers, in 1791 the British Parliament divided Quebec into Upper and Lower Canada, which later evolved into the provinces of English-speaking Ontario and French-speaking Quebec. The Maritime Provinces of the east were then separated into individual provinces.

The cultural differences between Francophone Canada and the rest of Canada have since erupted into serious political conflicts. The Francophone areas, mainly southern Quebec, argue that they are treated unfairly, since they have to learn English but the rest of the country is not required to learn French.

During the second half of the twentieth century, many people in Quebec supported a separatist movement that sought to break Quebec off from the rest of Canada into an independent country. In the 1980s, the separatist initiative lost in a public vote, but only by a small margin. The issue of Quebec's sovereignty continued to be raised in the public arena. In 1998, Canada's Supreme Court ruled that Quebec could not separate from the rest of Canada under international law but that the matter would have to be negotiated between Quebec and the rest of Canada if secession was to proceed.

French and English are the two official languages of the Canadian government as a whole, but the French people in Quebec, fearing that English was dominating the media, the Internet, and industry to such an extent that it was endangering their French culture, have declared French as the only official language of the province. To combat the encroachment of English, laws were enacted in Quebec requiring all public advertising to be in French, or if other languages are used, they must be half the size of the French letters. All businesses employing more than fifty employees are required to conduct all business in French. Immigrants who wish to be citizens of Quebec must learn French. All primary and secondary education takes place in French unless the child's parents were educated in English elsewhere in Canada. Civil servants dubbed the "language police" monitor and enforce the French language laws. A business found to be out of compliance with the language laws could be fined or shut down. Even though the official language of Quebec is French, since the national government takes place in both English and French, some services are still available in Quebec in English.

French Canada does not include the entire province of Quebec. Northern Quebec has traditionally been native Cree territory. When the initiative for Quebec to secede from Canada was being considered, an additional initiative was presented by the Cree to separate northern Quebec from the French region of Quebec. This would have caused the French separatist region of Quebec to lose major hydroelectric dams and important mineral and timber reserves. Other regions of Quebec that did not hold a French majority would then have promoted separating from Quebec and joining Ontario.

For Quebec to remain economically competitive, they would need the northern Cree region to remain part of the province to use the resources and energy found there. They would also need the businesses currently in Quebec to stay there, although many would probably have left if Quebec had become an independent state with French as the only language. When secession was being considered, it was found that businesses that employed more than fifty employees did not want to switch over and conduct all their business in French; they were conducting all their business in English to work with the global economic community. In 1994, the North American Free Trade Agreement (NAFTA) was established between Canada, the United States, and Mexico. The goal was to open up new lines of business operations between the three countries and to increase economic opportunities to better compete with the European Union and the Pacific Rim nations. What would Quebec do if it separated from Canada? Would Quebec be able to join the NAFTA agreement? What if the other NAFTA partners cut off Quebec? Quebec would have faced serious economic consequences if they had separated from Canada. A number of English-speaking businesses have already moved to Ontario, Canada's most populous province, to avoid changing to French. It appears Quebec will remain with the rest of Canada and work out any internal cultural issues.

Figure 4.4.244.4.24: Skyline of Toronto, One of the Most Diverse Cities in North America. Marcin Wichary – United – CC BY 2.0.

Other Ethnic Groups in Canada

Of course, we cannot forget the native groups who were displaced when the Europeans arrived. About 1.2 million people who identify themselves as Aboriginal live in Canada, or about 3.8 percent of the total population. They include North American Indians (also called First Nations), Métis (descendants of both Europeans and American Indians), and Inuit (inhabitants of the far Arctic north). Of those three groups, Inuit are the smallest, with only about fifty thousand remaining. These native people represent more than six hundred recognized groups and sixty-five language dialects, although only a handful of these languages are still spoken by a large enough core of people to remain viable languages for the long term.

Countless place names in Canada, just as in the United States, derive from native words. For example, the city of Quebec is built at the original head of navigation on the St. Lawrence River. Its name comes from an Algonquin word for "where the river narrows." The word *Canada* is likely derived from the St. Lawrence Iroquois word *Kanata*, meaning "village" or "settlement." While the French explorer Jacques Cartier was traveling up the St. Lawrence River in 1535, the indigenous peoples directed him to a specific village, the future Quebec City, using the word *Kanata*. He later used *Canada* to refer to the general area, and European mapmakers adopted it for the entire region.

Quebec is not the only place where devolutionary forces have been dividing cultural groups in Canada. In 1999, Nunavut officially broke from the Northwest Territory to become its own territory. Nunavut has only about thirty thousand people in an area larger than any other province or territory in Canada. It comprises about one-fifth of Canada's land area. Most of the people who had claimed the land before the Europeans arrived are Inuit. Iqaluit, the capital city of Nunavut, is on Baffin Island near Canada's east coast.

Canada has a great deal of ethnic diversity. One measure of this is the number of languages spoken there. One source estimates that there are about 145 languages spoken in Canada, including English and French. This reflects both the rich native heritage and the history of immigration from around the world. As of 2006, the foreign-born population was 6.2 million, or nearly 20 percent of Canada's population. There are few countries that match this level of immigration. Even the United States had only about a 12.5 percent foreign-born population in 2006. If Canada's current immigration rates continue, by 2017 more than 22 percent of Canada's population will be foreign born, which equals the high rate seen in the early twentieth century, during the golden age of Eastern European immigration to North America. During this period, people from all the countries of Eastern Europe immigrated to all parts of Canada (Statistics Canada).

Figure 4.25: Evidence of Eastern European Immigration This Ukrainian Greek Orthodox Church is located in the southern region of the Prairie Province of Manitoba. Photo by R. Berglee – CC BY-NC-SA.

The current surge of immigrants to Canada does not include many Europeans. Instead, these immigrants come from Asian countries, especially China and countries in South Asia such as India and Pakistan. If immigration patterns stay the same, by 2017 about one in five people in Canada will belong to a nonwhite minority group. According to the 2010 Canadian census, more than a million Chinese and more than a half million South Asians lived in Canada. Other groups increasing rapidly include people from the Middle East and Korea.

The Canadian Economy

Not surprisingly, Canada and the Unites States are each others' largest trading partners. More than 80 percent of Canadian exports go to the United States and 70 percent of imports to Canada come from the United States. Except for some natural-resource industries, most businesses are centered in Canadian cities to take advantage of the available labor force. Canada is rapidly moving toward a knowledge-based economy built on innovation and technology. Knowledge-intensive industries, such as biotechnology and information technology, are on the rise, and these are typically located in cities to facilitate partnering with universities and other researchers.

Although Canada is developing into a knowledge-based economy, the foundations of the Canadian economy have always been its abundant natural resources. Canada's primary industries have traditionally been agriculture, fishing, mining, fuel/energy, and logging/forestry. Success in tapping these natural resources for their economic benefit allowed the country to double in population since 1960 while the economy has increased sevenfold. The primary industries now make up less than 10 percent of the gross domestic product (GDP). Just as in the United States, the most dramatic structural change in the economy has been the rise of the service sector, which now employs about three-quarters of all Canadians and generates over 60 percent of the GDP. Canadian manufacturing has been a strong sector of the economy with close ties to United States and multinational corporations.

Figure 4.26: Bombardier CRJ900LR Jet Liner Taking Off from Heathrow Airport in London The CRJ jet aircraft series is manufactured in Canada. Wikimedia Commons – public domain.

Canada's economy is tightly tied to that of the United States. One of the best examples is how the Canadian economy fluctuates depending on whether the Canadian dollar was weak or strong compared with the US dollar. For example, in 2002, one American dollar was worth about $1.60 in Canada. For many years, the American dollar was much stronger on the world market than the Canadian dollar; therefore, Canadian goods and labor were less expensive for Americans than comparable US labor and goods. During the time of the weak Canadian dollar, many film and television industries moved to Canada to film television shows and movies, as it was less expensive to do so in Canada. Many popular television shows and movies have been shot in Canada, particularly in Toronto and Vancouver.

More recently, the value of the American dollar declined against other major currencies. The Canadian dollar remained strong, which meant that goods produced in Canada became much less affordable in the United States, causing the television and film industries to move back south of the border and Canadian exports to the United States to decline. In 2007, the Canadian dollar and the US dollar reached parity for the first time in thirty years. The two currencies continue to fluctuate with market values.

As mentioned earlier, NAFTA, the 1994 trilateral agreement between Canada, the United States, and Mexico, was one of the most significant economic events in North American history. For Canada, the agreement has meant more secure, stable access to US and Mexican markets. The agreement eliminated many tariffs; opened previously protected sectors in agriculture, energy, textiles, and automotive trade; and set specific rules for trade in services such as finance, transportation, and telecommunications. Perhaps most importantly for Canada, the agreement set rules for settling trade disputes.

The United States continues to exert its powerful influence on many countries in the world, but perhaps on none so strongly as Canada. Because of the geographical proximity of the two countries and the fact that the vast majority of Canadians live very close the US border, speak English as their first language, and share a great number of cultural similarities, American trends tend to be adopted by Canadians. Canadians differentiate themselves from Americans in legal issues, laws, and health care. For example, Canada has a health care system, funded by the provinces with financial help from the federal government, that provides free services to its citizens. Canadians often point to this difference as a one of the defining elements of their culture that is different from the United States.

Canada is a great consumer of American popular culture. Canadians listen to, watch, and read tremendous quantities of American music; television and movies; and news, books, and other literature—so much so that some Canadians believed Canadian culture was in danger of being extinguished. In response to these concerns, a law was passed and a watchdog agency created so that a certain percentage of all radio and television broadcasts emanating from Canadian radio and television stations had to originate in Canada or have significant Canadian content. Others were less worried about the impact of American pop culture on Canadian culture. This segment of the Canadian population felt that Canadians have long identified themselves in contrast to Americans; therefore, consuming American books, newspapers, television shows, and movies would only give Canadian a greater basis of comparison and thus strengthen the Canadian identity and perception of Canadian culture.

Key Takeaways

- Canada is a very large country with rich natural resources but a relatively small population that mostly lives in a narrow band in the southern part of the country.
- Canada's English and French bilingualism is part of its British and French colonial past. The French culture is dominant in Quebec, where the population has considered seceding from Canada and becoming an independent country.
- Canada's native population makes up less than 4 percent of the country's population but represents a great deal of cultural and linguistic diversity.
- More than one in five Canadians is an immigrant, and most of the recent immigrants come from non-Western countries, especially those in Asia.

Discussion and Study Questions

1. Where do most of the people live in Canada? Why do they live where they do?
2. What is Canada's capital city? What is Canada's largest city?
3. What is the difference in farming operations between the two main agricultural areas in Canada?
4. Where are large deposits of fossil fuels found in Canada? What are tar sands?
5. Where is the highest rainfall found in Canada? Where is the least amount of precipitation found in Canada? Explain this precipitation pattern.
6. Who were Canada's main colonizers? How does colonialism impact Canada today?
7. Where have devolutionary forces been active but not totally successful? Where have devolutionary activities already taken place in Canada?
8. What problems might Quebec experience if it were to secede from the rest of Canada?

9. Where is Nunavut? Why is it a separate territory? Why do you think its capital city in located on the eastern coast of the territory?
10. How are current Canadian immigrants impacting the cultural and ethnic makeup of the country?

Geography Exercise
Identify the following key places on a map:

Provinces and Territories

- Alberta
- British Columbia
- Manitoba
- New Brunswick
- Newfoundland
- Northwest Territories
- Nova Scotia
- Nunavut
- Ontario
- Prince Edward Island
- Quebec
- Saskatchewan
- Yukon Territory

Major Cities of Canada

- Calgary
- Charlottetown
- Edmonton
- Fredericton
- Halifax
- Iqaluit
- Montreal
- Ottawa
- Quebec City
- Regina
- St. Johns
- Toronto
- Vancouver
- Victoria
- Whitehorse
- Winnipeg
- Yellowknife

Locations and Features

- Arctic Ocean
- Baffin Island
- Canadian Shield
- Coastal Ranges
- Elsmere Island
- Hudson Bay
- Labrador
- Rocky Mountains
- St. Lawrence River

References

Province and Territory, "Population by Mother Tongue, by (2006 Census)," Statistics Canada, www40.statcan.ca/l01/cst01/DEMO11A-eng.htm.

Statistics Canada, http://www.statcan.gc.ca/start-debut-eng.html.

4.5. REGIONS OF THE UNITED STATES AND CANADA

The United States and Canada are vast countries with diverse regions, each offering unique landscapes, cultures, and climates. In the U.S., the Northeast is known for its historical significance, featuring cities like Boston and New York, along with a mix of urban and rural areas. The Southeast is characterized by its warm climate, rich music traditions, and Southern hospitality, with cities like New Orleans and Atlanta at its heart. The Midwest is often called the "breadbasket" due to its agricultural output, with charming towns and bustling cities like Chicago. The Southwest boasts arid landscapes, vibrant cultures influenced by Native American and Mexican traditions, and cities like Phoenix and Santa Fe. The West is famous for its stunning natural beauty, from the Rocky Mountains to the Pacific coast, with iconic cities such as Los Angeles and San Francisco.

Canada, on the other hand, features distinct regions like the Atlantic provinces, known for their maritime culture and stunning coastlines. Quebec, with its French heritage, offers a unique blend of European charm and North American vibrancy, especially in cities like Montreal and Quebec City. The Prairies are recognized for their vast farmland and strong Indigenous cultures, while British Columbia is celebrated for its breathtaking landscapes and outdoor activities, particularly around Vancouver. The North, encompassing territories like Yukon and Nunavut, is known for its rugged wilderness and rich Indigenous cultures.

Together, these regions showcase the rich tapestry of life across North America, each contributing to the continent's cultural and natural heritage.

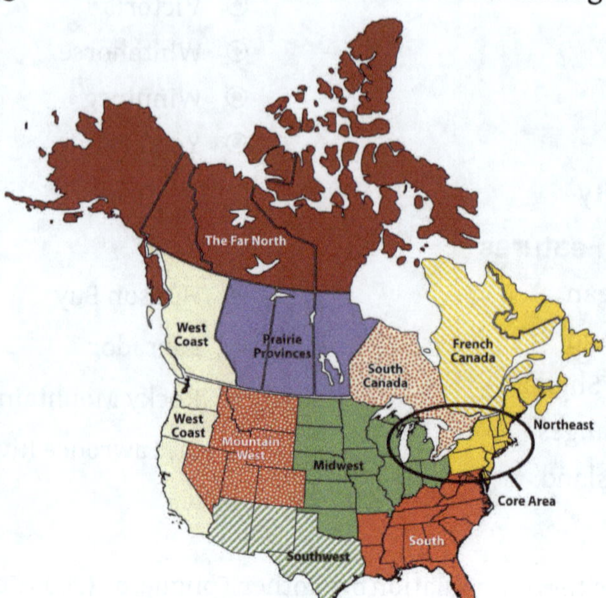

Figure 4.27: Main Regions of the United States and Canada

The Northeastern Core

The Northeastern Core includes the upper Midwest (Illinois, Indiana, Ohio, and Michigan); the mid-Atlantic states of Pennsylvania, New York, Maryland, and New Jersey plus northern Virginia; and the southern New England states of Connecticut, Rhode Island, and Massachusetts. It also includes southern Ontario, Canada's capital (Ottawa), and its largest city (Toronto). The physical environments of the Northeastern Core are quite diverse, including the northern Gulf-Atlantic Coastal Plain, the northern Appalachians, and the area surrounding the Great Lakes. This region, anchored by North America's largest metropolis, New York, is the economic heart of the United States and Canada and home to more than a third of each country's population. The megalopolis—the built-up area from Washington, DC, to Boston—is part of this region. The core region contains the Manufacturing/Rust Belt, which was once the main manufacturing region for North America but suffered decline with the advent of the information age. The core region hosts the headquarters of countless corporations, banks, financial markets (e.g., Wall Street), universities (from community colleges to the Ivy League), cultural institutions (e.g., Broadway, world-class museums, dance and music organizations), and even global organizations such as the United Nations.

Figure 4.28: Megalopolis

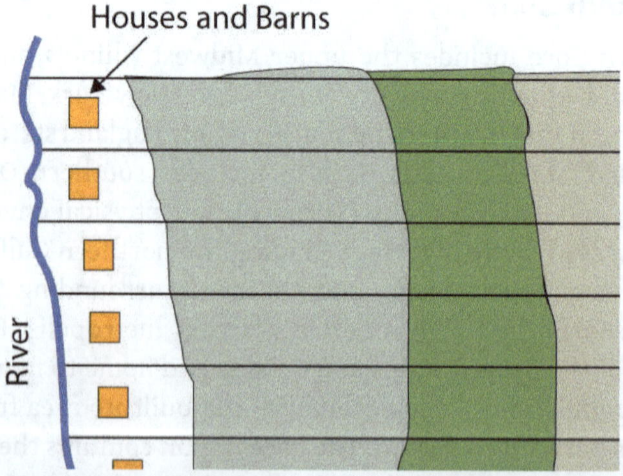

Figure 4.29: Downtown Philadelphia's city hall is an example of the urban qualities of the megalopolis that extends from Boston to Washington, DC.

Henk Sijgers – A direct line to city hall – CC BY-NC 2.0.

This large region includes geographic swaths of both wealth and economic suffering. Eight of the United States' ten wealthiest counties are in this region, most of them in the Washington, DC, area, and a number of billionaires live in New York City. Meanwhile, cities such as Detroit and Cleveland have suffered from deindustrialization and have experienced a major population decline since the 1950s. Detroit, for instance, has lost 61 percent of its population since 1950, and the decline continues. The city decreased in population by 25 percent just between 2000 and 2010. While manufacturing is not dead in the Northeastern Core, heavy industry has been in long-term decline.

New England and the Canadian Maritimes

New England and the Canadian Maritimes overlap with the Northeastern Core because its major city—Boston—is considered the northern edge of the megalopolis. South of Boston, the low-lying states of New England were the center of colonial settlement in the region and were the birthplace of America's Industrial Revolution. Southern New England began as an agricultural and fishing colony, and as industry developed in the nineteenth century, the region attracted European immigrants from Ireland, Italy, and elsewhere to work in its factories. The highly skilled workforce helped maintain a strong economy in southern New England, although there have been times of increased unemployment and economic hardship. Today the region has a more diverse economic base, including recreation and tourism, finance, telecommunications, and health care. The mountains of western New England have been particularly attractive for the development of ski resorts, and the coasts of New England are popular for summer vacationing.

As you move north from Boston, the terrain becomes more rugged and the soil less fertile. There are fewer economic reasons for people to live in northern New England, and the states of Vermont, New Hampshire, and Maine have always been less densely populated than the southern New England states. Maine is the least densely populated state in this region; about 90 percent of its land is forested, making it the most forested of the fifty United States. The vast Empty Quarter in western Maine consists of five million acres of privately owned forest and no permanent human inhabitants. Maine's leading economic activity is manufacturing, and the bulk of it is oriented around paper and other wood products.

Northern New England transitions to the even more rugged and remote uplands of the Canadian Maritimes: New Brunswick, Nova Scotia, Prince Edward Island, and Newfoundland and Labrador. The soil quality is quite poor, as glaciers removed most of it during the various ice ages. The region has a harsh climate and is removed from the major population centers of Canada and the eastern United States. The Maritime Provinces have always been oriented to the sea. Fishing and other sea-oriented businesses have historically been strong here, but in recent decades overfishing of the North Atlantic has caused a decline in the fishing economy. Tourism has been increasing as a source of revenue, especially in places such as Prince Edward Island, in which tourism is the dominant economic activity. Newfoundland, including Labrador, is Canada's poorest province.

French Canada

West of the Canadian Maritimes lies the province of Quebec, the heart of which is the St. Lawrence River valley, a lowland separating the Appalachian Mountains to the south from the inhospitable Canadian Shield to the north. As explained in Section 4.4.3 "British versus French Canada", France was the first European country to colonize the coastal regions of what is now Canada, the St. Lawrence River Valley, most of the land surrounding the Great Lakes, and the Ohio and Mississippi River valleys, south to the Gulf of Mexico. Although Great Britain obtained all that land from France in 1763 following the French and Indian War, enough French inhabitants occupied part of that territory that the region did not automatically become English speaking. The core of French Canada today is the St. Lawrence Valley from Montreal to the Atlantic coast and west of Montreal to Ottawa and north to the Hudson Bay. These French speakers, the descendants of the early French settlers, created a vibrant French-Canadian culture. About 21 percent of Canadians speak French as their mother tongue, including about 80 percent of Canadians living in the province of Quebec.

Throughout most of its history, the people of Quebec have been rural farmers, eking out a living on less-than-ideal land in a place with a short growing season. One unique characteristic of the farms in French Canada is their size and shape.

Early on, the farms were laid out as long lots, maximizing the number of farms that would have access to the transportation artery—usually a river, but sometimes a road. Each farm was about ten times longer than it was wide and had a small access point to the river, some fertile riverfront land, and a woodlot at the rear of the farm. This land-use pattern was common throughout French Canada and can even be seen today in the United States in former French colonies such as Tennessee and Louisiana.

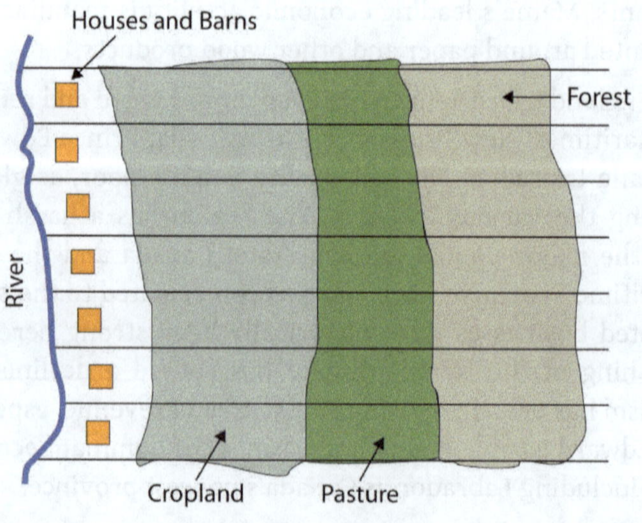

Figure 4.30: Long Lot Farms Typical in French Canada. Source: Image courtesy of E. Ratajeski.

Since then, Quebec's economy has developed to include a manufacturing sector (fueled by abundant hydroelectricity), tourism, and a variety of tertiary and quaternary industries. Montreal, Canada's second-largest metropolitan area with 3.9 million residents, is the largest French-speaking city in the Western hemisphere. It developed as the region's most important city in the mid-nineteenth century, as it controlled access through the St. Lawrence River and the Great Lakes. It became a diverse industrial center, with oil refineries, steel mills, flour and sugar refineries, and shop yards for railroad companies. Montreal attracted English speakers as well as the local **Francophones**, and at times in its history it has even had more English speakers than French speakers, despite being surrounded by a Francophone countryside. Most of northern Quebec is sparsely populated because of the lack of quality soil for agriculture, but a paper and pulp industry based on its forests has developed over the twentieth century, as well as hydroelectric power generation.

The American South

The South includes the entire southeastern portion of the United States from Kentucky south to Louisiana, east to Florida, and north to Virginia. The South consists of most of the Gulf-Atlantic Coastal Plain and the southern portion of the Appalachian Highlands.

Before the Civil War, the coastal plain was dedicated to plantation agriculture using African slave labor. Land not used for plantation crops such as tobacco, cotton, and rice was typically farmed by poor whites and later by poor blacks. Some were sharecroppers, while others farmed their own small plots, especially on the lesser-quality land in Appalachia. The South had little urbanization or industrialization at the time of the Civil War. Well into the twentieth century, the region remained rural and economically deprived.

Figure 4.31: Names often reflect the cultural region of their location. Business names with "Dixie" in them can be found throughout the South. The Dixie Grill restaurant is located in Morehead, Kentucky.

Photo by R. Berglee – CC BY-NC-SA.

Coal mining was a major source of employment in places such as West Virginia and eastern Kentucky for the first two-thirds of the twentieth century, but increased mechanization of mining methods, as well as new mining techniques such as mountaintop removal mining, decreased the number of miners needed, even as coal production increased.

The Appalachian South is perennially plagued by high unemployment, poverty, and difficult social conditions. Other areas of the rural South are also among the poorest in the nation, including the Mississippi Delta and the lower Mississippi River valley. Despite the continued swaths of poverty in the South, parts of the region have prospered in the past generation as Sun Belt migrants have moved to southern places such as Atlanta, Charlotte, Tampa, Miami, and dozens of smaller cities. This has fueled a period of urbanization and economic growth, and the newfound prosperity has helped integrate the South into the nation's economy.

The Midwest and Great Plains

The center of the continent contains a relatively level agricultural region: the Midwest and the Great Plains. This land includes some of the most fertile agricultural land in the entire world and has been dubbed America's breadbasket. The climate gets progressively more arid as you move to the west within this region, and the type of agriculture changes with the decrease in precipitation. Closer to the Rocky Mountains, the land is typically used for raising cattle, but enormous grain farms are found where water is available (especially through irrigation). The water for irrigation comes from the continent's largest aquifer, the Ogallala Aquifer. Water is often pumped to the surface using a system called center pivot irrigation. The heart of the spring wheat belt is North Dakota, and the crop is also common in eastern Montana and in Canada's Prairie Provinces of Alberta, Saskatchewan, and Manitoba. Winter wheat is common in Kansas and surrounding states. Farther to the east, where precipitation is more abundant, is the Corn Belt, focused in Iowa and Illinois.

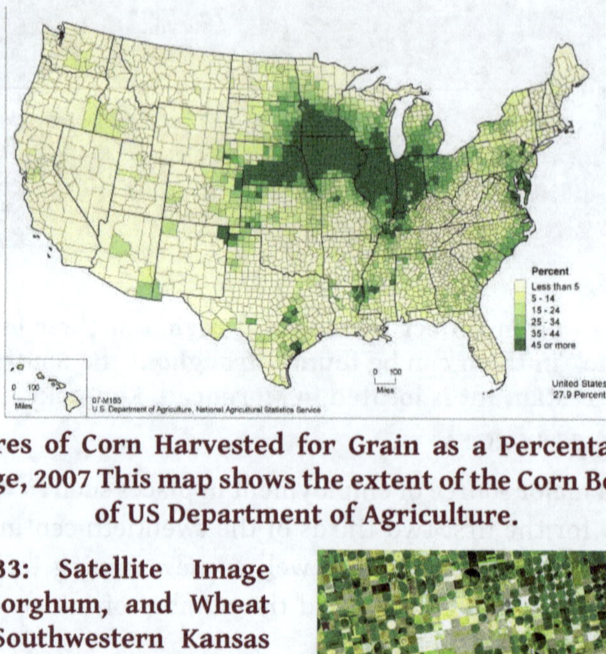

Figure 4.32: Acres of Corn Harvested for Grain as a Percentage of Harvested Cropland Acreage, 2007 This map shows the extent of the Corn Belt. Map courtesy of US Department of Agriculture.

Figure 4.33: Satellite Image of Corn, Sorghum, and Wheat Fields in Southwestern Kansas The circular fields are between a half-mile and mile in diameter, and are characteristic of fields that use center-pivot irrigation. Photo courtesy of NASA/GSFC/METI/ERSDAC/JAROS and U.S./Japan ASTER Science Team

The dominant city in this region is Chicago, which developed as a market town for the livestock and grain produced in the surrounding states and was linked to its hinterland through a complex network of rail lines. In fact, nearly all the major cities of this region developed as places for the buying, selling, and processing agricultural products. Today the Midwest and the Great Plains remain the most important food-producing areas in North America, although as agriculture has become increasingly mechanized and farms have gotten larger, the number of farmers has decreased. This region, especially the Great Plains, is experiencing a period of long-term population decline and aging.

The American Southwest

The states of Texas, New Mexico, and Arizona are considered the Southwest. The climate of the Southwest is more arid and receives a high amount of sunlight throughout the year. Desert conditions are integrated with higher elevations in the mountainous areas. Eastern Texas receives more rainfall from the Gulf of Mexico, and western Texas and the states of New Mexico and Arizona are quite arid and receive less rainfall. These conditions are more favorable to cattle ranching than to other agricultural activities. Large farming operations exist where water is available for irrigation. The warmer climate has been attractive for development and people emigrating from the colder regions of the north.

Figure 4.34: Typical Home in a Phoenix Suburb, Where Water Is a Valued Commodity There is no grass to mow, and cactus and palm trees are common. Photo by R. Berglee – CC BY-NC-SA.

The Southwest has a strong Hispanic heritage and was part of Spain's Mexican colony before England established colonies on the East Coast (the first Spanish settlers arrived in New Mexico in 1598). All three states have a large contingency of Hispanic residents, some of whom have descended from early pre–United States settlers, while others have come into the United States more recently across the long desert border between Mexico and the United States. According to the 2010 census, about 46 percent of New Mexico's population was Hispanic or Latino, and in Texas and Arizona the figures were 37 percent and 30 percent, respectively (US Census Bureau).

In Arizona, strong centrifugal forces have been in play because of its tough legal measures against illegal immigration. This issue exposes the social rift between the more European population of the state and the Hispanic immigrants. In 2008, the US Department of Homeland Security's Office of Immigration Statistics estimated that Arizona had one of the fastest-growing illegal immigrant populations in the country at 8.4 percent of the population (Hoefer, M., et. al., 2011). This was the second-highest percentage of undocumented residents in the country, after Nevada. In April 2010, the *New York Times* reported that Arizona's governor signed the nation's toughest bill on illegal immigration into law, designed to identify, prosecute, and deport illegal immigrants (Archibold, R. C., 2011). At the same time, Reuters reported that the when the bill was debated in the Arizona senate, the number of illegal immigrants was listed at 10 percent of the population (Reuters, 2011). Some opponents of the tough illegal immigration laws claimed that law enforcement officials would use racial profiling to target Hispanic residents, and the federal government took the state to court to halt its enforcement.

The Southwest also has a strong Native American presence, especially in New Mexico and Arizona. Twenty-one federally recognized tribes with more than 250,000 people (4.9 percent of the state's population) live in Arizona, and their reservations and traditional communities make up more than one-fourth of the state's land. The Navajo tribe is the largest in the United States, with more than 100,000 members in Arizona alone and others in surrounding states. The considerable Native American and Hispanic population in the Southwest means that non-Hispanic whites make up a minority of the population in New Mexico and Texas.

Figure 4.35: Navajo Dancer Performing the Eagle Dance One of the Pollen Trail Dancers, a Navajo group near Joseph City, Arizona, performs the Eagle Dance

on September 21, 2010, on the scenic South Rim of Grand Canyon National Park between Hopi House and Verkamp's Visitor Center. Grand Canyon National Park – Grand Canyon Eagle Dancer – CC BY 2.0.

The three Southwestern states have been recipients of Sun Belt migrations over the past few decades, as people have moved to the Southwest for tertiary-sector jobs and for the region's warm climate. The region is quite urbanized, and most of the new migrants are moving to cities. Three-fourths of Arizona's population live in the Phoenix or Tucson metropolitan area. The most populous metropolitan area in the Southwest is Dallas–Fort Worth, Texas, with 6.4 million residents in 2010, making it the fourth-largest metropolitan area in the United States. The Houston metropolitan area is not far behind with 6.1 million residents. The economy of Texas used to be based on oil and natural gas, but it has since become more diversified. Residents of these cities work in high-tech manufacturing, health care, business, and information. One of the most famous high-tech industries in Texas is space: Houston is home to the National Aeronautics and Space Administration's (NASA) Lyndon B. Johnson Space Center, where astronauts and thousands of others work in the space industry.

The Mountain West

From the Rocky Mountains to the Sierra Nevada and the Cascades and the **Intermontane** Basins and Plateaus in between, this part of North America has gone from the old Wild West to an area of rapid economic and population growth. The region encompasses western Colorado; western Wyoming; western Montana; Idaho; Nevada; Utah; the eastern portions of Washington, Oregon, and California; and the southern portion of the Canadian Rockies.

Figure 4.36: Rocky Mountains of Western Montana on US Highway 2. Photo by R. Berglee – CC BY-NC-SA.

The population of the Mountain West is growing much faster than the population of the United States as a whole. For example, Nevada's population grew 32.3 percent from 2000 to 2009, which is more than three times as much as the United States as a whole (9.1 percent). All US states in this region, except for Montana, grew at faster rates than the US average. Utah grew 24.7 percent, Idaho grew 19.5 percent, and Colorado grew 16.8 percent. What is fueling this growth? It is part of the larger pattern of Americans flocking to the Sun Belt, searching for an attractive climate and lifestyle. Jobs have been created in recreation (gambling, skiing), in high-tech firms, and in other tertiary sector industries. Many of the migrants come from southern California because the housing in the Mountain West is more affordable and the region is much less crowded. Nearly all the growth is occurring in urban and suburban areas.

However, the rapid growth of the West since 1990 has come at a cost. In some areas the large population is putting a strain on physical resources, such as water. Water is a hot-button political issue, particularly in the more arid states such as Nevada. Las Vegas, for example, is a desert city that gets 90 percent of its water from a Colorado River reservoir: Lake Mead. This water comes from snowmelt in the Rocky Mountains far to the east, and due to drought and high water demand, Lake Mead's water level has been dropping. If current patterns persist, Las Vegas will have a water crisis soon. The water shortage is happening even though Las Vegas has managed to reduce per capita water usage by raising prices for water and creating incentives to remove grass lawns. Las Vegas recycles 94 percent of all sewage water, which is the highest rate in the United States.

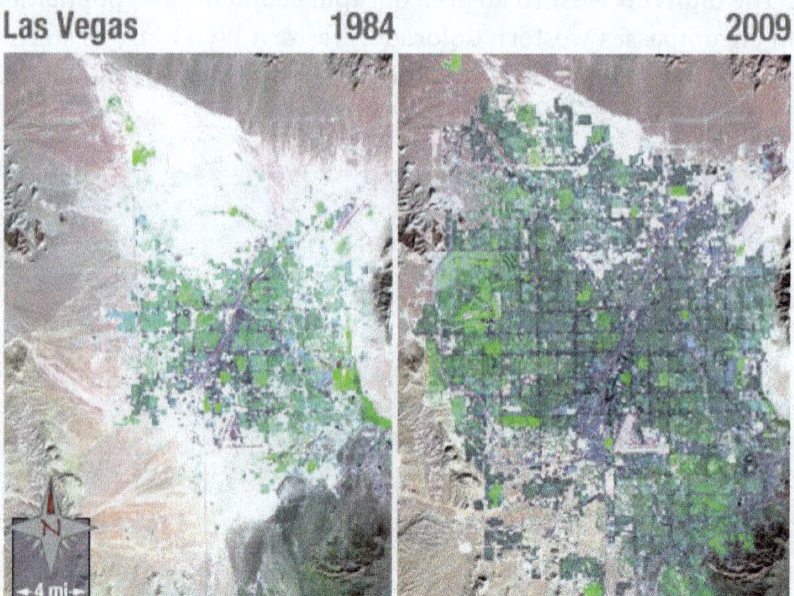

Figure 4.37: Urban Growth in Las Vegas, Nevada, from 1984 to 2009 Notice the dark purple of city streets and the bright green of irrigated vegetation. Images courtesy of NASA

As the West's population increased rapidly, the region's urban areas became part of the real estate bubble of the early 2000s. Real estate prices grew and the housing industry built one residential subdivision after another in the suburbs surrounding cities such as Las Vegas, Denver, and Salt Lake City. However, in the late 2000s the real estate market collapsed in many parts of the country, and Western cities were hard hit. As of early 2011, three of the five states with the highest foreclosure rates were in this region (Nevada, Utah, and Idaho), and the others were bordering states (Arizona and California).

The Pacific Coast

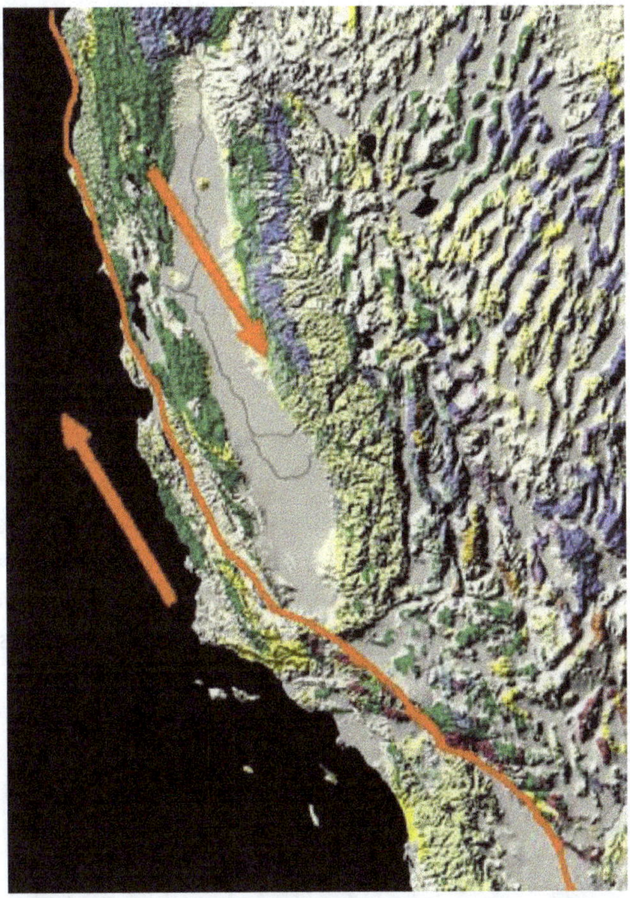

Figure 4.38: The San Andreas Fault The Pacific tectonic plate is moving northward relative to the North American plate. Image courtesy of USGS

The Pacific Coast includes the coastal portions of California, Oregon, and Washington, plus the southwestern portion of British Columbia in Canada. This region is typically thought of as two subregions: California and the Pacific Northwest. The two areas are quite different from each other in terms of climate and economy. However, both areas are part of the so-called Ring of Fire that encircles the Pacific

Ocean. The Ring of Fire is a zone of earthquakes and volcanic eruptions that occur near where the Pacific tectonic plate meets the surrounding plates. In the United States, two areas of concern are the San Andreas Fault in California and Mount St. Helens in Washington. The 1906 earthquake that destroyed San Francisco was a result of activity on the San Andreas Fault, and scientists predict that strong earthquakes will reappear along the fault in the future. Thousands of small earthquakes occur along the fault every year. Mount St. Helen's is a volcano in the Cascades that erupted in 1980, killing fifty-seven people and destroying hundreds of square miles of forest.

The Pacific Coast represents a large population center a continent away from what we consider the North American core. Most of the region's population is urban, and Los Angeles and its metropolitan area is by far the largest area of settlement. Twelve percent of the US population lives in California (thirty-seven million people), and the greater Los Angeles metropolitan area has nearly eighteen million people. Los Angeles is the second-largest US city after New York. Los Angeles is the quintessential automobile city. It developed into a major city in the mid-twentieth century at the time that automobile ownership had become common, and people who lived in the area tended to move to suburbs that were connected to each other by an extensive highway system. Los Angeles is a highly decentralized city, unlike cities in other parts of North America that formed during other transportation regimes.

The Pacific Coast region is also famous for its agriculture. California's Central Valley lies between the Coast Ranges to the west and the Sierra Nevada to the east and is among the most productive agricultural areas in the world. The irrigated farmland in the valley produces all types of nontropical crops and is the largest US producer of tomatoes, grapes, almonds, and other foods. When other parts of the country are still frozen in the winter months, the fields of the Central Valley are already producing bountiful harvests. California is also famous for its wine production, especially in Napa Valley near San Francisco.

Figure 4.39: The Pacific Ocean from Venice Beach in Los Angeles. Photo by R. Berglee – CC BY-NC-SA.

Besides agriculture, the economic base of the Pacific Coast is quite diverse and rich. If it were an independent country, California would be the world's sixth-largest economy. Los Angeles is considered the capital of the US entertainment industry, and other major industries include aerospace, manufacturing, and foreign trade. The port of Los Angeles is the busiest in North America, receiving shipments of goods from China and other Asian countries. Silicon Valley, near San Francisco, is a key area for high-tech research and Internet commerce. The Pacific Northwest is home to major corporations such as Boeing (whose headquarters recently moved to Chicago), Microsoft, and other famous companies such as Starbucks, Amazon.com, REI, T-Mobile, Costco, and Eddie Bauer. One of the richest Americans, Bill Gates, lives near Seattle.

Across the border to the north, Vancouver is Canada's third-largest metropolitan area with over two million residents. Vancouver is unlike any other city in North America. Nearly one-third of its residents are of Chinese origin, and more than half its population speak a language other than English at home. Vancouver began as a logging town but developed into its position as the Asian gateway to Canada because of its port, the busiest in Canada. Vancouver is a popular location for the film industry and is sometimes dubbed as "Hollywood North." It is also growing in the biotechnology and software industries.

The North

The North is the least densely populated of any region in North America due to its brutally cold winters, short growing season, and poor soils. It includes the boreal forests of the upper Great Lakes region and the Canadian Shield and the territory to the north of the tree line that extends beyond the Arctic Circle. Physically, this region is immense, including the state of Alaska plus most of Canada. The climate is similar to that of Russia: cold continental and arctic climates, arctic air masses swooping down from the north, and long winters. Most inhabitants of the northern portions of North America live in the forested areas rather than in the frozen Arctic.

Two groups of people live in this region. First are the native peoples who have always lived there. They are small in number and traditionally make a living by hunting and fishing. More recently, the native populations such as the Inuit and the First Nations in Canada subsist by combining wage employment with their traditional means of living off the land. American Indians or Alaskan natives make up about 15 percent of Alaska's population, for a total of roughly 106,000 people. In Canada's Northwest Territories, First Nations people make up just over half the population, but the total population is quite small—only about 41,000 in the entire territory. In Nunavut, the native population is about 85 percent of the total 30,000 residents, living in a territory the size of Western Europe.

Figure 4.40: Radar Station at Point Lay, Alaska. Wikimedia Commons – public domain.

The other residents are more recent immigrants who are there to exploit the land's natural resources. The economy is dominated by the primary economic sector: forestry, oil and natural gas extraction, and mining. In the Canadian Shield, metallic ores such as copper, gold, nickel, silver, and uranium are found in the rocks and diamond mines are in operation, as are mines producing rare earth elements used in computer screens, electric car batteries, and computer hard drives. These elements include metals such as cerium, terbium, dysprosium, and neodymium. Alaska is an oil-producing state, and the decision of whether to open additional areas of Alaska's Arctic to oil drilling remains controversial and uncertain.

Key Takeaways

- The economic core of North America has traditionally been in the US Northeast and its surrounding territory.
- As you move north from Boston into the rest of New England, the Canadian Maritimes, and Quebec, the economy is increasingly based on primary industries such as forestry and fishing.
- Agriculture and mining have been in decline in the South, while tertiary and quaternary industries have attracted new migrants to the region's urban and suburban areas.
- The Midwest and the Great Plains make up North America's breadbasket. The climate gets more arid as you move west, but through irrigation, agricultural productivity remains high.
- The Southwest is unique in its high proportion of Hispanic and Native American residents.

- The Mountain West is growing rapidly, especially in its urban and suburban areas. This is putting stress on the physical environment (particularly its water resources) and made the region susceptible to the real estate bubble collapse.
- The Pacific Coast region is prone to earthquakes and volcanic activity, yet it is home to the second-largest metropolitan area in the United States and is known for its rich, diverse economic base.
- The far northern stretches of North America are sparsely populated, with an economy based on primary industries such as forestry and extraction. The North, both Canada and Alaska, is also known for its large native populations.

Discussion and Study Questions

1. Where is the largest US megalopolis located? What region is it part of?
2. What is the purpose of long lots? Where would they be most prominent?
3. What have been some of the traditional economic activities in the American South? How is this changing?
4. Why is there a difference between where spring wheat and winter wheat is grown?
5. Name the main ethnic groups that are prominent in the American Southwest.
6. Why does the population continue to increase in the American West?
7. What makes the Pacific Coast region an unstable place for human development?
8. Why does the Pacific Coast region have both high agricultural production and large high-tech urban cities?
9. Why does the desert region of the American Southwest continue to attract a growing population?
10. What have been the main economic activities in the far north?

Geography Exercise
Identify the following key places on a map:

- Northwestern Core
- New England and the Canadian Maritimes
- French Canada
- The American South
- The Midwest and Great Plains
- The American Southwest

- The Mountain West
- The Pacific Coast
- The North

References

Archibold, R. C., "Arizona Enacts Stringent Law on Immigration," *New York Times*, April 23, 2010, accessed September 20, 2011, http://www.nytimes.com/2010/04/24/us/politics/24immig.html.

Hoefer, M., Nancy Rytina, and Bryan C. Baker, "Estimates of the Unauthorized Immigrant Population Residing in the United States: January 2008," Department of Homeland Security, accessed September 20, 2011, http://www.dhs.gov/xlibrary/assets/statistics/publications/ois_ill_pe_2008.pdf. Reuters, April 19, 2010, "Arizona Passes Tough Illegal Immigration Law," accessed September 20, 2011, www.reuters.com/article/2010/04/19/us-immigration-usa-arizona-idUSTRE63I6TU20100419. US Census Bureau, "State & County QuickFacts," quickfacts.census.gov/qfd/index.html.

4.6. CONCISE SUMMARY

The United States and Canada share numerous similarities, including expansive territory, history of European colonization, diverse immigrant populations, and elevated living levels. United States and Canada are increasingly diverging from their European roots as immigrants from non-European regions enter. The predominant group of immigrants in the United States originates from Latin America. The predominant group of immigrants to Canada originates from Asia. The United States and Canada both possess a limited indigenous population; however, in Canada, indigenous peoples have attained greater self-representation than in the United States, particularly since the establishment of Nunavut. Quebec, the Francophone core of Canada, has always endeavored to preserve its cultural distinctiveness while safeguarding its economic stability. Both nations are postindustrial, characterized by service- and information-driven economies. The United States constitutes the largest economy globally and has a historical precedent of disseminating its culture, ideologies, and military strength throughout. North America comprises diverse regions characterized by unique cultural and physical attributes. Every region possesses majority and minority populations determined by immigration or indigenous ancestry. Economic situations differ among regions. The Sun Belt is increasingly drawing a multitude of information-centric high-tech companies.

5 Chapter

MIDDLE AMERICA

5.1. INTRODUCTION: AN OVERVIEW

Middle America often evokes images of small towns, sprawling farmlands, and a strong sense of community. It's a region characterized by its rich cultural tapestry, where traditions blend with modernity. The heartbeat of Middle America can be felt in its bustling county fairs, local diners, and friendly gatherings. Here, values like hard work, family, and neighborliness thrive, even as economic challenges and changing demographics shape the landscape. This area plays a crucial role in the broader narrative of the United States, reflecting both the nation's heartland spirit and the complexities of contemporary life.

Figure 5.1 Middle America: Caribbean, Mexico, and Central America

Physical Geography

Middle America has various types of physical landscapes, including volcanic islands and mountain ranges. Tectonic action at the edge of the Caribbean Plate has brought about volcanic activity, creating many of the islands of the region as volcanoes rose above the ocean surface. The island of Montserrat is one such example. The volcano on this island has continued to erupt in recent years, showering the island with dust and ash and making habitation difficult. Many of the other low-lying islands, such as the Bahamas, were formed by coral reefs rising above the ocean surface. Tectonic plate activity not only has created volcanic islands but also is a constant source of earthquakes that continue to be a problem for the Caribbean community.

The republics of Central America extend from Mexico to Colombia and form the final connection between North America and South America. The **Isthmus of Panama**, the narrowest point between the Caribbean Sea and the Pacific Ocean, serves as a **land bridge** between the continents. The backbone of Central America is mountainous, with many volcanoes located within its ranges. Much of the Caribbean and all of Central America are located south of the Tropic of Cancer and are dominated by tropical type A climates. The mountainous areas have varied climates, with cooler climates located at higher elevations. Mexico has extensive mountainous areas with two main ranges in the north and highlands in the south. There are no landlocked countries in this realm, and coastal areas have been exploited for fishing and tourism development.

Rimland and Mainland

Using a regional approach to the geography of a realm helps us compare and contrast a place's features and characteristics. Location and the physical differences explain the division of Middle America into two geographic areas according to occupational activities and colonial dynamics: the rimland, which includes the Caribbean islands and the Caribbean coastal areas of Central America, and the mainland, which includes the interior of Mexico and Central America.

Colonialism thrived in the rimland because it consists mainly of islands and coastal areas that were accessible to European ships. Ships could easily sail into a cove or bay to make port and claim the island for their home country. After an island or coastal area was claimed, there was unimpeded transformation of the area through plantation agriculture. On a plantation, local individuals were subjugated as servants or slaves. The land was planted with a single crop—usually sugarcane, tobacco, cotton, or fruit—grown for export profits. Most of these crops were not native to the Americas but were brought in during colonial times. European diseases killed vast numbers of local Amerindian laborers, so slaves were brought from Africa to do the work.

Plantation agriculture in the rimland was successful because of the import of technology, slave labor, and raw materials, as well as the export of the harvest to Europe for profit.

Plantation agriculture changed the rimland. The local groups were diminished because of disease and colonial subjugation, and by the 1800s most of the population was of African descent. Native food crops for consumption gave way to cash crops for export. Marginal lands were plowed up and placed into the plantation system. The labor was usually seasonal: there was a high demand for labor at peak planting and harvest times. Plantations were generally owned by wealthy Europeans who may or may not have actually lived there.

The mainland, consisting of Mexico and the interior of Central America, diverged from the rimland in terms of both colonial dynamics and agricultural production. The interior lacked the easy access to the sea that the rimland enjoyed. As a result, the hacienda style of land use developed. This Spanish innovation was aimed at land acquisition for social prestige and a comfortable lifestyle. Export profits were not the driving force behind the operation, though they may have existed. The indigenous workers, who were poorly paid if at all, were allowed to live on the haciendas, working their own plots for subsistence. African slaves were not prominent in the mainland.

In the mainland, European colonialists would enter an area and stake claims to large portions of the land, often as much as thousands or even in the millions of acres. Haciendas would eventually become the main landholding structure in the mainland of Mexico and many other regions of Middle America. In the hacienda system, the Amerindian people lost ownership of the land to the European colonial masters. Land ownership or the control of land has been a common point of conflict throughout the Americas where land transferred from a local indigenous ownership to a colonial European ownership.

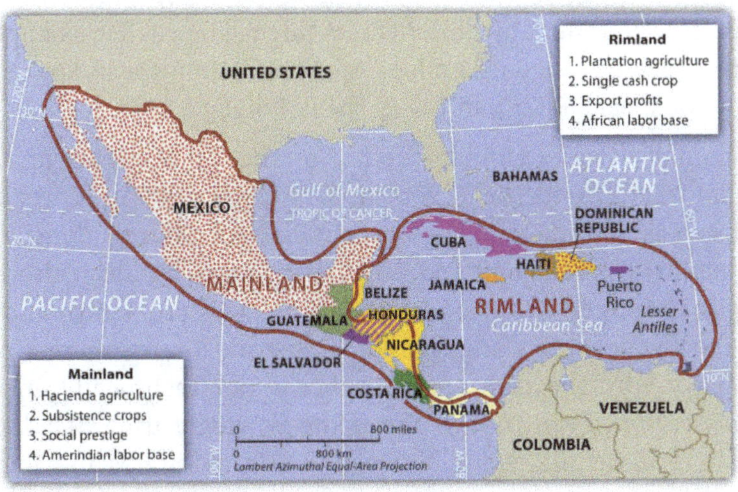

Figure 5.2: Mainland and Rimland Characteristics of Middle America Based on Colonial-Era Economic Activities

The rimland was more accessible to European ships, and the mainland was more isolated from European activity.

The plantation and hacienda eras are in the past. The abolition of slavery in the later 1800s and the cultural revolutions that occurred on the mainland challenged the plantation and hacienda systems and brought about land reform. Plantations were transformed into either multiple private plots or large corporate farms. The hacienda system was broken up, and most of the hacienda land was given back to the people, often in the form of an ejidos system, in which the community owns the land but individuals can profit from it by sharing its resources. The ejidos system has created its own set of problems, and many of the communally owned lands are being transferred to private owners.

The agricultural systems changed Middle America by altering both the systems of land use and the ethnicity of the population. The Caribbean Basin changed in ethnicity from being entirely Amerindian, to being dominated by European colonizers, to having an African majority population. The mainland experienced the mixing of European culture with the Amerindian culture to form various types of mestizo groups with Hispanic, Latino, or Chicano identities.

The European Invasion

Though the southern region of the Americas has commonly been referred to as "Latin America," this is a misnomer because Latin has never been the lingua franca of any of the countries in the Americas. What, then, is the connection between the southern region of the Americas and Latin? To understand this connection, the reader needs to bring to mind the dominant languages as well as the origin of the colonizers of the region called "Latin America." Keep in mind that the name of a given country does not always reflect its lingua franca. For example, people in Mexico do not speak a language called "Mexican"; they speak Spanish. Likewise, Brazilians do not speak "Brazilian"; they speak Portuguese.

European colonialism had an immense effect on the rest of the world. Among other things, colonialism diffused the European languages and the Christian religion. Latin Mass has been a tradition in the Roman Catholic Church. Consider the Latin-based Romance language group and how European colonialism altered language and religion in the Americas. The Romance languages of Spanish and Portuguese are now the most widely used languages in Middle and South America, respectively. This is precisely why the term *Latin America* is not technically an appropriate name for this region, even though the name is widely used. *Middle America* is a more accurate term for the region between the United States and South America, and *South America* is the appropriate name for the southern continent in spite of the connection to Latin-based languages.

European colonialism impacted Middle America in more ways than language and religion. Before Christopher Columbus arrived from Europe, the Americas did not have animals such as horses, donkeys, sheep, chickens, and domesticated cattle.

This meant there were no large draft animals for plowing fields or carrying heavy burdens. The concept of the wheel, which was so prominent in Europe, was not found in use in the Americas. Food crops were also different: the potato was an American food crop, as were corn, squash, beans, chili peppers, and tobacco. Europeans brought other food crops—either from Europe itself or from its colonies—such as coffee, wheat, barley, rice, citrus fruits, and sugarcane. Besides food crops, building methods, agricultural practices, and even diseases were exchanged.

The Spanish invasion of Middle America following Columbus had devastating consequences for the indigenous populations. It has been estimated that fifteen to twenty million people lived in Middle America when the Europeans arrived, but after a century of European colonialism, only about 2.5 million remained. Few of the indigenous peoples—such as the Arawak and the Carib on the islands of the Caribbean and the Maya and Aztec on the mainland—had immunities to European diseases such as measles, mumps, smallpox, and influenza. Through warfare, disease, and enslavement, the local populations were decimated. Only a small number of people still claim Amerindian heritage in the Caribbean Basin, and some argue that these few are not indigenous to the Caribbean but are descendants of slaves brought from South America by European colonialists.

Columbus landed with his three ships on the island of Hispaniola in 1492. Hispaniola is now divided into the countries of Haiti and the Dominican Republic. With the advantage of metal armor, weapons, and other advanced technology, the Spanish invaders quickly dominated the local people. Since Europe was going through a period of competition, warfare, and technological advancements, the same mind-set carried forward to the New World. Indigenous people were most often made servants of the Europeans, and resistance resulted in conflict, war, and often death. The Spanish soldiers, explorers, or adventurers called conquistadors were looking for profits and quick gain and ardently sought gold, silver, and precious gems. This quest for gain pitted the European invaders against the local groups. The Roman Catholic religion was brought over from Europe and at times was zealously pushed on the local "heathens" with a "repent or perish" method of conversion.

Many of the Caribbean islands have declared independence, but some remain crown colonies of their European colonizers with varying degrees of autonomy. Mexico achieved independence from Spain by 1821, and most Central American republics also gained independence in the 1820s. In 1823, the United States implemented the Monroe Doctrine, designed to deter the former European colonial powers from engaging in continued political activity in the Americas. US intervention has continued in various places in spite of the reduction in European activity in the region. In 1898, the United States engaged Spain in the Spanish-American War, in which Spain lost its colonies of Cuba, Puerto Rico, and others to the United

States. Puerto Rico continues to be under US jurisdiction and is not an independent country.

The Maya and the Aztec

Though the region of Mexico has been inhabited for thousands of years, one of the earliest cultures to develop into a civilization with large cities was the Olmec, which was believed to be the precursor to the later Mayan Empire. The Olmec flourished in the south-central regions of Mexico from 1200 BCE to about 400 BCE. Anthropologists call this region of Mexico and northern Central America Mesoamerica. It is considered to be the region's cultural hearth because it was home to early human civilizations. The Maya established a vast civilization after the Olmec, and Mayan stone structures remain as major tourist attractions. The classical era of the Mayan civilization lasted from 300 to 900 CE and was centered in the Yucatán Peninsula region of Mexico, Belize, and Central America. Guatemala was once a large part of this vast empire, and Mayan ruins are found as far south as Honduras. During the classical era, the Maya built some of the most magnificent cities and stone pyramids in the Western Hemisphere. The city-states of the empire functioned through a sophisticated religious hierarchy. The Mayan civilization made advancements in mathematics, astronomy, engineering, and architecture. They developed an accurate calendar based on the seasons and the solar system. The extent of their immense knowledge is still being discovered. The descendants of the Maya people still exist today, but their empire does not.

Figure 5.3: Mayan Site of Uxmal in the Yucatán Region of Mexico The classical Mayan era lasted from 300 to 900 CE. Many magnificent cities were built with stone and remain today as major tourist attractions. Photo by R. Berglee – CC BY-NC-SA.

The Toltec, who controlled central Mexico briefly, came to power after the classical Mayan era. They also took control of portions of the old Mayan Empire from the north. The Aztec federation replaced the Toltec and Maya as the dominant civilization in southern Mexico. The Aztec, who expanded outward from their base in central Mexico, built the largest and greatest city in the Americas of the time, Tenochtitlán, with an estimated population of one hundred thousand. Tenochtitlán was located at the present site of Mexico City, and it was from there that the Aztec expanded into the south and east to create an expansive empire. The Aztec federation was a regional power that subjugated other groups and extracted taxes and tributes from them. Though they borrowed ideas and innovations from earlier groups such as the Maya, they made great strides in agriculture and urban development. The Aztec rose to dominance in the fourteenth century and were still in power when the Europeans arrived.

Spanish Conquest of 1519–21

After the voyages of Columbus, the Spanish conquistadors came to the New World in search of gold, riches, and profits, bringing their Roman Catholic religion with them. Zealous church members sought to convert the "heathens" to their religion. One such conquistador was Hernán Cortés, who, with his 508 soldiers, landed on the shores of the Yucatán in 1519. They made their way west toward the Aztec Empire. The wealth and power of the Aztecs attracted conquistadors such as Cortés, whose goal was to conquer. Even with metal armor, steel swords, sixteen horses, and a few cannons, Cortés and his men did not challenge the Aztecs directly. The Aztec leader Montezuma II originally thought Cortés and his men were legendary "White Gods" returning to recover the empire. Cortés defeated the Aztecs by uniting the people that the Aztecs had subjugated and joining with them to fight the Aztecs. The Spanish conquest of the Aztec federation was complete by 1521.

As mentioned, the Spanish invasion of Middle America had devastating consequences for the indigenous populations. It is estimated that there were between fifteen and twenty-five million Amerindians in Middle America before the Europeans arrived. After a century of European colonialism, there were only about 2.5 million left (Module 01). Cortés defeated the Amerindian people by killing off the learned classes of the religious clergy, priestly orders, and those in authority. The local peasants and workers survived. The Spanish destroyed the knowledge base of the Maya and Aztec people. Their knowledge of astronomy, their advanced calendar, and their engineering technology were lost. Only through anthropology, archaeology, and the relearning of the culture can we fully understand the expanse of these early empires. The local Amerindian descendants of the Maya and the Aztec still live in the region, and there are dozens of other Amerindian groups in Mexico with their own languages, histories, and cultures.

The Spanish Colonial City

As the Spanish established urban centers in the New World, they structured each town after the Spanish pattern, with a plaza in the center. Around the plaza on one side was the church (Roman Catholic, of course). On the other sides of the plaza were government offices and stores. Residential homes filled in around them. This pattern can still be seen in almost all the cities built by the Spanish in Middle and South America. The Catholic Church not only was located in the center of town but also was a supreme cultural force shaping and molding the Amerindian societies conquered by the Spanish.

In Spain, the cultural norm was to develop urban centers wherever administration or military support was needed. Spanish colonizers followed a similar pattern in laying out the new urban centers in their colonies. Extending out from the city center (where the town plaza, government buildings, and church were located) was a commercial district that was the backbone of this model. Expanding out on each side of the spine was a wealthy residential district for the upper social classes, complete with office complexes, shopping districts, and upper-scale markets.

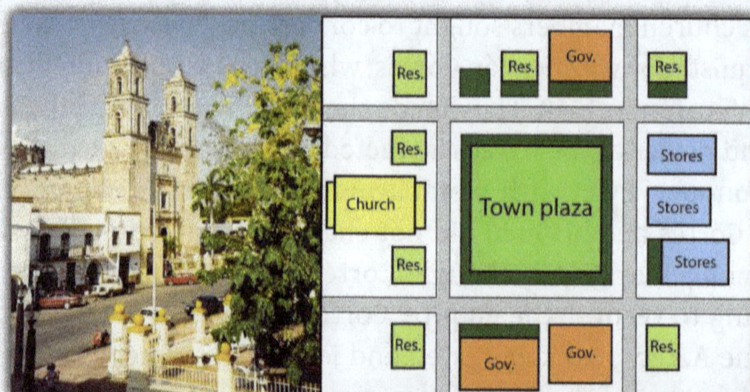

Figure 5.4: Catholic Cathedral across from a Plaza in the Yucatán City of Valladolid (*left*); Model of a Spanish Colonial Urban Pattern (*right*) The Spanish colonial urban pattern had a plaza in the center of the city with government buildings around the square and a Catholic church on one side. Photo by R. Berglee – CC BY-NC-SA.

Surrounding the central business district (CBD) and the spine of most cities in Middle and South America are concentric zones of residential districts for the lower, working, and middle classes and the poor. The first zone, the zone of maturity, has well-established middle-class residential neighborhoods with city services. The second concentric zone, the zone of transition (in situ accretion), has poorer working-class districts mixed with areas with makeshift housing and without city services. The outer zone, the zone of periphery, is where the expansion of the city occurs, with makeshift housing and squatter settlements. This zone has little or no city services and functions on an informal economy. This outer zone often branches into the city, with slums known as favelas or barrios that provide the working poor access to the city without its benefits. Impoverished immigrants that arrive in the city from the rural areas often end up in the city's outer periphery to eke out a living in some of the worst living conditions in the world.

Cities in this Spanish model grow by having the outer ring progress to the point where eventually solid construction takes hold and city services are extended to accommodate the residents. When this ring reaches maturity, a new ring of squatter settlements emerges to form a new outer ring of the city. The development dynamic is repeated, and the city continues to expand outward. The urban centers of Middle and South America are expanding at rapid rates. It is difficult to provide public services to the outer limits of many of the cities. The barrios or favelas become isolated communities, often complete with crime bosses and gang activities that replace municipal security.

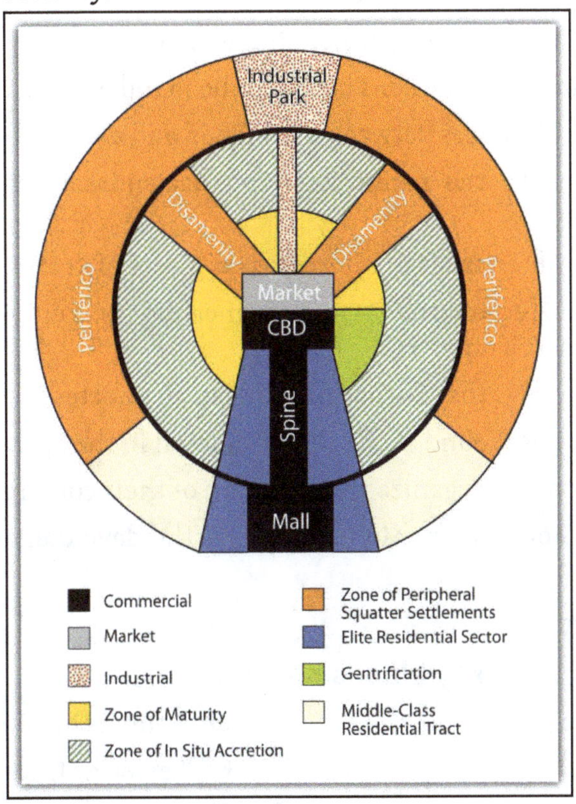

Figure 5.5: Spanish-American City Structure According to the Ford-Griffin Model

Key Takeaways

- Haciendas were located chiefly in the mainland and plantations were located mainly in the rimland.

- Both the hacienda and the plantation structures of agriculture altered the ethnic makeup of their respective regions. The rimland had an African labor base, and the mainland had an Amerindian labor base.
- In their quest for wealth, Spanish conquistadors destroyed the Aztec Empire and colonized the Middle American mainland. Much historical knowledge was lost with the demise of the learned class of the Aztec Empire.
- Europeans introduced many new food crops and domesticated animals to the Americas and in turn brought newly discovered agricultural products from America back to Europe.
- The Spanish introduced the same style of urban planning to the Americas that was common in Spain. Many cities in Middle and South America were patterned after Spanish models.

Discussion and Study Questions

1. What are the three main regions of Middle America?
2. What are the main distinctions between the mainland and the rimland?
3. What are the differences between a hacienda and a plantation?
4. What happened to the plantations and haciendas established during the colonial era?
5. Why is Middle America often referred to as a part of "Latin America"?
6. Who were the Aztec and the Maya, and when did their empires flourish? What happened to these empires?
7. What are some ways that European colonialism affected this realm?
8. What features were found at the center of town in the Spanish urban model?
9. How did the Spanish organize the structure of their colonial cities?
10. How does the Ford-Griffin Model illustrate the development of the Spanish-American city?

Geography Exercise

Identify the following key places on a map:

- Atlantic Ocean
- Bahamas
- Baja Peninsula
- Caribbean Sea
- Central America
- Greater Antilles
- Gulf of Mexico
- Isthmus of Panama
- Lesser Antilles
- Mainland
- Pacific Ocean
- Rimland
- Yucatán Peninsula

References

"Module 01: Demographic Catastrophe—What Happened to the Native Population after 1492?," http://www.dhr.history.vt.edu/modules/us/mod01_pop/context.html.

5.2 MEXICO

Physical Characteristics

Mexico is the eighth-largest country in the world and is about one-fifth the size of the United States. Bordered to the north by the United States, Mexico stretches south to Central America, where it is bordered by Guatemala and Belize. One of Mexico's prominent geographical features is the world's longest peninsula, the 775-mile-long Baja California Peninsula, which lies between the Pacific Ocean and the Gulf of California (also known as the Sea of Cortez). The Baja California Peninsula includes a series of mountain ranges called the Peninsular Ranges.

The Tropic of Cancer cuts across Mexico, dividing it into two different climatic zones: a temperate zone to the north and a tropical zone to the south. In the northern temperate zone, temperatures can be hot in the summer, often rising above 80 °F, but considerably cooler in the winter. By contrast, temperatures vary very little from season to season in the tropical zone, with average temperatures hovering very close to 80 °F year-round. Temperatures in the south tend to vary as a function of elevation.

Mexico is characterized by a great variety of climates, including areas with hot humid, temperate humid, and arid climates. There are mountainous regions, foothills, plateaus, deserts, and coastal plains, all with their own climatic conditions. For example, in the northern desert portions of the country, summer and winter temperatures are extreme. Temperatures in the Sonoran and Chihuahuan Deserts exceed 110 °F, while in the mountainous areas snow can be seen at higher elevations throughout the year.

Two major mountain ranges extend north and south along Mexico's coastlines and are actually extensions of southwestern US ranges. The Sierra Madre Occidental and the Sierra Madre Oriental run roughly parallel to each other. The Sierra Madre Occidental, an extension of the Sierra Nevada range, runs about 3,107 miles along the west coast, with peaks higher than 9,843 feet. The Sierra Madre Oriental is an extension of the Rocky Mountains and runs 808 miles along the east coast. Between these two mountain ranges lies a group of broad plateaus, including the Mexican Plateau, or Mexican Altiplano (a wide valley between mountain ranges). The central portions, with their rolling hills and broad valleys, include fertile farms and productive ranch land. The Mexican Altiplano is divided into northern and southern sections, with the northern section dominated by Mexico's most expansive desert, the Chihuahuan Desert.

Another prominent mountain range is the Cordillera Neovolcánica range, which as its name suggests, is a range of volcanoes that runs nearly 620 miles east to west across the central and southern portion of the country. Geologically speaking, this range represents the dividing line between North and Central America. The peaks of the Cordillera Neovolcánica can reach higher than 16,404 feet in height and are snow covered year-round.

Copper Canyon, in the northern Mexican state of Chihuahua, is about seven times larger than the Grand Canyon. Copper Canyon was formed by six rivers flowing through a series of twenty different canyons. Besides covering a larger area than the Grand Canyon, at its deepest point, Copper Canyon is 1,462 feet deeper than the Grand Canyon.

Though sandy beaches often come to mind when thinking about Mexico, the mountainous regions are home to pine-oak forests. More than a quarter of Mexico's landmass is covered in forests; as a result, timber is an important natural resource. Mexico ranks fourth in the world for biodiversity; it has the world's largest number of reptile species, ranks second for mammals, and ranks fourth for the number of amphibian and plant species. It is estimated that more than 10 percent of the world's species live here. Forest depletion is a key environmental concern, but timber remains an important natural resource. The loss of natural habitat coincides with depletion of natural resources and an increase in population.

Mexico is home to a range of volcanoes, some of which are active. Popocatépetl and Ixtaccíhuatl ("Smoking Warrior" and "White Lady," respectively, in the Náhuatl language—a primary language of the indigenous peoples in central Mexico) occasionally send out puffs of smoke clearly visible in Mexico City, reminding the city's inhabitants that eruption is a possibility. Popocatépetl is one of the most active volcanoes in Mexico, erupting fifteen times since the arrival of the Spanish in 1519 CE. This volcano is close enough to populations to threaten millions of people.

Figure 5.2.65.2.6: Major Volcanoes of Mexico. Source: Map courtesy of USGS/ Cascades Volcano Observatory, http://vulcan.wr.usgs.gov/Volcanoes/Mexico/ Maps/map_mexico_volcanoes.html.

Three tectonic plates underlie Mexico, making it one of the most seismically active regions on earth. In 1985, an earthquake centered off Mexico's Pacific coast killed more than ten thousand people in Mexico City and did significant damage to the city's infrastructure.

Many of Mexico's natural resources lie beneath the surface. Mexico is rich in natural resources and has robust mining industries that tap large deposits of silver, copper, gold, lead, and zinc. Mexico also has a sizable supply of salt, fluorite, iron, manganese, sulfur, phosphate, tungsten, molybdenum, and gypsum. Natural gas and petroleum also make the list of Mexico's natural resources and are important export products to the United States. There has been some concern about declining petroleum resources; however, new reserves are being found offshore in the Gulf of Mexico.

Though only about 13 percent of Mexico's land area is cultivated, favorable climatic conditions mean that food products are also an important natural resource both for export and for the feeding the country's sizable population. Tomatoes, maize (corn), vanilla, avocado, beans, cotton, coffee, sugarcane, and fruit are harvested in sizable quantities. Of these, coffee, cotton, sugarcane, tomatoes, and fruit are primarily grown for export, with most products bound for the United States.

Mexico has very pronounced wet and dry seasons. Most of the country receives rain between June and mid-October, with July being the wettest month. Much less rain occurs during the other months: February is usually the driest month. More importantly, Mexico lies in the middle of the hurricane belt, and all regions of both coasts are at risk for these storms between June and November. Hurricanes along the Pacific coast are much less frequent and less violent than those along Mexico's Gulf and Caribbean coasts. Hurricanes can cause extensive damage to infrastructure along the coasts where major tourist resorts are located. Mexico's extensive and beautiful coastlines provide an important boon to the nation's tourism industry.

The Core versus the Periphery

In the past few decades, the Mexican economy has slowly become less centralized and more focused on the private sector. The Mexican economy is a mix of modern industry, agriculture, and tourism. Current estimates indicate that the service sector makes up about 60 percent of the economy, followed by the industrial sector at 33 percent. Agriculture represents just above 4 percent. Per-capita income in Mexico is about one-third of what it is in the United States. The Mexican labor force is estimated at forty-six million individuals; 14 percent of the labor force work in agriculture, 23 percent in the industrial sector, and 62 percent in the service sector.

Mexico is an example of a country with a clear core-periphery spatial relationship. Mexico City and its surrounding metropolitan centers represent the county's core:

the center of activity, industry, wealth, and power. Industries and manufacturing have been traditionally located in this region. The core region has most of the country's 110 million people (as of 2010). Mexico's population is about 77 percent urban, with the largest urban areas found in the core region.

Figure 5.7: Mexican Economic Core Area Centered on Urban Areas around Mexico City

The periphery is the northern region, including the border area, and the southern region, including the Yucatán Peninsula.

Figure 5.8: Mexico City on a Clear Day with the Ridge of the Mountains Visible in the Background A day with extensive air pollution will restrict the view of the horizon. Matthew Rutledge – Wide view of Polanco – CC BY 2.0.

Mexico City is one of the largest cities in the world and anchors the core region of Mexico. In 2010, the official population of Mexico City was about eighteen million, but unofficial population estimates can reach thirty million. The actual population of Mexico City is unknown because of the hundreds of slums that surround the city on the slopes of the central valley. Mexico City is growing at a rate of more than one thousand people per day through a combination of the number of births and the number migrants. The lure of opportunities and advantages still pulls migrants to the city in search of a better life. Higher populations tax the resources in rural areas, where jobs and opportunities are hard to find. This push-pull relationship creates a strong rural-to-urban shift in Mexico. This same trend is found throughout the developing world.

Mexico City is a historic and vibrant city, but is not without problems. At higher than seven thousand feet in elevation, it is located between two mountain ranges. Air pollution is severe and is augmented by frequent air inversions that trap pollution over the city. To reduce air pollution, people are only allowed to drive their vehicles on certain days according to odd or even license plate numbers. Older vehicles that do not pass emission standards are banned. Fresh water is in short supply, and wastewater from sewage is discharged into lakes down the valley. Amerindians who live by these lakes or on the islands have to deal with the pollution. Because about four to five million inhabitants of Mexico City have no utilities, human waste buildup has become a challenge. Fresh water is pumped into the city through pipelines from across the mountains. Leakage and inadequate maintenance cause a large percentage of the water to be lost before it can be used in the city. Water is also drawn from underground aquifers beneath the city, which has caused parts of the city to sink as much as two feet, causing serious structural damage to historic buildings.

Figure 5.9: Mayan Home in the Rural Village of Yachachen in the Yucatán Peninsula This is located in a peripheral region of Mexico. Photo by R. Berglee – CC BY-NC-SA.

One of Mexico's largest peripheral regions lies to the south, along the country's border with Guatemala. It includes the state of Chiapas and most of the Yucatán Peninsula and is primarily inhabited by Amerindians of Mayan ancestry. As is typical of peripheral regions, little political or economic power is held by the residents, who find themselves at the lowest end of the social and economic order. The highland region of Chiapas and the Yucatán Peninsula are primarily agricultural regions with few industries. However, tourism has changed the northern Yucatán: Cancún has developed into a major tourist destination, and Mayan ruins in the region attract thousands of tourists each season. Unless the local population can benefit from tourism, there are few other opportunities for employment in this part of Mexico.

Mexican Social Order

The early European control of the land, the economy, and the political system created conflict for the people of Mexico. The country has experienced domination followed by revolution at various times, starting with colonial domination, then economic domination, and lastly political domination. In each historic cycle, revolution and conflict were followed by change. The result was a mixing and acculturation of the Europeans and the Amerindians, which created the current mestizo mainstream society. Mestizos make up about 60 percent of the current population, Europeans make up about 9 percent, and Amerindians make up about 30 percent. More than sixty indigenous languages spoken by Amerindian groups are recognized in Mexico. At least seventeen indigenous languages are spoken by more than one hundred thousand people or more in Mexico, most of them living in the southern part of the country.

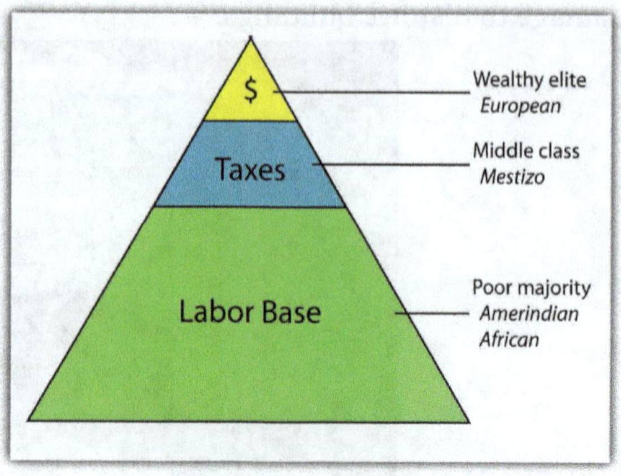

Figure 5.10: Socioeconomic Classes in Mexico and Most of Latin America

Mexican society is regionally and ethnically diverse, with sharp socioeconomic divisions. Many rural communities have strong ties with their regions and are often referred to as *patrias chicas* ("small homelands"), which helps to perpetuate the cultural diversity. The large number of indigenous languages and customs, especially in the southern parts of Mexico, further emphasize cultural diversity. *Idigenismo* ("pride in the indigenous heritage") has been a unifying theme of Mexico since the 1930s. However, daily life in Mexico can be dramatically different according to socioeconomic class, gender, ethnicity, rural or urban settlements, and other cultural differences. A peasant farmer in the rainforests of the Yucatán will lead a very different life than a museum curator in Mexico City or a lower-middle-class auto factory worker in Monterrey.

The current social status of Mexican society can be illustrated by a pyramid shape (see Figure 5.10 "Socioeconomic Classes in Mexico and Most of Latin America"). Those of European descent are at the top of the pyramid and control a higher percentage of the wealth and power even though they are a minority of the population. The small middle class is largely mestizo, including managers, business people, and professionals. The working poor make up most of the population at the bottom of the pyramid. The lower class contains the highest percentage of people of Amerindian descent or, in the case of the Caribbean, African descent.

The most desirable type of social structure is illustrated by a diamond shape: in the middle is a large, employed middle class that can pay most of the taxes and purchase consumer goods that help bolster the economy. The narrow top is made up of the richest, and the narrow bottom is made up of the poorest (see Figure 5.11 "A More Ideal Socioeconomic Class Structure with a Large Middle-Class Tax Base"). Unfortunately, this optimal type of social structure does not always materialize in the manner hoped for. As an example, a goal of the economic planners of the United States has been to create a wide social profile. Unfortunately, in recent decades the US middle class has been declining, and the wealthy class and the working-class poor have increased. In Mexico, about 40 percent of the population lives in poverty.

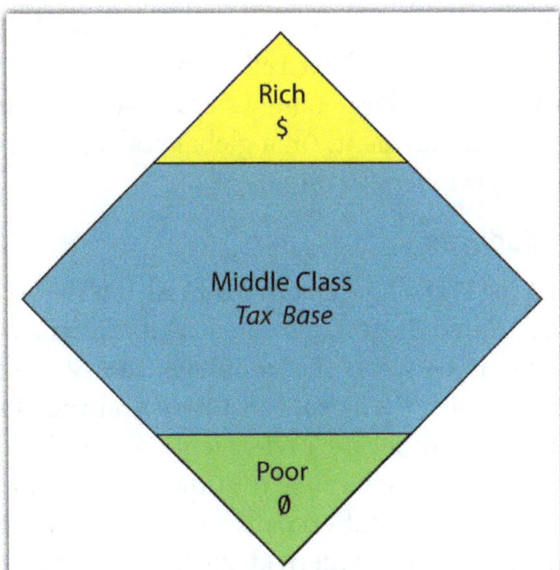

Figure 5.11: A More Ideal Socioeconomic Class Structure with a Large Middle-Class Tax Base

Over the course of the past century, the people of Mexico have been working through a demographic transition. As the rural regions of Mexico continued to have a high fertility rate, death rates declined, and the country's population grew exponentially. In 1970, the population of Mexico was about fifty million. By the year 2000, it had doubled to more than one hundred million. However, the population estimate for 2010 was just greater than 110 million. As Mexico urbanizes and industrializes, family size and fertility rates have been in decline, and population growth has slowed.

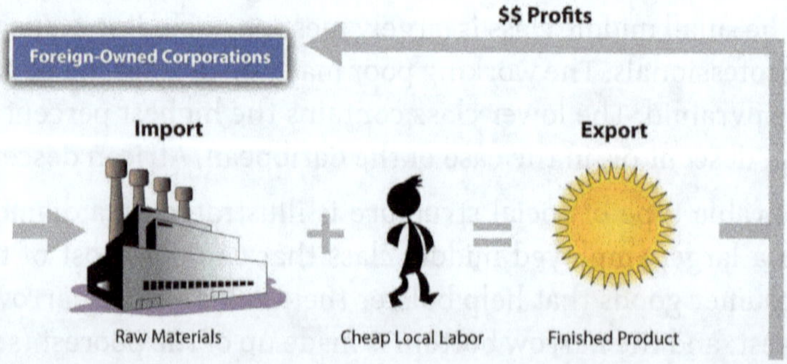

Figure 5.12: Population Pyramids for Mexico in 1980 and 2010 Population pyramids for Mexico in 1980 and again in 2010. The 1980 pyramid indicates rapid population growth. The 2010 pyramid illustrates a slight decline in the past few years. Source: Data courtesy of US Census Bureau International Programs.

Rural-to-urban shift has increased the population of Mexico City, which is considered the primate city of the country. Rural Amerindian groups in the isolated and remote mountainous regions of Mexico have historically been self-sufficient for their daily needs and have relied on the land for their livelihoods. In the past few decades, however, large family sizes have forced many young people to look to the cities for employment. On a global scale, people in many countries are migrating from the periphery to the core.

NAFTA and Maquiladoras

The North American Free Trade Agreement (NAFTA), is a 1994 economic agreement between Canada, the United States, and Mexico that eliminated or reduced the tariffs, taxes, and quotas between the countries to create the world's largest trading bloc to compete with the European Union and the global economy. This theoretically allows more corporate investments across borders and increases foreign ownership of business facilities. It stimulated a shift in the location of industrial activity and in the migration patterns of people in Mexico. Capitalizing on the old industrial locations of northern Mexico, such as Monterrey, corporations started to relocate manufacturing plants from the United States to the Mexican side of the border to take advantage of Mexico's low-cost labor. The aspect of cheaper labor was a benefit understood to bolster corporate profits and reduce product costs. The United States is one of the world's largest consumer markets, so these manufacturing plants, called maquiladoras (also known as *maquilas*), could benefit both countries.

Maquiladoras are foreign-owned factories that import most of the raw materials or components needed for the products they manufacture, assemble, or process with local cheap labor, and then they export the finished product for profit. US corporations own more than half the maquiladoras in Mexico, and about 80 percent of the finished goods are exported back to the United States. Although most maquiladoras are located near the US-Mexican border, additional factories are located around Monterrey and other cities with easy access to the United States. A major trade corridor is developing between Monterrey and Dallas/Ft. Worth, which acts as a doorway to the US markets.

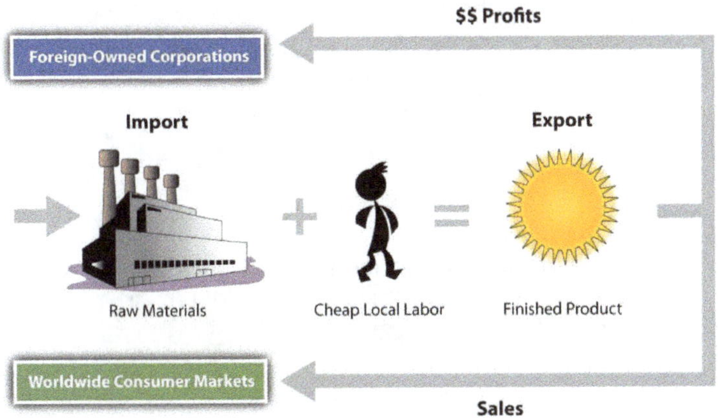

Figure 5.13: Import/Export System of a Maquiladora Operation

Thousands of maquiladoras flourish along the US-Mexican border, although the Mexican government has also promoted maquiladoras in other parts of Mexico. Maquiladoras provide jobs for workers in Mexico and provide cheaper goods for US consumers. However, this system has inherent problems. Labor unions in the United States complain that the high-paying industrial jobs that support the US middle class are being lost to cheap Mexican labor. Labor laws in Mexico are less rigorous than US laws, allowing for longer work hours and fewer benefits for maquiladora employees. In addition, pollution standards in Mexico are not as restrictive as those in the United States, giving rise to environmental concerns. The central US-Mexican border region has dry or arid type B climates with fresh water in short supply, and water is in high demand in industrial processes. With the rapid increase in employment along the border, many of the people who work in the factories do not have adequate housing or utilities. Extensive slum areas have grown around maquiladora zones, which have little law enforcement, high crime, and few services.

The US-Mexican border region has become a strong pull factor, enticing poor people who seek greater opportunities and advantages to move from Mexico City and other southern regions of Mexico to the border region to look for work. When they do not find work, they are tempted to cross the US border illegally. The United States is considered a land of opportunity and attracts immigrants—both legal and illegal—from Mexico. For political and economic reasons, the main US political parties have been hesitant to seriously address the problem of the millions of illegal immigrants.

Figure 5.14: US-Mexican Border

Opportunities and advantages drive the push-pull of migrants searching for improved economic conditions.

It is not only US corporations that have taken advantage of the explosion in the number of maquiladoras in Mexico. European and Japanese companies have also muscled in on a share of the market. Capitalism thrives on cheap labor and accessible raw materials. With much of this industrial activity located in the northern sector of Mexico, it becomes easier to understand the difficult issues that confront Mexico's southernmost state of Chiapas, where there is little benefit from this growth of economic activity.

In 1995, Chile was considered a possible addition to the countries participating in NAFTA. US congressional differences, however, have prevented Chile from being accepted as a full member. As a result, Chile remains a "silent partner" and conducts business according to similar rules. Agreements with Chile block Asian goods from making their way into the United States through Chile and Mexico. The

United States, Mexico, and Canada all have full-fledged independent free-trade agreements with Chile.

The **Free Trade Area of the Americas (FTAA)** is a plan to integrate the entire Western Hemisphere into one giant trading bloc. The same concerns that the European Union faced regarding currency, language, and law confront this proposal. A new currency called the Eagle was proposed as early as the 1990s to replace the Canadian dollar, the US dollar, and the Mexican peso. In later years, a currency called the Amero was proposed for the same purpose, but its implementation is unlikely. Any change in the US dollar would affect a great number of countries: Puerto Rico (a territory of the United States) and the countries of Ecuador, Panama, and El Salvador already use the US dollar as their standard currency. A one-currency solution might become a more viable option if the US dollar were to crash or significantly lose its value in the world marketplace.

A goal of NAFTA is to exploit cheap labor until the Mexican economy rises to a level similar to that of the United States and Canada, equalizing migration patterns and eventually bringing about a situation in which the border checkpoints between the countries could be eliminated, as they have been within the European Union. Through the development of a larger middle class in Mexico, the three main countries of NAFTA would have similar standards of living. Mexico has a long way to go to arrive at this status but is making progress at the expense of the United States and Canada. Corruption, organized crime, and drug wars have made progress in Mexico more difficult.

Critics of NAFTA claim that the term *free trade* really means corporate trade. NAFTA is also viewed as a component of globalization in the form of corporate colonialism, which only benefits those wealthy enough to hold investments at the corporate level. The exploitation of cheap labor has caused undue immigration across the US-Mexican border, bringing millions of illegal workers into the United States. The Mexican government has not adequately addressed Mexico's economic conditions to provide jobs and opportunities for the people or to use the wealth held or controlled by the elite minority to enhance economic opportunities for the middle- and lower-class majority.

European Colonialism (1492/1945–present): Exploitation of a country for physical control and economic gain.

Corporate Colonialism (1991–present): Exploitation of a country for economic gain and not physical control.

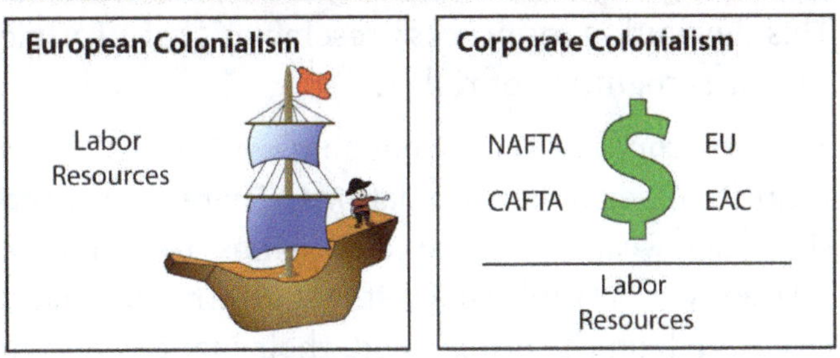

Figure 5.15: Labor and Resources in Globalization

Slave labor was prominent during European colonial era, whereas cheap labor is the target in neocolonial activity—that is, corporate colonialism.

Chiapas and NAFTA

The state of Chiapas in Mexico has an unequal distribution of wealth, a situation evident in most core-peripheral spatial relationships. Located in the rural highlands of Mexico and inhabited by a minority group that holds to the Mayan language and traditions, Chiapas has few economic opportunities for its people. Wealthy landowners and the ruling elite who have long held power have routinely taken advantage of peasant farmers. The aristocracy uses the best land and pays low wages to local workers. Medical care, education, and government assistance have been slow in coming to this region and its people.

In the past few decades, various Amerindian groups have organized in the rural areas of Mexico in an attempt to counter the power of the political elite. In Chiapas a group calling itself the Zapatista National Liberation Army (ZNLA) organized to coordinate an offensive against the Mexican government in various towns in the region. The ZNLA was organized to coincide with the implementation of NAFTA among the United States, Canada, and Mexico in January of 1994. As each country claimed benefits from this agreement, the peripheral region of Chiapas sought to receive their share of those benefits. The Mexican military was quick to react to the ZNLA offensive and rapidly drove them out of the towns they had occupied. The publicity and the international press coverage assisted the ZNLA in getting their message out to the rest of the world.

Since 1994, the ZNLA's guerilla forces have used their familiarity with the mountains for sanctuary and have faced off against the Mexican military when negotiations with the federal government have broken down. The ZNLA want greater recognition of their rights and their heritage and more autonomy over their region and lands. This devolutionary process resembles that of various European regions desiring similar recognition of rights.

Similar conflicts are ongoing in other rural states of Mexico with majority Amerindian populations. There is a direct relationship between social status and wealth and skin color in most regions of Mexico. The skin tone is directly related to a person's social status. On the one hand, Aztec and Mayan heritage is celebrated; on the other hand, their identity and darker skin relegates them to a lower socioeconomic status.

Illegal Drug Trafficking in Mexico

The illegal drug trade is a multibillion-dollar industry, and Mexico has traditionally been the transitional area or stop-off point between the South American drug producing areas and entrance into US markets. Cocaine, marijuana, and more recently heroin were produced in the Andes Mountains of South America and shipped north to the United States. Colombian cartels were once the main controllers of illegal drugs in the Western Hemisphere, but in recent decades, organized crime units in Mexico have muscled in on the control of drugs coming through Mexico, making deals with their South American counterparts to become the main traffickers of drugs into the United States, and the influence and power of Mexican drug cartels has increased immensely since the demise of the Colombian cartels in the 1990s. Enormous profits fuel the competition for control. Just as the United States has declared a war on drugs and has used its Drug Enforcement Administration (DEA) as a main arm in combating the industry, Mexico has had to address its own issues in the illegal drug trade.

Illegal drug income flowing into Mexico has become a major part of the economy in specific areas. Drug kingpins have used their economic power to buy off local police forces and silence opposition. They have also been known to provide poor neighborhoods with funding for services that would normally be designated as government obligations. These actions have often provided a mixed reaction within the population in local areas. The drug cartels have become an integrated part of the fabric of Mexico.

In an attempt to combat the situation, the Mexican government has been engaged in its own internal war against the illegal drug trade. The battles between the drug cartels and the Mexican government have created a serious internal conflict in the country, killing thousands of innocent bystanders in the cross fire. Armed conflicts between rival cartels or local gangs have increased the violence that has been intensifying since 2000. Mexican cities near the US border have experienced increased incidences of major drug-related murders and gang violence. Higher volumes of firearms trafficking from the United States and abroad into Mexico have been fueling the armed conflicts. Military and police casualties have increased, and the number of drug-related shootings are on the rise.

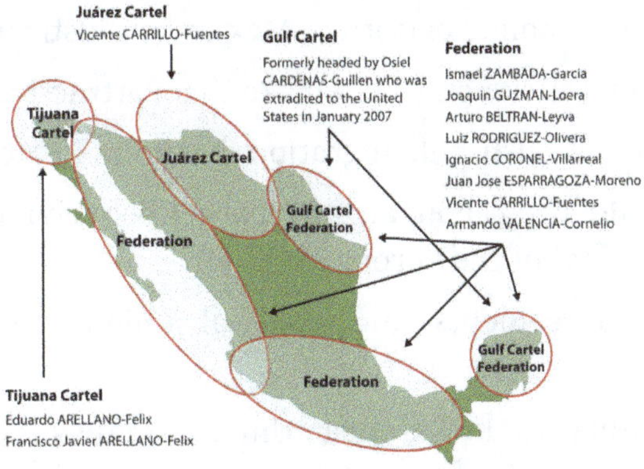

Figure 5.2.165.2.16: Influence of Major Drug Cartels in Mexico. Map courtesy of US Congress, Committee on Foreign Relations.

Cartels have been known to use jet airliners, semitrucks, and even submarines in their attempts to ship illegal drugs into the United States. Large tunnels have been found beneath the US-Mexican border that were used to smuggle drugs. Intimidation and corruption have been standard practices used by drug traffickers to protect their interests. Bribes, payoffs, and corruption have been difficult to battle in a country with a high percentage of the population living in poor conditions.

Key Takeaways

- Mexico possesses extensive natural resources that provide for a wide range of biodiversity and economic activities.
- Mexico portrays a clear core-periphery spatial relationship. Mexico City and its urban neighbors anchor the core, while the northern border region and states such as Chiapas represent the rural periphery.
- People of European heritage continue to hold positions of power and privilege in Mexico's socioeconomic class structure. Amerindian populations exist at the lowest level with the fewest economic opportunities.
- Economic reforms that coincide with NAFTA have greatly enhanced the industrial capacity of Mexico and helped integrate the country into the global economy.
- Mexican drug traffickers have become the major controllers of illegal drugs entering the United States from the south. Drug cartels in Mexico reap enormous profits and have become a major problem for the Mexican government and the country.

Discussion and Study Questions

1. What are the main physical features of Mexico?
2. How does the core-periphery spatial relationship apply to Mexico?
3. How can the socioeconomic pattern of Mexico be illustrated?
4. How are maquiladoras structured to make them attractive to corporations?
5. What are the current push-pull migration dynamics in Mexico?
6. What change would bring about an open border between the United States and Mexico similar to that of the European Union?
7. How are trade agreements and political unions reshaping our global community?
8. How is NAFTA similar to the European Union, and how is it different?
9. How is Chiapas caught in the politics between the core and the periphery?
10. How has the illegal drug trade affected Mexico?

Geography Exercise
Identify the following key places on a map:

- Baja California Peninsula
- Chiapas
- Chihuahuan Desert
- Cordillera Neovolcánica
- Mexican Altiplano
- Mexico City
- Monterrey
- Sierra Madre Occidental
- Sierra Madre Oriental
- Yucatán Peninsula

5.3. CENTRAL AMERICA

Physical Environment

Central America is a land bridge connecting the North and South American continents, with the Pacific Ocean to its west and the Caribbean Sea to its east. A central mountain chain dominates the interior from Mexico to Panama. The coastal plains of Central America have tropical and humid type A climates. In the highland interior, the climate changes with elevation. As one travels up the mountainsides, the temperature cools. Only Belize is located away from this interior mountain chain. Its rich soils and cooler climate have attracted more people to live in the mountainous regions than along the coast.

Hurricanes, tropical storms, earthquakes, and volcanic activity produce recurring environmental problems for Central America. In 1998, Hurricane Mitch swept through the region, devastating Nicaragua and El Salvador, which had already been devastated by civil wars in previous years.

The volcanic activity along the central mountain chain over time has provided rich volcanic soils in the mountain region, which has attracted people to work the land for agriculture. Central America has traditionally been a rural peripheral economic area in which most of the people have worked the land.

Family size has been larger than average, and rural-to-urban shift dominates the migration patterns as the region urbanizes and industrializes. Natural disasters, poverty, large families, and a lack of economic opportunities have made life difficult in much of Central America.

Altitudinal Zonation

High mountains ranges run the length of Central and South America. The Andes Mountains of South America are the longest mountain chain in the world, and a large section of this mountain range is in the tropics. Tropical regions usually have humid type A climates. What is significant in Latin America is that while the climate at the base of the Andes may be type A, the different zones of climate and corresponding human activity vary as one moves up the mountain in elevation. Mountains have different climates at the base than at the summit. Type H highland climates describe mountainous areas that exhibit different climate types at varying degrees of elevation.

Human activity varies with elevation, and the activities can be categorized into zones according to altitudinal zonation. Each zone has its own type of vegetation and agricultural activity suited to the climate found at that elevation. For every thousand-foot increase in elevation, temperature drops 3.5 ºF. In the tropical areas of Latin America, there are five established temperature-altitude zones. Elevation zones may vary depending on a particular location's distance from the equator.

1. Tierra caliente (*hot land*): Sea level to 2,500 feet
2. Tierra templada (*temperate land*): 2,500 to 6,000 feet
3. Tierra fria (*cold land*): 6,000 to 12,000 feet
4. Tierra helada (*frozen land*): 12,000 to 15,000 feet
5. Tierra nevada (*snowy land*): Above 15,000 feet

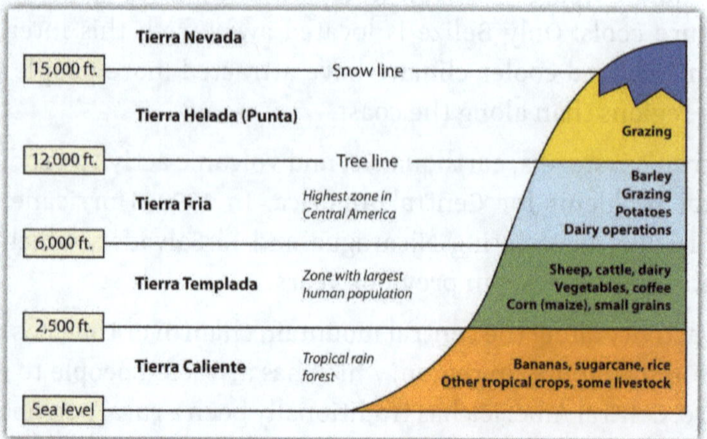

Figure 5.17: Altitudinal Zonation System in Latin America

Tierra Caliente (*Hot Land*): Sea Level to 2,500 Feet

From sea level to 2,500 feet are the humid tropical lowlands found on the coastal plains. The coastal plains on the west coast of Middle America are quite narrow, but they are wider along the Caribbean coast. Vegetation includes tropical rain forests and tropical commercial plantations. Food crops include bananas, manioc, sweet potatoes, yams, corn, beans, and rice. Livestock are raised at this level, and sugarcane is an important cash crop. Tropical diseases are most common, and large human populations are not commonly attracted to this zone.

Tierra Templada (*Temperate Land*): 2,501 to 6,000 Feet

From 2,500 to 6,000 feet is a zone with cooler temperatures than at sea level. This is the most populated zone of Latin America. Four of the seven capitals of the Central American republics are found in this zone. Just as temperate climates attract human activity, this zone provides a pleasant environment for habitation. The best coffee is grown at these elevations, and most other food crops can be grown here, including wheat and small grains.

Tierra Fria (*Cold Land*): 6,001 to 12,000 Feet

From 6,000 to 12,000 feet is the highest zone found in Middle America. This zone is usually the limit of the tree line; few trees grow north of this zone. The shorter growing season and cooler temperatures found at these elevations are still adequate for growing agricultural crops of wheat, barley, potatoes, or corn. Livestock can graze and be raised on the grasslands. The Inca Empire of the Andes Mountains in South America flourished in this zone.

Tierra Helada (*Frozen Land*): 12,001 to 15,000 Feet

Some classify this as the “Puna” zone. At this elevation, there are no trees. The only human activity is the raising of livestock such as sheep or llama on any short grasses available in the highland meadows. Snow and cold dominate the zone. Central America does not have a *tierra helada* zone, but it is found in the higher Andes Mountain Ranges of South America.

Tierra Nevada (*Snowy Land*): Above 15,000 Feet

There is little human activity above 15,000 feet. Permanent snow and ice is found here, and little vegetation is available. Many classification systems combine this zone with the *tierra helada* zone.

European Colonialism

Amerindian groups dominated Central America before the European colonial powers arrived. The Maya are still prominent in the north and make up about half the population of Guatemala. Other Amerindian groups are encountered farther south, and many still speak their indigenous languages and hold to traditional cultural customs. People of European stock or upper-class mestizos now control political and economic power in Central America. Indigenous Amerindian groups find themselves on the lower rung of the socioeconomic ladder.

Figure 5.18: Central American Republics

During colonial times, the Spanish conquistadors dominated Central America with the exception of the area of Belize, which was a British colony called British Honduras until 1981. Guatemala, El Salvador, Honduras, Nicaragua, and Costa Rica were Spanish colonies and became independent of Spain in the 1820s. Panama was a part of Colombia and was not independent until the United States prompted an independence movement in 1903 to develop the Panama Canal. As is usually the case with colonialism, the main religion and the lingua franca of the Central American states are those of the European colonizers, in this case Roman Catholicism and Spanish. In some locations, the language and religion take on variant forms that mix the traditional with the European to create a unique local cultural environment.

People and Population

About 50 percent of the people of Central America live in rural areas, and because the economy is agriculturally based, family size has traditionally been large. Until the 1990s, family size averaged as high as six children. As the pressures of the postindustrial age have influenced Central America, average family size has been decreasing and is now about half that of the pre-1990s and is declining. For example, the World Bank reports that in Nicaragua the average woman has 2.68 children during her lifetime (The World Bank).

Rural-to-urban shift is common, and as the region experiences more urbanization and industrialization, family size will decrease even more.

During the twentieth century, much of Central America experienced development similar to stage 2 of the index of economic development. An influx of light industry and manufacturing firms seeking cheap labor has pushed many areas into stage 3 development. The primate cities and main urban centers are feeling the impact of this shift.

Over the years, larger family sizes have created populations with a higher percentage of young people and a lower percentage of older people. Cities are often overwhelmed with young migrants from the countryside with few or no places to live. Rapid urbanization places a strain on urban areas because services, infrastructure, and housing cannot keep pace with population growth. Slums with self-constructed housing districts emerge around the existing urban infrastructure. The United States has also become a destination for people looking for opportunities or advantages not found in these cities.

CAFTA and Neocolonialism

Just as Canada, the United States, and Mexico signed the North American Free Trade Agreement (NAFTA) into law in 1994, the United States and five Central American states signed the Central American Free Trade Agreement (CAFTA) in 2006. The agreement was signed by trade representatives from El Salvador, Honduras, Nicaragua, Guatemala, and the United States. The CAFTA-DR agreement, which includes the Dominican Republic, was ratified in 2007. In 2010, Costa Rica's legislature approved a measure to join the agreement. CAFTA is supported by the same forces that advocated neocolonialism in other regions of the world.

CAFTA's purpose is to reduce trade barriers between the United States and Central America, thus affecting labor, human rights, and the flow of wealth. During negotiations for CAFTA, US political forces cited CAFTA as a top priority and argued that it would help move forward the possibility of the larger Free Trade Area of the Americas (FTAA), which would create a single market for the Americas.

Countries gain national wealth in the three main ways: by growing it, extracting it, or manufacturing it. These methods, however, contribute to a nation's wealth only if the wealth stays within the country. With free-trade agreements such as NAFTA and CAFTA, the wealth gained from manufacturing, which has the highest value-added profits, does not stay in the country of production. Instead, the profits are carted off to the foreign corporation that controls the industrial factory. Multinational corporations see Central American countries as profitable sites for industrial; they can exploit cheap labor sources and at the same time provide jobs for local people. These advantages should result in lower product costs for consumers.

There have been protest marches and anti-CAFTA activities in many Central American countries. Costa Rica, one of the most stable countries in the region, had problems passing the agreement because of voter opposition. One of the primary arguments opponents to CAFTA make is that the wealth generated by the exploitation of the available cheap labor will not stay in Central America; instead, it will be removed by the wealthy core nations, just as European colonialism removed the wealth generated by the conquistadors and shipped it back to Europe. Those who oppose CAFTA and corporate colonialism also cite the following arguments:

- A popular argument against CAFTA is that "free trade" is the same as corporate trade. Expanding corporate-controlled free trade makes the global south more dependent on the global north, and the corporations reap the profits.
- CAFTA promotes corporate colonialism or neocolonialism. The "have" countries dominate and take advantage of the "have-not" countries to an ever-greater extent. The small countries of Central America cannot compete with large US corporations, which pressure and influence political systems to provide advantages and opportunities to exploit the smaller, weaker nations of Central America.
- CAFTA diminishes the power of Central American countries to regulate their own economies and protect their own citizens; that is, concentration of power in the hands of corporations with strong ties to those in political power allows the elite to maintain control over a country's economy.
- CAFTA forces small developing countries with no chance of competing successfully against the United States to open their markets to powerful US corporations.
- CAFTA leads to further privatization of social services, decreases public access to basic services, and gives corporations more money and control.
- CAFTA forces competition for the lowest wages and lowest production costs, which drives wages down in the United States and keeps them down throughout Central America while at the same time providing huge profits to multinational corporations.

Figure 5.19: Protest against CAFTA in Central America The banner reads, "For the sovereignty of the people...we demand the repeal of CAFTA (Central American Free Trade Agreement)." Laura – For the sovereignty of the people... – CC BY 2.0.

Supporters of CAFTA claim that it provides jobs, infrastructure, and opportunities to the developing countries of Central America. In return, cheap consumer goods are available to the people. The globalized economy is a mixed game: on the one hand, consumer goods are inexpensive to purchase; on the other hand, the world's wealth flows into the hands of a few people at the top and is not always shared with most of the people who contribute to it.

The Republics: Diverse Political Geography

Central American countries might share similar climate patterns, but they do not share similar political or economic dynamics. The political geography of the region is diverse and ranges from a history of total civil war to peace and stability. The growing pains of each country as it competes and engages in the global economy often cause turmoil and conflict. Each state has found a different path, but each has dealt with similar issues with varying degrees of success. Barriers to progress range from political corruption to gang violence. Stability has come to the communities that have found new avenues of gaining wealth and creating a higher standard of living.

Guatemala

Figure 5.20: Woman Selling Fruit in a Guatemalan Market. John Barrie – Antigua Market – CC BY 2.0.

In the late 1900s, Guatemala, El Salvador, and Nicaragua experienced devastating civil wars that divided their people and destroyed their economies. In the Mayan state of Guatemala, the 1960–96 civil war was fought between the right-wing Ladinos (urbanized mestizos and Maya) and the left-wing rural Amerindian Mayan majority. The genesis of this war was democratically elected president Jacobo Arbenz's social reforms, which conflicted with the interests of the US-based United Fruit Company. In 1954, US-backed forces, funded by the Central Intelligence Agency (CIA), overthrew Arbenz and laid the groundwork for civil unrest for the next four decades. Right-wing and left-wing death squads terrorized the country until the latter 1990s, when the Catholic Church brokered a peace accord. The poor and devastated country is now moving forward on its path to recovery.

El Salvador

In the coffee republic of El Salvador, the civil war of 1979–92 was fought between the government-backed wealthy land-owning elite and the peasants who worked the land and lived in poverty. A few powerful families owned almost the entire country. Coffee is a major export crop for El Salvador, a country with a mild climate at its higher elevations. Arabica coffee grows well at these elevations. To protect their economic interests, US coffee companies backed the wealthy elite in El Salvador and lobbied the support of the US government. US military advisors and CIA support aided El Salvador's government forces. At the same time, the peasants of El Salvador were soliciting support from Nicaragua and Cuba, which were backed by the Soviet Union.

After the civil war devastated the country and killed an estimated seventy-five thousand people, a peace agreement that included land reform was finally reached in 1992. El Salvador is a small country about the size of the US state of New Jersey with a population of more than six million people. The war devastated this rural mountainous country and forced more than three hundred thousand people to become refugees in other countries. Many migrated north to the United States. Recovery from the war has been difficult and has been hampered by natural disasters such as hurricanes and earthquakes.

Nicaragua

At the same time that civil wars were going on in Guatemala and El Salvador, there was conflict in Nicaragua. After US marines occupied the country from 1926 to 1933, the US-backed Somoza family took power and remained there for decades. By 1978, violent opposition to governmental manipulation and corruption engulfed the country. An estimated fifty thousand people died in a bitter civil war that ousted the Somoza regime and brought the Marxist Sandinista government to power in 1979.

Nicaraguan aid to leftist rebels in El Salvador caused the United States to sponsor anti-Sandinista contra (short for counterrevolutionary) guerrillas through much of the 1980s and to bring about a second Nicaraguan civil war. In 1982, the US Congress blocked direct US aid to the contra forces through the Boland Amendment. Covert activity by CIA operatives continued to fund the contra forces by selling surplus US arms to Iran, brokered through Israel. In spite of a US embargo against Iran and animosity between Israel and Iran, the deals went through with hopes of negotiating the release of US hostages in Lebanon. The profits from these illegal covert arms sales were funneled into support for the contra forces in Nicaragua, and the scandal, known as the Iran-Contra Affair, has become a standard reference for US intervention in Central America.

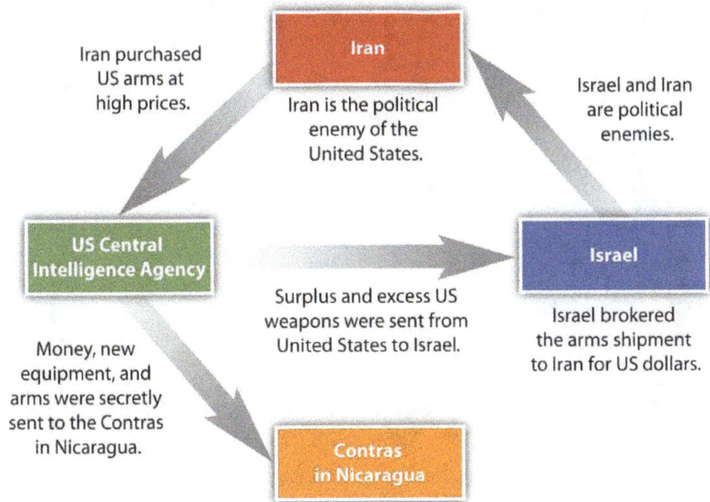

Figure 5.21: Dynamics of the Iran-Contra Affair

In 1990, at the end of the Sandinista-Contra War, democratic elections were carried out. Regardless of the Iran-Contra Affair, the US-backed candidate defeated the Sandinista incumbent. The civil war between the Sandinistas and the contras cost an estimated thirty thousand lives. The country's infrastructure and economy were both in shambles after this era. Despite this history, the people of Nicaragua have worked hard to move forward. Increasing stability in the past decade has improved the country's potential for economic opportunities and has prompted the country to promote tourism and work to increase employment opportunities for its people.

Honduras

Honduras has not experienced civil war, even though it is located in the midst of three troubled neighbors. It is considered a banana republic. American fruit companies have dominated the economy of this poor country and have supported the buildup of arms to ensure its stability. The term *banana republic* applies here only in the manner in which the region was dominated by foreign companies that grew bananas for export. Often the fruit companies would buy up large tracts of land and employ (for low wages) those displaced from the land to help grow the bananas. There have been incidences in history when US fruit companies involved themselves in the political affairs of Central American countries to gain an economic advantage. Foreign fruit companies have monopolized the market in Central America to extract higher profits and control economic regulations. At the present time, international corporations have started to invest in places such as Honduras to capitalize on the country's cheap labor pool and relatively stable economic and political conditions.

Figure 5.22: Three Banana Republics of Central America: Guatemala, Nicaragua, and Honduras

All three have been dominated by US corporate interests.

Mike Mozart – Chiquita Bananas Minions Movie Stickers – CC BY 2.0.

Costa Rica

If there is a bright spot in Central America, it is the democratic and peaceful Costa Rica, which does not have an army. The stable, democratically elected government and growing economy has earned the country the nickname the *Switzerland of Central America*. Multinational companies have been moving here to take advantage of the stable economic conditions, low labor costs, and supportive environment for its employees. The California-based Intel Corporation has a large microchip-manufacturing site in Costa Rica, which contributes heavily to the country's economy. The tropical climate and stable economy of Costa Rica also attract US tourists and people looking for a place to live after retirement. Costa Rica has borrowed heavily to finance social programs, education, and infrastructure and relies on tourism, outside forces, and economic development to help pay the bills.

Belize

Figure 5.23: Geographer Dr. David Meyer examines a sugarcane stalk in Belize. Belize portrays traits of a rimland state, complete with plantation agriculture and African influence.

Photo by R. Berglee – CC BY-NC-SA.

At the northern end of Central America is the former British colony of Belize, which in gained independence in 1981. Belize borders the Caribbean Sea and has a hot, tropical type A climate. It is small in size—about the size of El Salvador—and in population, with only about three hundred thousand people. Belize's lingua franca is English, but Spanish is increasing in usage because of immigration. It has the longest coral reef in the Western Hemisphere and has been promoting ecotourism as a means of economic development to capitalize on this aspect. After hurricanes ravaged the coastal Belize City, the country shifted its capital forty-five miles inland to Belmopan as a protective measure. Belmopan is a small, centrally located city with only about ten thousand people. It is called a forward capital, a term used to describe a capital city of a country that has been moved to better serve or protect the country's interests.

Panama and the Panama Canal

During the 1880s, the region of Panama was part of South America and was controlled by colonial Colombia, which was formerly colonized by Spain. To travel from the Atlantic Ocean to the Pacific Ocean, ships had to sail around the southern tip of South America, which was time consuming and difficult to negotiate in some places due to ocean currents.

France made an agreement with Colombia to purchase a strip of land in Panama ten miles wide and about fifty miles long to build a canal. The French had experience in building the Suez Canal between the Red Sea and the Mediterranean and applied their skills in Panama. The tropical climate and swampy terrain, however, quickly defeated the French workers with malaria, yellow fever, and other tropical diseases.

In the United States, there was an increasing need to shorten the shipping distance between California and New York. Before the United States took over the canal project after the French abandoned it, Panama was separated from Colombia in a brief civil war and declared independent in 1903.

Understanding the problems that the French had encountered, the United States first sent civil engineers and medical professionals to Panama to drain the swamps and apply tons of chemicals such as the insecticide DDT to eradicate the mosquito population. These chemicals were later found to be toxic to humans but worked well in eliminating the mosquito problem. The Panama Canal was finally completed by the United States and opened for business in 1914 after tremendous difficulties had been overcome.

Many workers were imported from the Caribbean to help build the canal, which changed the ethnic makeup of Panama's population. About 14 percent of the population of Panama has West Indian ancestry, and many of the laborers were of African descent. The difference in ethnicity caused an early layering of society, with those from the Caribbean finding themselves at the lower end of the socioeconomic scale.

Figure 5.24: Locks on the Panama Canal Almost fifteen thousand vessels pass through the Panama Canal every year. Robbie – Pedro Miguel Locks, Panama Canal – CC BY-NC-ND 2.0.

The Panama Canal is a marvel of engineering. An interior waterway was dammed up to create the artificial Lake Gatún at eighty-five feet above sea level. This large inland lake provides a freshwater channel extending most of the way across the Isthmus of Panama. Canal channels on each end of Lake Gatún connect it with the sea. Locks raise and lower ships from sea level to the eighty-five-foot water level of the canal and the lake. Gravity provides fresh water from Lake Gatún to fill the locks that raise and lower ships. As ships travel through the locks, the fresh water is eventually emptied into the sea. Rainfall is critical to resupply the water in Lake Gatún to keep the water channel constant and to keep the canal locks in operation. The canal channel has to be dredged periodically to keep it from silting in. In recent years, deforestation has reduced the number of trees around the lake, resulting in more silt entering the lake bed. A program to replant trees has been implemented to secure the lake and restore the natural conditions.

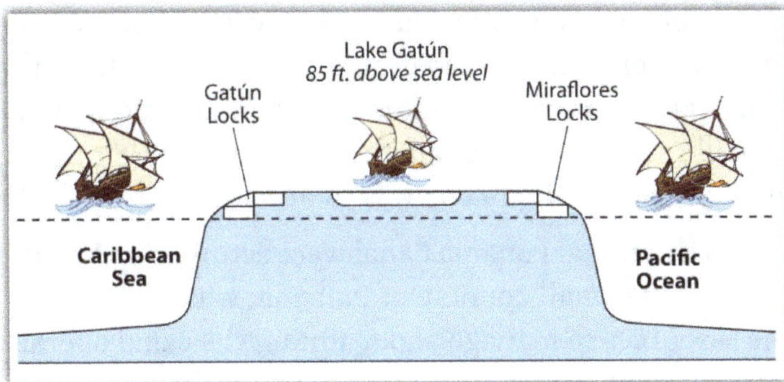

Figure 5.25: Panama Canal System

The water for the locks is from Lake Gatún, which is eighty-five feet above sea level.

Recently, the politics of the Panama Canal have become more of an issue than the operation of the canal itself. In 1977, US president Jimmy Carter entered into an agreement with Panamanian president Omar Torrijos to return the canal to the government of Panama. Under this agreement, both the Panama Canal Zone and the actual canal were to be returned to Panama by the end of 1999. Many Americans opposed the return of the canal to Panama. President Ronald Reagan campaigned on this position. The United States had military installations in the Canal Zone and had used this area as a training ground for the Vietnam War and other military missions. The United States operated the School of the Americas (SOA) in the Canal Zone, which was a place to train counterinsurgents and military personnel from other countries. The SOA was moved to Fort Benning, Georgia, in 1984 and was renamed the Western Hemisphere Institute for Security Cooperation (WHINSEC) in 2001.

One of the early graduates of the SOA was a young Panamanian officer by the name of Manuel Noriega, who was placed on the CIA payroll in 1967. He was an important figure, helping with the US war against Nicaragua and generally serving US interests in the region in spite of the fact he was a known drug dealer. In May 1989, Noriega was elected president of Panama and became less supportive of US interests in the region. In December 1989, the United States invaded Panama and captured Noriega. He was sentenced to forty years in a US prison for drug trafficking and held as a political prisoner. Even after Noriega's arrest, the United States was not allowed to retain use of the Canal Zone for military purposes, which was a major reason for the US presence in Panama. The Panama Canal Zone was an excellent geographical location for US military operations because it provided an excellent base to monitor military activity in South America. US military planes could fly from US bases to Panama without refueling, and the planes could then fly out of Panama to monitor activity in South America.

One of President Carter's arguments for the return of the canal to Panama was that after the US military had supported the war with Colombia to make Panama independent in 1903, there had been no proper authorization from the Panamanian people to cede the Canal Zone to the United States. International law ruled that the Canal Zone was still sovereign Panamanian territory. The US military claimed the reason for remaining in the Canal Zone was to provide security for the canal.

The Canal Zone and the actual Panama Canal were returned to Panama in 2000. The question arises, does the small country of Panama, with only about three million people, have the resources to manage and maintain the canal operations? To assist in economic development, Panama has established a free-trade zone next to the canal to entice international commerce. Originally established in 1948, the free-trade zone has become one of the largest of its kind in the world. Panama City has also become a hub of international banking with the dubious claim of being a main money-laundering center for Colombian drug money. Panama is striving to be a main economic center for the region, which would advance economic globalization and trade for Panama.

Key Takeaways

- Central America shares a similar climate type and physical features. It has enormous potential for tourism development. The political history of the region is quite diverse, with each republic experiencing different political and economic conditions.
- High population growth and rapid rural-to-urban shift has created higher unemployment rates and fewer economic opportunities. CAFTA was implemented to help multinational corporations tap into the cheap labor pool.
- The United States has had a major impact on this region both politically and economically. The United States has intervened in civil wars and invaded Panama. US companies have dominated much of the region's fruit and coffee production. Most recently, the United States has supported industrial activities and the implementation of CAFTA.
- The Panama Canal has been a vital transportation link between the Atlantic and Pacific Oceans for the United States and the world. The Panama Canal Zone was a valuable strategic location for the operation of US military

exercises. The United States was a major controller of the Panama Canal until the year 2000, when the canal and the Canal Zone was returned to Panama.

Discussion and Study Questions

1. How does altitudinal zonation influence human activity? What are the main altitudinal zones?
2. What geographic aspects do the Central American republics share?
3. How are the Central American republics different in their political histories?
4. How does one define a banana republic or a coffee republic? Where are they found in Central America?
5. What countries are included in CAFTA-DR? Why would anyone oppose this agreement?
6. What was the Iran-Contra Affair? How does this portray the politics of the region?
7. How does Costa Rica attract US business ventures?
8. Who started building the Panama Canal? Who completed it? Who controls it today?
9. How did the building of the Panama Canal change Panama's ethnic makeup?
10. Why was the Panama Canal so important to the United States?

Geography Exercise
Identify the following key places on a map:

- Forward capital of Belmopan
- Panama Canal

References

The World Bank, "Fertility Rate, Total (Births per Woman)," http://data.worldbank.org/indicator/SP.DYN.TFRT.IN.

5.4. THE CARIBBEAN

The areas of Middle America and South America, encompassing the Caribbean, exhibit analogous colonial histories of incursion, supremacy, and evolution by external European powers. The Caribbean Basin is typically categorized into the Greater Antilles and the Lesser Antilles, denoting the larger and smaller islands, respectively. The Greater Antilles comprises the four major islands of Cuba, Jamaica, Hispaniola, and Puerto Rico. The Lesser Antilles are located in the eastern and southern regions. The Bahamas are geographically situated in the Atlantic

Ocean, rather than the Caribbean Sea; yet, they are commonly connected with the Caribbean region and frequently linked to the Lesser Antilles. Middle America can be categorized into two geographic regions based on occupational activities and colonial influences. The rimland encompasses the Caribbean islands and the Caribbean coastal regions of Central America. The mainland comprises the interior regions of Mexico and Central America.

Figure 5.26: Caribbean Regions of the Greater Antilles, the Lesser Antilles, and the Bahamas

Many of the Caribbean islands experience the rain shadow effect. Jamaica has as much as a twenty-inch difference in rainfall between the north side and south side of the island because most of the rain falls on the north side, where the prevailing winds hit the island. The Blue Mountains in the eastern part of the island provide a rain shadow effect. Puerto Rico has a tropical rain forest on the northeastern part of the island, which receives a large amount of rainfall. The rain shadow effect creates semidesert conditions on the southwestern side of Puerto Rico because the southwestern side receives little rainfall. Low elevation islands such as the Bahamas do not receive as much rain because they are not high enough to affect the precipitation patterns of rain clouds.

European Colonialism in the Caribbean

The Spanish were not the only Europeans to take advantage of colonial expansion in the Caribbean: the English, French, Dutch, and other Europeans followed. Most of the European colonial countries were located on the west coast of Europe, which had a seafaring heritage. This included smaller countries such as Denmark, Sweden, and Belgium. The Caribbean Basin became an active region for European ships to enter and vie for possession of each island.

Many of the Caribbean islands changed hands several times before finally being secured as established colonies (see Table 5.1 "Historical Caribbean Colonizers"). The cultural traits of each of the European colonizers were injected into the fabric of the islands they colonized; thus, the languages, religions, and economic activities of the colonized islands reflected those of the European colonizers rather than those of the native people who had inhabited the islands originally. The four main colonial powers in the Caribbean were the Spanish, English, Dutch, and French. Other countries that held possession of various islands at different times were Portugal, Sweden, and Denmark. The United States became a colonial power when they gained Cuba and Puerto Rico as a result of the Spanish-American War. The US Virgin Islands were purchased from Denmark in 1918. Sweden controlled the island of St. Barthelemy from 1784 to 1878 before trading it back to the French, who had been the original colonizer. Portugal originally colonized Barbados before abandoning it to the British.

Table 5.1 Historical Caribbean Colonizers

Colonizer	European colonies
Spain	Cuba, Dominican Republic, Puerto Rico
British	Bahamas, Jamaica, Cayman Islands, Turks and Caicos Islands, Antigua, Dominica, St. Lucia, St. Vincent, Grenada, Barbados, Virgin Islands, Trinidad and Tobago, Montserrat, Anguilla, St. Kitts and Nevis
Dutch	Curacao, Bonaire, Aruba, St. Eustatius, Saba and Sint Maarten (south half)
French	Haiti, Guadeloupe, Martinique, St. Martin (north half), St. Barthelemy
United States	Puerto Rico, Virgin Islands, Cuba

Colonialism drastically altered the ethnic makeup of the Caribbean; Amerindians were virtually eliminated after the arrival of Africans, Europeans, and Asians. The current social hierarchy of the Caribbean can be illustrated by the pyramid-shaped graphic that was used to illustrate social hierarchy in Mexico (Figure 5.10 "Socioeconomic Classes in Mexico and Most of Latin America"). Those of European descent are at the top of the pyramid and control a higher percentage of the wealth and power even though they are a minority of the population. In the Caribbean, the middle class includes mulattos, or people with both African and European heritage, many of which include managers, businesspeople, and professionals. In some countries, such as Haiti, the minority mulatto segment of the population makes up the power base and holds political and economic advantage over the rest of the country while the working poor at the bottom of the pyramid make up most of the population. In the Caribbean, the lower economic class contains the highest percentage of people of African heritage.

Not only was colonialism the vehicle that brought many Africans to the Caribbean through the slave trade, but it brought many people from Asia to the Caribbean as well. Once slavery became illegal, the colonial powers brought indentured laborers to the Caribbean from their Asian colonies. Cuba was the destination for over one hundred thousand Chinese workers, so Havana can claim the first Chinatown in the Western Hemisphere. Laborers from the British colonies of India and other parts of South Asia arrived by ship in various British colonies in the Caribbean. At the present time, about 40 percent of the population of Trinidad can claim South Asian heritage and a large number follow the Hindu faith.

The Greater Antilles

Cuba: A Rimland Experience

The largest island in the Greater Antilles is Cuba, which was transformed by the power of colonialism, the transition to plantation agriculture, and a socialist revolution. The island country of Cuba is slightly larger than the US state of Kentucky, but it has more than eleven million people, while Kentucky has just over 4.2 million. The elongated island has the Sierra Maestra mountains on its eastern end, the Escambray Mountains in the center, and the Western Karst region in the west, near Viñales. Low hills and fertile valleys cover more than half the island. The pristine waters of the Caribbean that surround the island make for some of the most attractive tourism locations in the Caribbean region.

It has been estimated that as many as one hundred thousand Amerindians inhabited Cuba when Christopher Columbus first landed on the island in 1492. Except for brief control by the British, the island was a Spanish colony until 1898.

Plantation agriculture was established, and slaves provided the labor. History indicates that more than eight hundred thousand African slaves were brought to Cuba between 1800 and 1870. Slave labor was in high demand on the extensive sugar plantations that dominated the island's economy. The African influence can still be witnessed today in the main religion of the island, Santeria, which is an overlay of African-based spirits on top of Catholic saints. At the present time, an estimated 70 percent of Cubans practice some aspect of Santeria.

With the defeat of Spain in the Spanish-American War, the United States gained possession of the Spanish possessions of Cuba, Puerto Rico, Guam, the Philippines, and various other islands and thus became a colonial power. Cuba technically became independent in 1902 but remained under US influence for decades. Sugar plantations and the sugar industry came to be owned and operated by US interests, and wealthy Americans bought up large haciendas (large estates), farmland, and family estates, as well as industrial and business operations. Organized crime syndicates operated many of the nightclubs and casinos in Havana. As long as government leaders supported US interests, things went well with business as usual.

Figure 5.27: US "Colonial" Influence in Cuba The old capitol building in Havana, a replica of the US Capitol, was built by the United States during their control of Cuba. The building is a tourist area and no longer used for the government. The old US cars in the photo were made before the Cuban Revolution (1958) but are still used and make up about half the motor vehicles in Havana. Photo by R. Berglee – CC BY-NC-SA.

The Cuban Revolution

In January of 1934, with the encouragement of the US government, Fulgencio Batista led a coup that took control of the Cuban government. Fidel Castro, once a prisoner under Batista and having fled to Mexico in exile for a number of years, returned to Cuba to start a revolution. Joining him were his brother Raúl Castro and revolutionaries such as Che Guevara, an Argentinean doctor turned comrade-in-arms. Starting in the remote and rugged Sierra Maestras in the east, Castro rallied the support of the Cuban people. By the end of 1958, the Cuban Revolution brought down the US-backed Batista government. Castro gained power and had the support of most of the Cuban population.

Figure 5.28: Fidel Castro (*left*); Billboard in Havana Promoting the Virtues of Revolutionaries Antonio Maceo and Che Guevara (*right*). Wikimedia Commons – CC BY 3.0; R. Berglee – CC BY-NC-SA.

Castro worked to recover Cuba for Cubans. The government cleared rampant gambling from the island, forcing organized crime operations to shut down or move back to the United States. Castro nationalized all foreign landholdings and the sugar plantations, as well as all the utilities, port facilities, and other industries. Foreign ownership of land and businesses in Cuba was forbidden. Large estates, once owned by rich US families, were taken over and recovered for Cuban purposes.

The US Embargo Era

Castro's policy of seizing (nationalizing) businesses and property raised concerns in the United States. As a result, US president Dwight D. Eisenhower severed diplomatic relations with Cuba in 1960 and issued an executive order implementing a partial trade embargo to prohibit the importation of Cuban goods. Later presidents implemented a full-scale embargo, restricting travel and trade with Cuba. In March 1960, the Central Intelligence Agency trained Cuban exiles for an invasion of Cuba at the Bay of Pigs, on the south side of the island. This failed invasion attempt only resulted in consolidating the Cuban people's support for Castro and his socialist government.

To deter any further US plans of invading or destabilizing Cuba, Castro sought economic and military assistance from the Soviet Union. Soviet premier Nikita Khrushchev agreed to secretly send missiles armed with nuclear weapons capable of hitting targets within the United States. In September 1962, US spy planes identified the missile sites. On October 22, 1962, President John F. Kennedy announced a naval blockade of the island and informed Khrushchev that any Soviet ship crossing the blockade would be sunk. At the last second, the two leaders resolved this dispute (called the Cuban Missile Crisis) before it erupted into a potential nuclear war. Khrushchev recalled the ships and agreed to dismantle the Cuban missile sites. In return, the United States agreed not to invade Cuba and to remove US missiles from sites in Turkey that were aimed at the Soviet Union.

The collapse of the Soviet Union in 1991 caused a downturn in Cuba's economy. With the loss of Soviet aid, the 1990s were a harsh time for Cubans, a period of transition. Castro turned to tourism and foreign investment to shore up his failing economy. Tensions between the United States and Cuba did not improve. In 1996, the United States strengthened the trade embargo with the Helms-Burton Act. At the turn of the twenty-first century, Cuba emerged as the lone Communist state in the Americas. Castro was the longest-governing leader of any country in the world. He never kept his promises of holding free elections; instead, he cracked down on dissent and suppressed free speech. He turned over power to his brother Raúl in 2006.

A Post-Castro Cuba

With Fidel Castro no longer in power, Cuba's future looks more positive but difficult. The island has natural resources, a great climate, and an excellent location but is also struggling economically. Cuba has a high literacy rate and has standardized health care, though medical supplies are often in short supply. The Cubans who live in dire poverty look to the future for relief. Personal freedoms have been marginal, and reforms are slowly taking place in the post-Fidel era. As the largest island in the Caribbean, Cuba has the potential to become an economic power for the region. There is vast US interest in regaining US dominance of the Cuban economy, and corporate colonialists would like to exploit Cuba's economic potential. Keeping corporate colonialism out is what Fidel's socialist experiment worked so hard to achieve, even at the expense of depriving the Cuban people of civil rights and economic reforms.

Cuba today is in transition from a socialist to a more capitalist economy and relies on outside sources for energy and food. In 2008, the average wage in Cuba was about twenty dollars per month. There was almost total employment, and everyone was on an equal footing in regard to free health care, education, and housing. At the same time, the underground informal economy was thriving and was pushing the formal economy to make changes. That same year, Raúl Castro declared that workers with different skills and occupations could earn wages at varying levels. Cubans are now allowed to have cell phones and computers, though Internet access has been restricted by the Cuban government. More goods and money are being brought in from Cuban family members who live in the United States. The US government grants general licenses to allow a number of categories of people to travel to Cuba. It is only a matter of time before full travel restrictions are lifted.

Figure 5.29: Dump Truck Taxi Cubans use all available resources and opportunities to get by. These people are catching a ride on a dump truck to get where they want to go. Photo by R. Berglee – CC BY-NC-SA.

Cuba is counting on tourism for an added economic boost. With some of the finest beaches and the clearest waters in the Caribbean, Cuba is a magnet for tourists and water sports enthusiasts. Its countryside is full of wonders and scenic areas. The beautiful **Viñales Valley** in western Cuba has been listed as a UNESCO World Heritage Site for its outstanding karst landscape and traditional agriculture as well as for its architecture, crafts, and music. Karst topography is made up of soluble rock, such as limestone, which in the Viñales Valley results in unusual bread loaf–shaped hills that create a scenic landscape attractive for tourism. This region is also one of Cuba's best tobacco-growing areas and has great potential for economic development. Cuba is gearing up for an increase in tourism when travel restrictions are lifted by the United States. There is already a focus on improving tourism services to people traveling there from China, Australia, Japan, and other countries. Millions more from the United States are expected to travel to Cuba once the travel restrictions are lifted. The Cuban economy is banking on tourism to forge a path to a more prosperous future.

Figure 5.30: Viñales Valley in Western Cuba In 2008, Hurricanes Gustav and Ike devastated the tobacco crops, but the region is recovering and is a major tourist area. Photo by R. Berglee – CC BY-NC-SA.

The Commonwealth of Puerto Rico

Populated for centuries by Amerindian peoples, the island of Puerto Rico was claimed by the Spanish Crown in 1493, following Columbus's second voyage to the Americas. In 1898, after four hundred years of colonial rule, during which the indigenous population was nearly exterminated and African slave labor was introduced, Puerto Rico was ceded to the United States as a result of the Spanish-

American War. Puerto Ricans were granted US citizenship in 1917. Popularly elected governors have served since 1948. In 1952, a constitution was enacted providing for internal self-government. In elections held in 1967, 1993, and 1998, Puerto Rican voters chose to retain the commonwealth status, although they were almost evenly split between total independence and becoming a US state.

Puerto Rico is the smallest of the four islands of the Greater Antilles and is only slightly larger than the US state of Delaware. Puerto Rico's population is about four million, similar to the population of Kentucky or Oregon. As US citizens, Puerto Ricans have no travel or employment restrictions anywhere in the United States, and about one million Puerto Ricans live in New York City alone. The commonwealth arrangement allows Puerto Ricans to be US citizens without paying federal income taxes, but they cannot vote in US presidential elections. The Puerto Rican Federal Relations Act governs the island and awards it considerable autonomy.

Figure 5.31: US Government Building in San Juan, Puerto Rico, with Both US and Puerto Rican Flags. Rafael Torres – Puerto Rico Capitol Building – CC BY-NC-ND 2.0.

Puerto Rico has one of the most dynamic economies in the Caribbean Basin; still, about 60 percent of its population lives below the poverty line. A diverse industrial sector has far surpassed agriculture as the primary area of economic activity. Encouraged by duty-free access to the United States and by tax incentives, US firms have invested heavily in Puerto Rico since the 1950s, even though US minimum wage laws apply. Sugar production has lost out to dairy production and other livestock products as the main source of income in the agricultural sector. Tourism has traditionally been an important source of income, with estimated arrivals of more than five million tourists a year. San Juan is the number one port for cruise ships in the Caribbean outside Miami. The US government also subsidizes Puerto Rico's economy with financial aid.

The future of Puerto Rico as a political unit remains unclear. Some in Puerto Rico want total independence, and others would like to become the fifty-first US state; the commonwealth status is a compromise. Puerto Rico is not an independent country as a result of colonialism. Many of the islands and colonies in the Caribbean Basin have experienced dynamics similar to Puerto Rico in that they are still under the political jurisdiction of a country that colonized it.

Hispaniola: The Dominican Republic and Haiti

Sharing the island of Hispaniola are the two countries of Haiti and the Dominican Republic. The island became a possession of Spain under European colonialism after it was visited by Columbus in 1492 and 1493. The Tiano-Arawak people were nonviolent and welcomed the Europeans, who in turn pressed them into servitude and slavery. French buccaneers settled on the western portion of Hispaniola and started growing tobacco and agricultural crops. France and Spain finally agreed to divide the island into two colonies: the western side would be French, and the eastern side would be Spanish.

The Dominican Republic holds the largest share of Hispaniola. A former Spanish colony, the Dominican Republic has weathered the storms of history to become a relatively stable democratic country. It is not, of course, without its problems. The Dominican Republic has long been viewed primarily as an exporter of sugar, coffee, and tobacco, but in recent years the service sector has overtaken agriculture as the economy's largest employer. The mountainous interior and the coastal beaches are attractive to the tourism market, and tourism remains the main source of economic income. The economy is highly dependent on the United States, which is the destination for nearly 60 percent of its exports. Remittances from workers in the United States sent back to their families on the island contribute much to the economy. The country suffers from marked income inequality; the poorest half of the population receives less than one-fifth of the gross domestic product (GDP), while the richest 10 percent enjoys nearly 40 percent of GDP. High unemployment and underemployment remains an important long-term challenge. The Central American-Dominican Republic Free Trade Agreement (CAFTA-DR) came into play in March 2007, boosting investment and exports and reducing losses to the Asian garment industry. In addition, the global economic downturn has not helped the Dominican Republic.

Plantation agriculture thrived in Haiti during the colonial era, producing sugar, coffee, and other cash crops. The local labor pool was insufficient to expand plantation operations, so French colonists brought in thousands of African slaves to work the plantations, and people of African descent soon outnumbered

Europeans. Haiti became one of the most profitable French colonies in the world with some of the highest sugar production of the time. A slave revolt that began in 1792 finally defeated the French forces, and Haiti became an independent country in 1804. It was the first country ever to be ruled by former slaves. However, the transition to a fully functional free state was difficult. Racked by corruption and political conflicts, few presidents in the first hundred years ever served a full term in office.

The United States occupied Haiti from 1915 to 1934 in an attempt to instill a US presence and bring some sense of stability. From 1957 to 1986, Dr. François "Papa Doc" Duvalier and then his son Jean-Claude "Baby Doc" Duvalier controlled the government. They created a private army and terrorist death squads known as Tonton Macoutes. Many Haitians fled to the United States or Canada, especially to French-speaking Quebec. After the Duvalier era, a Catholic priest by the name of Jean-Bertrand Aristide won the presidency through democratic elections only to be deposed of by a military coup a few years later. Haiti has had a difficult time finding political and economic stability.

Haiti is the poorest nation in the Western Hemisphere, and many Haitians live in dire poverty with few employment opportunities. An elite upper-class minority controls the bulk of the nation's wealth. Many people in Haiti have sought comfort in Voodou (Vodoo), a religious practice steeped in African beliefs brought over with the slave trade. Often misunderstood by outsiders, Vodou's its main objective is to bring good health and well-being.

Figure 5.32: UN peacekeeping troops patrol the streets of Port au Prince after Haiti's 2010 earthquake.

Haiti's January 12, 2010, earthquake was a major setback for such a poor country that was already in need of aid and support. The 7.0 magnitude earthquake, with many aftershocks, struck Haiti about fifteen miles from Port-Au-Prince, resulting in as many as two hundred thousand deaths. More than two million were immediately homeless, and about a million more were in need of aid. The lack of building standards in Haiti contributed to the collapse of structures and the devastation. Food and other aid were brought in by many international agencies and other countries to address the situation. Efforts continue to address the needs of the Haitian people to stabilize the situation and provide support and future opportunities.

Jamaica

The tropical island of Jamaica is physically smaller than the US state of Connecticut. In 2010, it had a population of about 2.8 million. Jamaica was settled by the Spanish early in the sixteenth century. The Taino Indians, who had inhabited Jamaica for centuries, were gradually eradicated and replaced by African slaves. When England seized the island from Spain in 1655, it established a plantation economy based on sugar, cocoa, and coffee. Two hundred years later (1834), the abolition of slavery freed a quarter million slaves, many of whom became small farmers. Jamaica gradually obtained independence from Britain, with full independence achieved in 1962. Sugar, cocoa, and coffee production continue on the island, of which more than half is mountainous. The **Blue Mountains** of eastern Jamaica are known for their Blue Mountain coffee production.

Deteriorating economic conditions during the 1970s led to recurrent violence as rival gangs affiliated with the major political parties evolved into powerful organized crime networks involved in international drug smuggling and money laundering. Violent crime, drug trafficking, and poverty pose significant challenges to the country. Nonetheless, many rural and resort compounds remain relatively safe and contribute substantially to the tourism sector. The beautiful beaches and lush interior make for an attractive destination for cruise ships and other tourists.

Tourism and Economic Activity in the Rimland

The physical geography of the Caribbean region makes it a prime location for tourism. Its beautiful coastal waters and warm tropical climate draw in tourists from all over North America and the world. Tourism is the number one means of economic income for many places in the Caribbean Basin, and the tourist industry has experienced enormous growth in the last few decades. Tourism is a major component of efforts by leaders of the islands of the Lesser Antilles to achieve economic development for their people. In the last decade, there has been strong growth in the number of cruise ships operating in the Caribbean. Cruise ships from the southern coasts of the United States ply their trade around the islands and coastal regions. San Juan receives the largest number of cruise ship travelers, but other areas well known to tourists include the Bahamas, St. Martin, and the Virgin Islands. Even the poorest country in the Caribbean, Haiti, has tried to attract cruise ships to its ports. The western Caribbean itinerary includes the Cayman Islands, Jamaica, and Mexico or Central American ports. The main restriction on cruise ship travel is the hurricane season, from June to November.

Figure 5.33: Carnival *Victory* Cruise Ship in San Juan Harbor Large cruise ships in the Caribbean can hold up to four thousand passengers and crew members. The major cruise lines do not operate in the Caribbean during hurricane season. Photo by R. Berglee – CC BY-NC-SA.

One might reasonably think that the economic benefits of tourism would be entirely positive. However, this is not necessarily the case. Even though tourism has become a vital economic component of the Caribbean Basin, in the long term, tourism creates many problems. Large cruise ships and pleasure crafts can overtax the environment; there have been occasions where there were actually more tourists than citizens on an island. An increase in tourist activity brings with it an increase in environmental pollution.

Most people in the Caribbean Basin live below the poverty line, and the investment in tourism infrastructure, such as exclusive hotels and five-star resorts, takes away resources that could be allocated to schools, roads, medical clinics, and housing. However, without the income from tourism, there would be no money for infrastructure. Tourism attracts people who can afford to travel. Most of the jobs in the hotels, ports, and restaurants where wealthy tourists visit employ people from poorer communities at low wages. The disparity between the rich tourist and the poor worker creates strong centrifugal cultural dynamics. The gap between the level of affluence and the level of poverty is wide in the Caribbean. In the model of how countries gain wealth, tourism is a mixed-profit situation. Local businesses in the Caribbean do gain income from tourists who spend their money there; however, the big money is in the cruise ship lines and the resort hotels, which are mainly owned by international corporations or the local wealthy elite.

There is little activity in the Caribbean Basin in the manufacturing sector. Although US firms have invested in Puerto Rico and the Dominican Republic has experienced growth in light industries and information technologies, the remaining islands have had little industrial growth. Unless an island state has natural resources such as oil or minerals, as is the case with Trinidad and Tobago, there are few opportunities other than tourism to bolster the economy.

Agricultural products have been traditionally a large part of the economic activity of the islands of the Lesser Antilles. Grenada, for example, is known for its nutmeg and other spices. Bananas, sugar, and other fruit and food crops have also been export products. The problem has been that the profit margins on the products are low and prices are subject to international markets, which fluctuate widely. With an increasing population and few opportunities or advantages, countries such as Haiti suffer from poverty and unemployment. Being an island, there are few methods of expanding the economy. People often try to migrate to another county in search of employment and a more hopeful future.

Offshore Banking

Other methods of gaining wealth in the Caribbean include offshore banking and financial services. Various islands have established themselves as banking centers where one can set up financial accounts that are outside the jurisdiction of other countries. These offshore accounts provide tax havens for individuals or corporations that wish to evade taxes in their home countries. Many of these island banking centers do not share account information with tax agencies or government offices of other countries. Places such as the Cayman Islands have a worldwide reputation for professional financial services for offshore banking. As a result, the people of the Cayman Islands have a high standard of living with much national wealth. The Bahamas and the Turks and Caicos have also established offshore financial centers, and other Caribbean Islands are working to increase their visibility with similar services. All are hoping to gain income from this low-impact and high-income enterprise. The United States and other countries have made attempts to pressure these islands to share the financial account information of people evading taxes. If successful, there may be less of an incentive for individuals and businesses to use offshore accounts to shelter income from taxes or for those involved in illegal money-laundering schemes to hide money in the Caribbean.

The wide level of diversity in the Caribbean has not made it easy for outside entities to provide support and assistance for common economic goals. The Caribbean Community (CARICOM) is an organization with fifteen Caribbean members established to promote economic integration and cooperation in the region. It hopes to coordinate foreign policy for the region and ensure that the benefits of integration are equitably shared. CARICOM is an attempt to compete with other trade organizations such as the North American Free Trade Agreement (NAFTA) and CAFTA-DR.

Caribbean Music

The Caribbean culture includes many varieties of music that have developed on several islands. The region is a breeding ground for innovative music and rhythms that emulate the cultural traditions and attitudes of the people. Every island has its own traditional festivals that include parades, music, and dance. The music scene reveals the uniqueness of the Caribbean. Listening to the myriad of sounds generated from the different islands opens a window to the assortment of cultural backgrounds found in the Caribbean.

The Caribbean is full of local musical variations, and many types of music are found only on a single island or two. For example, the Dominican Republic has meringue music, Dominica has bouyon, and Haiti has its festive rara music.

The Cuban influence on music is vividly evident in the spicy salsa tunes that have emanated from the Caribbean over the years. African and Spanish cultural influences have helped shape the salsa music that has evolved from Cuba and neighboring islands of Puerto Rico and the Greater Antilles. With a strong percussion component, snappy brass additions, and rhythmic guitar sections, salsa has become popular throughout many parts of Latin America. Other genres of music coming out of Cuba include rumba, habanera, son, and timba, to name a few.

Calypso music comes from a mix of African influences on the island of Trinidad. Calypso began taking shape at the beginning of the twentieth century and gained popularity through Carnival and other Caribbean festivals. It has evolved to incorporate the steel pan and other musical instruments. A commercial version of calypso became popular with Harry Belafonte's version of the Jamaican folk tune of "Day-O," known as the "Banana Boat Song"; however, Belafonte's 1956 album *Calypso* had more of a Jamaican mento musical style and he was not from Trinidad. Mento is a more folksy rural style of acoustical music that influenced other forms of music such as reggae in Jamaica.

The pan (steel pan) is the national instrument of Trinidad and Tobago, where it was created. The pan was originally made from fifty-five-gallon oil drums. The bottom is cut off at various levels to provide different sounds. The top is then shaped into a chromatically pitched percussion instrument. It is struck with a pair of straight sticks. An entire family of pans has been developed and can be assembled into a steel pan orchestra. The instrument has become popular outside the Caribbean as well.

Figure 5.34 Steel Pans from Trinidad

Reggae music started coming out of Jamaica in the 1960s with the music of such artists as Bob Marley and the Wailers, which featured Peter Tosh and Bunny Wailer. The rhythmic style with offbeat accents evolved from earlier genres to become a standard of Jamaican music. Musicians from the Beatles to Eric Clapton have used reggae rhythms. Outside Jamaica, reggae has hit the charts thanks to groups such as UB40. Reggae music has often been associated with the Rastafarian movement or Rasta, which is based on a religious ideology including the beliefs that former Ethiopian emperor Haile Selassie was God incarnate and the biblical Zion was in Africa.

Key Takeaways

- Colonialism created a high level of ethnic, linguistic, and economic diversity in the Caribbean. The main shifts were the demise of indigenous groups and the introduction of African slaves. The African influence can be witnessed in the religions of Santeria in Cuba, Vodoo (Voodou) in Haiti, and Rasta in Jamaica.
- The Caribbean Basin faces many challenges, including natural elements such as hurricanes, earthquakes, and volcanic activity. Economic conditions are often hampered by environmental degradation, corruption, organized crime, or the lack of employment opportunities.
- The Cuban Revolution led by Fidel Castro created a socialist state that nationalized foreign-owned assets and brought about a trade embargo by the United States. Cuba lost its aid from the Soviet Union after the USSR's collapse in 1991 and has been increasing its focus on tourism and capitalistic reforms.
- Tourism can bring added economic income for an island country, but it also shifts to the service sector resources that are needed for infrastructure and services. A high percentage of tourism income goes to external corporations.
- The diversity of the Caribbean is evident in the wide range of musical types generated from the islands. Individual islands are known for certain types of music; salsa, reggae, and calypso are examples.

Discussion and Study Questions

1. Which islands make up the Greater Antilles? Where are the Lesser Antilles?
2. Which European countries were the main colonizers of the Caribbean?
3. How is Cuba a good example of a rimland country?
4. Why does the United States still have an economic embargo against Cuba?
5. How is the political system in Cuba different from that of the United States?

6. Why is Puerto Rico a commonwealth of the United States?
7. How is Haiti different from its neighbor, the Dominican Republic?
8. What are the positive and negative perspectives on the Caribbean tourism industry?
9. Besides tourism, what other methods do people in the Caribbean islands use to generate wealth?
10. How do differences in musical styles tell a story of Caribbean culture?

Geography Exercise

Identify the following key places on a map:

- The Bahamas
- Blue Mountains
- Greater Antilles
- Lesser Antilles
- Sierra Maestra
- Viñales Valley

5.5. TROPICAL CYCLONES (HURRICANES)

North and south of the equator, above the oceans, a weather phenomena known as a tropical cyclone can form, significantly transforming the physical and cultural landscape upon making landfall. This meteorological phenomenon is referred to as a hurricane in the North Atlantic Ocean and the Caribbean Sea. In the North Pacific Ocean, this weather phenomenon is referred to as a typhoon. In the Indian Ocean and South Pacific Ocean regions, it is referred to as a tropical cyclone or just a cyclone. All these storms are classified as tropical due to their frequent development between the Tropics of Cancer and Capricorn.

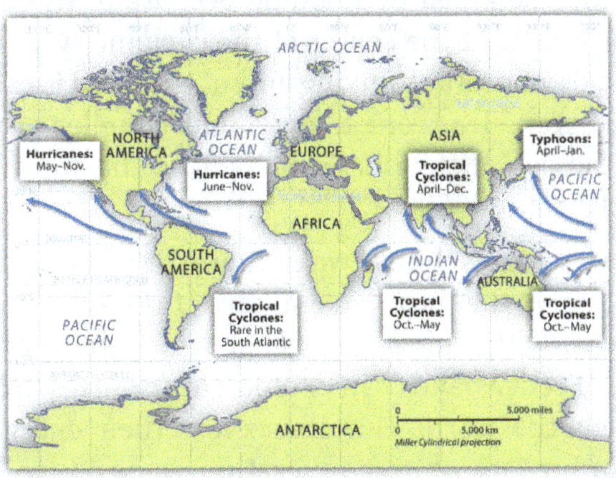

Figure 5.35: Cyclones, Hurricanes, and Typhoons and Their Respective Locations around the World

Hurricanes form over sea temperatures over 80 ºF. As the air warms, it ascends swiftly, inducing the influx of air to substitute the ascending air, thereby generating vigorous wind currents and stormy conditions. The swiftly ascending humid air subsequently cools and condenses, leading to intense precipitation and a downdraft of cooler air. The Earth's rotation induces a cyclonic pattern in storm systems. Tropical storms north of the equator rotate counterclockwise. Tropical storms south of the equator spin in a clockwise direction. Hurricanes originate as tropical depressions, characterized by wind speeds ranging from twenty-five to thirty-eight miles per hour. Cyclonic activity and elevated temperatures sustain the system. A storm is classified as a tropical storm when its sustained winds range from thirty-nine to seventy-three miles per hour. Tropical depressions are assigned numbers, whereas tropical storms receive names. A storm is categorized as a hurricane when winds attain a sustained speed of seventy-four miles per hour.

Hurricane Dynamics

Hot air rises. A water temperature of at least 80 ºF can sustain rising air in the development of a tropical depression. These storms continue to be driven by the release of the latent heat of condensation, which occurs when moist air is carried upward and its water vapor condenses. This heat is distributed within the storm to energize it. As the system gains strength, a full-scale hurricane can develop. Rising warm air creates a low-pressure area that draws in air from the surface. This action pushes water toward the center, creating what is called a storm surge. Storm surges can average five to twenty feet or more depending on the category of the hurricane. Cyclonic rotation is created by rotation of the earth in a process called the Coriolis effect. The Coriolis effect is less prominent along the equator, so tropical cyclones usually do not develop within five degrees north or south of the equator.

When a hurricane makes landfall (comes ashore), the storm surge causes extensive flooding. More people are killed by flooding because of the storm surge than by any other hurricane effect. At the center of the cyclonic system is the hurricane's eye, where there is a downdraft of sinking air but the wind is calm and there are no clouds. The eye can extend from one to one hundred miles or more. Many people who have been in the eye of the hurricane believe the storm has passed, but in reality they are in the center of it.

Bordering the eye of a storm is the eye wall, where the strongest winds and heaviest rainfall are found. This is the most violent part of the hurricane. Beyond the eye wall are feeder bands, with thunderstorms and rain showers that spiral inward toward the eye wall. Feeder bands can extend out for many miles and increase as the heat engine feeds the storm. Hurricanes lose their energy when they move over land because of the lack of heat generation. Once on land, the storm system breaks down. Rainfall and winds can continue, but with decreased intensity.

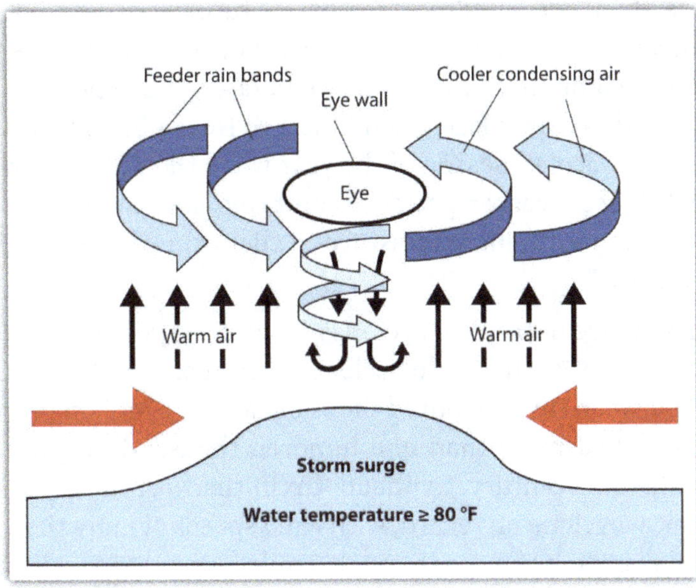

Figure 5.36: Dynamics of Hurricane Components

Centuries ago, the Spanish used the term *hurakan*, an indigenous word for "evil spirits" or "devil wind," to name the storms that sank their ships in the Caribbean. Hurricanes are rated according to sustained wind speed using the Saffir-Simpson Scale. This scale rates a hurricane according to five categories (see Figure 5.37 "Saffir-Simpson Hurricane Scale"). Category 1 hurricanes have sustained wind speeds of at least seventy-four miles per hour and can inflict heavy damage to buildings, roofs, windows, and the environment. Category 5 hurricanes have sustained winds of more than 155 miles per hour and destroy everything in their paths. Hurricanes can also spawn tornadoes, which increase their potential for destruction.

Category	Wind Speeds	Storm Surge
1	74–95 mph	4–5 ft
2	96–110 mph	6–8 ft
3	111–130 mph	9–12 ft
4	131–155 mph	13–18 ft
5	>155 mph	>18 ft

Figure 5.37: Saffir-Simpson Hurricane Scale

Annually, more than one hundred tropical disturbances develop in the North Atlantic, but only about ten make it to a tropical storm status and five to six become hurricanes. Only two or three hit the United States in a typical year. Hurricane season for the North Atlantic lasts from June 1 to November 30. Tropical cyclones develop during the warmest season of the year when the water temperature is the highest. Though these weather patterns can bring enormous devastation to the landscape, they also redistribute moisture in the form of rain and help regulate global temperatures.

The devastating nature of tropical cyclones is the main concern when forecasting a potential storm. In 1970, the Bholo cyclone hit the coast of Bangladesh, resulting in the death of between three hundred thousand and one million people. A number of cyclones that killed more than one hundred thousand people each have hit Bangladesh in the past century. Typhoon Tip in the Northwest Pacific in 1979 is the largest tropical cyclone on record, with wind speeds of more than 190 miles per hour and a total diameter of more than 1,350 miles—equal to the distance from the Mexican border to the Canadian border in the United States. Typhoons can be, on average, twice the size of hurricanes.

Figure 5.38: Hurricane Katrina at Various Stages of Its Development across the Gulf of Mexico

Hurricane Camille was the strongest US hurricane on record at landfall, with sustained winds of 190 miles per hour and wind gusts of up to 210 miles per hour. Camille hit the US Mississippi coast in 1969 as a category 5 hurricane. It devastated everything in its path, killing 259 people. Hurricane Katrina in 2005 was one of the most costly storms to impact the United States. Katrina started out as a tropical depression while in the Bahamas. The storm reached a category 5 hurricane as it passed through the Gulf of Mexico but diminished in strength when making landfall in Louisiana, with sustained winds of 125 miles per hour (a strong category 3 hurricane). Katrina caused widespread devastation along the central Gulf Coast and devastated the city of New Orleans. At least 1,836 people lost their lives, and the cleanup cost an estimated $100 billion.

Since records were started in 1851 for hurricanes in the Atlantic Basin, there have been thirty-two hurricanes that reached category 5 in the region. A few of them have reached all the way to the Central American coast. Hurricane Mitch hit the coast of Central America in 1998 and dumped over seventy-five inches of rain across the countries of Honduras, El Salvador, Nicaragua, and Guatemala. Devastating winds and heavy rain caused the deaths of up to twenty thousand people. Destructive category 5 hurricanes Edith and Felix made landfall in Nicaragua in 1971 and 2007, respectively. The Yucatán Peninsula and the coast of Mexico have also witnessed a number of devastating category 5 hurricanes.

The Caribbean Basin is located in the path of many hurricanes developing out of the Cape Verde region of the North Atlantic. For example, 2008 was a particularly devastating hurricane season, with sixteen tropical storms and eight full-scale hurricanes, five of which caused massive devastation. Three category 4 hurricanes (Ike, Gustav, and Paloma) cut through the northern Caribbean to hit the Greater Antilles.

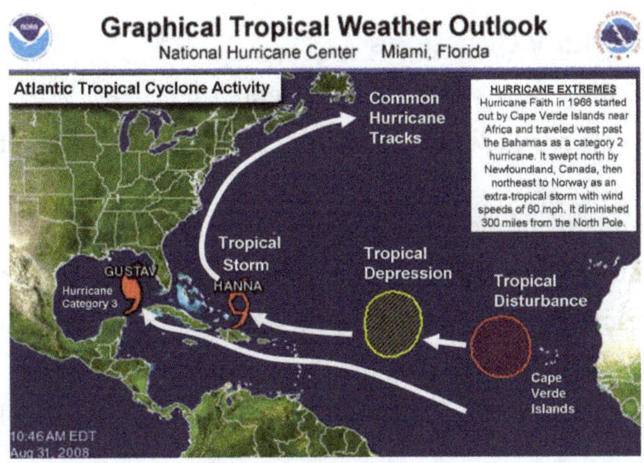

Figure 5.39: Direct Path of Hurricane Activity This photo indicates hurricane Gustav and tropical storm Hanna, as well as an existing tropical depression (which became Hurricane Ike) and a tropical disturbance. Tropical Storm Hanna later developed into a full-scale hurricane. Map courtesy of National Hurrican Center.

The most devastating was Ike, which ripped through the Caribbean, across the entire length of Cuba, and then on to the Gulf Coast of Louisiana and Texas. Ike's immense size contributed to the fact that it was the third most costly hurricane on record. Ike caused an estimated $7.3 billion in damage to Cuba and more than $29 billion in damage in the United States. Hurricane Gustav made landfall in Hispaniola and Jamaica before increasing in strength and causing about $3.1 billion in damage to Cuba. In November of 2008, Hurricane Paloma made landfall in Cuba and caused an additional $300 million in damage to the island. Many of the other Caribbean islands were also devastated by the hurricanes that hit the region in 2008.

Key Takeaways

- Tropical cyclones occur in the tropical regions over warm ocean water. In the North Atlantic, they are called hurricanes; in the North Pacific, they are called typhoons; and in the Indian Ocean, they are called cyclones.
- Hurricanes start as tropical depressions with wind speeds of at least twenty-five miles per hour. As wind speeds increase to thirty-nine miles per hour, the disturbances are called tropical storms and are named. When wind speeds reach seventy-four miles per hour, they become hurricanes.
- Rising air pulls water to the center of the storm, creating a storm surge, the most dangerous feature of the storm because of the immense flooding it can cause when reaching land.
- Hurricane season is between June 1 and November 30. Cruise ships do not usually operate in the Caribbean during this time.

Discussion and Study Questions

1. Why do tropical cyclones form near the equator?
2. What are the stages of weather patterns that build up to a tropical cyclone (hurricane)?
3. What are the main components of a hurricane?
4. What part of the hurricane usually causes the most damage or loss of life?
5. How are hurricanes classified? What are the main categories of a hurricane?
6. How many tropical disturbances develop in the North Atlantic each year? How many develop into full-scale hurricanes each year? How many hurricanes usually hit the United States each year?
7. Why is it often more difficult for the Caribbean islands to recover from a hurricane than the United States?
8. What path do hurricanes usually follow in the North Atlantic?
9. Where do cyclones and typhoons develop other than the North Atlantic?
10. When is the main hurricane season in the North Atlantic? How does the hurricane season impact tourism in the Caribbean?

5.6. CONCISE SUMMARY

The Caribbean, Mexico, and Central America constitute the region of Middle America. Two categories of development patterns arose under European colonialism. The rimland, characterized by plantation agriculture, predominated in the Caribbean and coastal areas. The mainland, characterized by its haciendas, predominated Mexico and the interior territories of Central America.European colonialism devastated the Amerindian inhabitants of the Caribbean and subjugated the Aztec Empire on the continent. Colonialism transformed food production, construction techniques, urbanization, language, and religion in the region.African slave labor became prevalent in the Caribbean, significantly changing the ethnic composition of numerous islands. Amerindians constitute the majority of the lower working class on the mainland. A small contingent of affluent Europeans remains at the apex of the socioeconomic hierarchy. The majority of Mexico's population is of mestizo descent.Mexico has evolved from a Spanish colony to a participant in the North American Free Trade Agreement (NAFTA). Trade links have facilitated the industrialization of Mexico's economy and generated jobs, particularly in the thriving maquiladoras. Mexico possesses abundant natural resources yet continues to face challenges in offering economic opportunities for its whole populace. Wealth and power are governed by an elite minority of European descent. Numerous geographic ideas and principles can be utilized in this context: rural-to-urban migration, core-periphery spatial dynamics, altitudinal zonation, and the influence of climatic variations on human settlement.The increase in population and the scarcity of employment possibilities have exacerbated poverty levels in numerous regions. A significant discrepancy exists between the income levels of the affluent and the impoverished. Haiti is among the most impoverished nations in the Western Hemisphere. The United States has significantly influenced this region, both politically and economically. The US military has intervened in numerous locations to safeguard its interests. American corporations have predominated the economy of the region. The United States has recently endorsed industrial endeavors and the execution of free-trade agreements to capitalize on inexpensive labor. Natural disasters such as earthquakes, volcanoes, and hurricanes persist in causing havoc and damage to human operations throughout Middle America. Additional environmental concerns, like deforestation and soil deterioration, have emerged as significant difficulties.

Central America is a heterogeneous and fragmented region, with each country, island, or republic exhibiting distinct geographical characteristics. The diverse musical forms that have originated from the region exemplify cultural diversity. Tourism is a significant economic industry that exerts varied effects on the local context. Each segment of the Middle American region has endeavored to enhance its tourism appeal to strengthen its economy. The global economy has compelled the political entities of the area to collaborate more closely to promote their economic interests. Trade organizations like NAFTA, the Central American-Dominican Republic Free Trade Agreement (CAFTA-DR), and the Caribbean Community (CARICOM) aim to enhance economic integration. Certain individuals contend that transnational firms are poised to gain the most from free-trade agreements.

Tropical cyclones are meteorological phenomena that manifest in tropical regions above warm oceanic waters. In the North Atlantic, wind speeds of seventy-four miles per hour are classified as hurricanes. The hurricane season spans from June 1 to November 30, during which cruise ships typically refrain from operating in the Caribbean. Intense winds and storm surges have resulted in significant floods and destruction to the human environment.

Chapter 6 SOUTH AMERICA

6.1. INTRODUCTION: AN OVERVIEW

South America exhibits a vast array of physical features, ranging from the towering Andes Mountains to the tropical forests of the Amazon Basin. This collection of physical attributes provides several resources, enabling individuals to participate in economic activities, accumulate wealth, and meet their requirements. The extensive Andes mountain range possesses mineral wealth that has been exploited since antiquity. Precious metals have been extracted from the mountains, providing significant opportunity for those fortunate enough to receive their wealth. Fossil fuels have been discovered in significant quantities in the remote northern areas of Venezuela and Colombia. The Amazon Basin has served as a source of hardwood timber and, more recently, significant mineral resources. This region hosts some of the largest iron ore mines globally. The expansive plains of Brazil and the fertile soils of the Pampas facilitate extensive agricultural enterprises that supply food goods for both the continent and the globe. The arid Atacama region in northern Chile contains some of the world's greatest copper reserves. The different climate zones facilitate the development of a large array of species. The harsh physical terrain of South America has generated both obstacles and possibilities for its inhabitants.

Prior to the advent of European colonialism, numerous indigenous communities established themselves as states or empires. The Inca Empire was the most extensive empire in existence when the Europeans came. The early empires constructed monumental edifices necessitating the transportation of enormous cut stones, each weighing several tons, accomplished without contemporary equipment or technology. European colonialism transformed the continent in multiple ways. The Europeans not only vanquished and subjugated indigenous Amerindian groups like the Inca, but also engaged in cultural exchanges that transformed the lifestyles of numerous South Americans.

Figure 6.1 South America: Political Map of the Countries and Various Physical Regions

Colonialism established numerous contemporary national borders and shaped trade interactions with the newly formed colonies. The plantation system and the importation of African slaves significantly altered the ethnic composition of the population residing along the eastern coast. Following the abolition of slavery, indentured servitude facilitated the influx of laborers from Asia to bolster the workforce.

Indentured servants were typically impoverished individuals who consented to labor for a specified duration, generally less than seven years, in exchange for essential provisions such as shelter, sustenance, and transportation or attire. These persons often did not get a salary but may have gotten a one-time payment upon the conclusion of the contracted service. In advantageous circumstances, indentured servants were regarded as family and acquired significant experience and vocational skills for their future sustenance. Numerous circumstances were significantly less advantageous and resembled a variant of slavery, wherein individuals did not endure their servitude but perished due to disease, severe conditions, or death.

The physical environment and the impact of colonization both contributed to the cultural characteristics of South American regions. The continent can be segmented into regions based on ethnic majorities shaped by early colonial growth. The amalgamation of ethnic groups from Europe, Africa, and Asia, along with the indigenous people, has resulted in a varied cultural mosaic. For instance, the majority of the population of Guyana and Suriname is of Asian heritage, whereas most individuals in Argentina and Uruguay are of European ancestry. of Peru and Bolivia, the predominant demographic is Amerindian, while numerous regions along the eastern coast of Brazil are primarily inhabited by individuals of African descent. A significant proportion of the South American population have a mixed ethnic heritage. Comprehending the cultural geography of South America enhances the understanding of the region's human development trends.

The contemporary economic advancement of South America has facilitated its integration into the global economy. The degrees of economic growth differ significantly within the domain. Distinct core-peripheral spatial patterns are evident across multiple regions of the continent, and the rural-to-urban transition has been pronounced in numerous areas. The rural areas in the Andes or the interior experience a deficiency of economic assistance required to improve their infrastructure. Simultaneously, urban regions are experiencing tremendous expansion and are fully integrated with global markets and cutting-edge technologies. The majority of major cities are situated in coastal areas. This trend of urbanization primarily stems from colonial endeavors and their impact. The nations of South America are collaborating to enhance trade and commercial activity. Trade agreements and economic unions have emerged as conventional strategies for establishing corporate relationships to augment economic prospects.

Physical Geography

The far-reaching Andes Mountains and the massive Amazon River system dominate South America's physical geography. The five-thousand-mile-long Andes Mountain chain extends along the entire western region of the continent from Venezuela to southern Chile. The Andes are the longest mountain chain on Earth and the highest in the Americas. The Andes Mountain range has more than thirty peaks that reach at least twenty thousand feet in elevation, many of which are active volcanoes. The Andes has provided isolation to the **Inca Empire**, mineral wealth to those with the means for extraction, and a barrier to travelers crossing the continent. The Andes' minerals include gold, silver, tin, and other precious metals. Mining became a major industry in the colonial era and continues to the present.

At the core of the continent is the mighty Amazon River, which is more than two thousand miles long and has an enormous drainage basin in the largest tropical rain forest in the world. The Amazon's many tributaries are larger than many other world rivers. Flowing parallel to the equator, the Amazon is in contention with the Nile as the longest river in the world. In 2007, a dispute arose over the actual length of the Amazon River. Brazilian scientists using satellite imagery and teams on the ground using a global positioning system (GPS) calculated a new length for the Amazon. Using similar methods, the Nile River also had a new length calculated. Both were determined to be longer than originally measured. At this time, the actual length of each river is openly disputed. The most accurate statement is that the Amazon River has the greatest discharge of any river in the world, and it is potentially also the longest river in the world.

The Amazon carries about a fifth of all river water in the world. The Amazon and its many tributaries drain the entire interior region of the continent, covering 40 percent of South America. During the rainy season, the Amazon River can be more than one hundred miles wide. No bridges span the Amazon River. Its source is a glacial stream located high in the Peruvian Andes, about 150 miles from the Pacific Ocean.

The Amazon's extended tributaries—such as the **Rio Negro**, the **Madeira**, and the **Xingu**—move massive amounts of water through the Amazon Basin and are major rivers in their own right. The Amazon has more than 1,100 tributaries; a dozen are more than one thousand miles long. Hydroelectric dams are located on the tributaries to produce electricity for the region's fast-growing development. South America has additional large rivers that drain the continent, including the **Orinoco**, which flows through Venezuela; the **Sao Francisco**, which flows through southeast Brazil; and the **Paraguay** and the **Paraná** Rivers, which flow south from Brazil into the **Rio de la Plata** between Argentina and Uruguay.

Lake Titicaca rests in the middle of the **Altiplano Region** of the Central Andes on the border between Peru and Bolivia. The Altiplano Region is a wide basin between two main Andean mountain ranges. The word altiplano means "valley" in Spanish. There are a substantial number of altiplanos in South America. They provide for agricultural production and human habitation. The Altiplano Region has been home to ancient civilizations that date back to early human settlements. Lake Titicaca is a large freshwater lake about 120 miles long and 50 miles wide. The surface is at an elevation of about twelve thousand feet above sea level, and the lake is more than nine hundred feet deep in some areas. Usually at such high elevations, the temperature would dip below freezing and restrict agriculture. However, the large lake acts as a solar energy collector by absorbing energy from the sun during the day and giving off that energy in the form of heat during the night. The energy redistribution allows for a moderate temperature around the lake that is conducive to growing food crops. With abundant fresh water and the ability to grow food and catch fish, the Altiplano Region has supported human habitation for thousands of years. On the Bolivian side of the lake is the ancient stone site of Tiahuanaco, which has had some artifacts dated to 1500 BCE or earlier. People continue to live around and on Lake Titicaca and maintain a subsistence lifestyle.

Across the Andes Mountains from the Altiplano Region is the Atacama Desert. The Atacama is one of the driest places on Earth: in some parts, no rain has fallen in recorded history. In normal circumstances, the Atacama would be a desolate region without human activity, but that is not the case. Some of the world's largest copper reserves are found here. Nitrates, which are used in fertilizers, are also found in large quantities. Mining the Atacama has brought enormous wealth to people fortunate enough to be on the receiving end of the profits. The rain shadow effect is responsible for the extraordinary dryness of the Atacama. The Andes are quite high at this latitude, and the winds blow in rain clouds from the east. When the clouds reach the mountains, they ascend in elevation, releasing their precipitation without ever reaching the western side of the Andes.

Figure 6.2: Lake Titicaca with Traditional Reed Boat Made byAmerindian Locals Lake Titicaca is the highest-elevation navigable lake in the world. Benedict Adam – Peru – CC BY 2.0.

South America has large agricultural plateaus east of the Andes, such as the **Mato Grosso Plateau**, which includes a portion of the great cerrado agricultural region of central Brazil. The Cerrado is a vast plain that has been developed for agriculture and produces enormous harvests of soybeans and grain crops. Bordering the Cerrado to the southeast are the **Brazilian Highlands**, an extensive coffee-growing region. The **Pampas** in eastern Argentina, Uruguay, and southernmost Brazil is another excellent agricultural region with good soils and adequate rainfall. Farming, cattle ranching, and even vineyards can be found here, making the Pampas the breadbasket of the Southern Cone. To the south of the Pampas is the lengthy expanse of **Patagonia**, which covers the southern portion of Argentina east of the Andes. Patagonia is a prairie grassland region that does not receive a very much rainfall because of the rain shadow effect of the Andes to the west. The main activities in Patagonia are the raising of cattle and other livestock. The region is starting to attract attention for the extraction of natural resources such as oil, natural gas, and valuable minerals.

To the northern part of the continent in Venezuela and Colombia, sandwiched between the Andes Mountains and the **Guiana Highlands,** is a grassland region with scrub forests called the Llanos. The human population is small because of the remoteness of the region along the Orinoco River basin. The Guiana Highlands of southeast Venezuela and the Guianas are an isolated set of mountainous plateaus mixed with rugged landscapes and tropical climates. Angel Falls, the highest waterfall in the world, with a free fall of more than 2,647 feet and a total drop of about 3,212 feet (more than half a mile), is located here. To the northwest of the Llanos and the Guiana Highlands in Venezuela is **Lake Maracaibo**, a large inland lake open to the Caribbean Sea. A coastal lake, Maracaibo rests atop vast oil reserves that provide economic wealth for Venezuela.

European Colonialism

South America's colonial legacy shaped its early cultural landscape. The indigenous people, with their empires and local groups, were no match for the Iberian invaders who brought European colonialism to the continent. South America was colonized exclusively by two main Iberian powers: Spain colonized the western part of the South America, and Portugal colonized the east coast of what is present-day Brazil. The only region that was not colonized by those two powers was the small region of the **Guianas**, which was colonized by Great Britain, the Netherlands, and France.

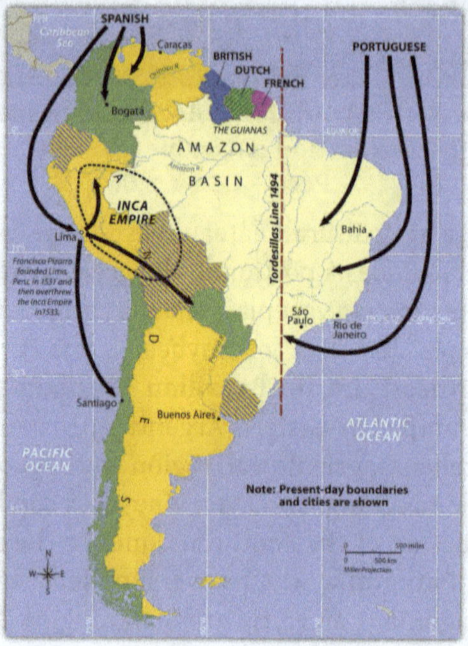

Figure 6.3: Colonial Activity in South America

Everything changed with the invasion of the Iberian colonizers. The underlying tenets of culture, religion, and economics of the local indigenous people were disrupted and forced to change. It is no mystery why the two dominant languages of South America are Spanish and Portuguese and why Roman Catholicism is the realm's dominant religion.

Colonialism also was responsible for transporting food crops such as the potato, which originated in the Peruvian Andes, to the European dinner table. Today, coffee is a main export of Colombia, Brazil, and other countries in the tropics. Coffee was not native to South America but originated in Ethiopia and was transferred by colonial activity. The same is true of sugarcane, bananas, and citrus fruits; oranges were not native to South America, but today Brazil is the number one exporter of orange juice.

Colonialism was driven by the desire for profit from the quick sale of products such as gold or silver, and there was a ready market for goods not found in Europe, such as tobacco, corn, exotic animals, and tropical woods.

Plantation agriculture introduced by the Europeans led to a high demand for manual laborers. When the local populations could not meet the labor demand, millions of African workers were brought through the slave trade. These African slaves introduced their own unique customs and traditions, altering the culture and demographics of the Western Hemisphere. The current indigenous Amerindian population, a fraction of what it was before the Europeans arrived, makes up only a small percentage of South America's total population. Europeans colonizers generally took the best land and controlled the economic trade of the region. The acculturation in South America is directly related to the European colonial experience.

The Inca Empire and Francisco Pizarro

Not long after Hernán Cortés conquered the Aztec Empire of Mexico in 1521, a young Spanish conquistador named **Francisco Pizarro**, stationed in what is now Panama, heard rumors of silver and gold found among the South American people. He led several sailing excursions along the west coast of South America. In 1531 CE, he founded the port city of Lima, Peru. Since 1200 CE, the Inca had ruled a large empire extending out from central Peru, which included the high-elevation Altiplano Region around Lake Titicaca. The Inca Empire dominated an area from Ecuador to Northern Chile. The Inca were not the most populous people but were a ruling class who controlled other subjugated groups. Pizarro, with fewer than two hundred men and two dozen horses, met up with the Inca armies and managed to defeat them in a series of military maneuvers. The Inca leader was captured by the Spanish in 1533. Two years later, in 1535, the Inca Empire collapsed.

Figure 6.4: Colonialism in South America

The two main colonizers in South America were Spain and Portugal. The Spanish conquistador Francisco Pizarro defeated the Inca Empire.

The Inca Empire was significant thanks to the high volume of gold and silver found in that region of the Andes. The story is told that Pizarro placed the captured Inca leader in a small room and told him that if the room was filled with gold to the top of the Inca leader's head, Pizarro would let the Inca leader go free. Gold was summoned from the people and the countryside, and the room was filled. The conquistadors forced the Inca leader to convert to Catholicism and be baptized and then killed him. The gold was melted down and transported back to Spain. The mineral wealth of the Andes made the conquistadors rich. Lima was once one of the wealthiest cities in the world. Europeans continued to dominate and exploit the mining of minerals in Peru and Bolivia throughout the colonial era. European elites or a Mestizo ruling class has dominated or controlled the local Amerindian groups in the Andes since colonial times.

Figure 6.5: The Jesuit Church of La Compañia de Jesús on the Plaza de Armas in Cuzco, Peru The original church was constructed in 1571 on the site of the ancient Incan palace. The earthquake of 1650 caused severe damage to the building, so it had to be reconstructed in about 1688. Many Catholic cathedrals in Latin America were built with stones from ancient sites. This cathedral is claimed to be the Western Hemisphere's most ideal example of colonial baroque architecture. James Preston – La Compania de Jesus, Cusco – CC BY 2.0.

Many Amerindian groups inhabited this region before the Inca Empire, such as the people who built and lived in the ancient city of Tiahuanaco, which could have bordered Lake Titicaca during an earlier climate period, when the lake was much larger. Clearly, humans have lived in South America for thousands of years. There is speculation that travel between South America and the Mediterranean region occurred earlier than current historic records indicate. Many of the ruins in the western region of South America have not been thoroughly excavated or examined by archaeologists, and the size and scope of many of the stone structures stand as testimony to the advancements in engineering and technology employed in an era that, according to historical records, had only primitive tools.

The Iberian Division of the Continent

The Spanish conquistadors were not the only European invaders to colonize South America. Colonial influence—which forced a change in languages, religion, and economics—also came from the small European kingdom of Portugal. Portuguese ships sailed along the eastern coast of South America and laid claim to the region for the king. The Portuguese did not find large gold or silver reserves, but they coveted the land for the expansion of their empire. Soon the Spanish and the Portuguese were fighting over the same parts of South America. In 1494, the issue was brought before the Roman Catholic Church. The **Tordesillas Line** was drawn on a map to divide South America into the Spanish west and the Portuguese east. The region that is now Brazil became the largest Portuguese colonial possession in South America, a center for plantation agriculture similar to that in the Caribbean. For this reason, a large African population lives in Brazil, and most of the people in Brazil speak Portuguese and are Roman Catholics.

Independence did not come for the Spanish colonies until 1816 and 1818, when Chile and Argentina broke away in an independence movement in the south. Simón Bolívar led liberation movements in the north. By 1824, the Spanish were defeated in South America. Brazil did not gain independence from Portugal until 1822, when the prince of Portugal declared an independent Brazil and made himself Brazil's first emperor. It was not until 1889 that a true republic was declared and empire was abolished.

Colonialism and the Guianas

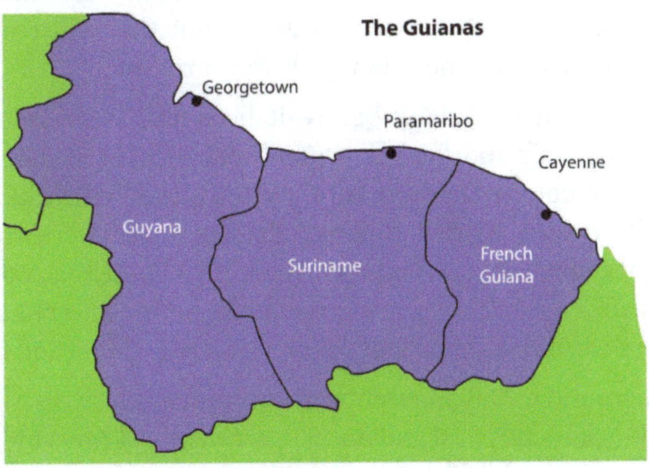

Figure 6.6: The Guianas of South America

The Guianas in the northeast were the only European colonies in South America that were not under Spanish or Portuguese control. The British, Dutch, and French all held claims to different parts of the Guianas. French Guiana remains a colony (department) of France to this day. The transition from colony to independent state has required persistence, time, and patience. Guyana and Suriname only just received their independence in the latter part of the twentieth century.

The coastal location of the Guianas provided European colonialists with an excellent site for plantation agriculture. Coastal cities had easy access to the Atlantic trade system. The active slave trade in the Atlantic brought African slaves to the Guianas to work the plantations. When slavery was abolished, indentured servants were brought in from other parts of the world that were colonized by the same European powers. Colonialism and plantation agriculture entirely changed the cultural dynamics of the Guianas. Port cities along the coast are the dominant urban centers. The Guianas follow a pattern similar to that of the rimland of the Caribbean and are included with the Caribbean in many studies. At the present time, bauxite (aluminum ore) mining and oil deposits along the coast provide modest income, and many people make their living in subsistence agriculture.

Guyana

Guyana and its neighbors have a tropical type A climate. The main interior regions are covered with dense forests. Some deforestation has taken place in Guyana; even so, forests cover more than 80 percent of Guyana's land area with a diverse range of forests, ranging from dry evergreen forests to tropical rain forests, and with thousands of species of plants that are only found in this region. Guyana's natural habitats and biodiversity are unparalleled, which is attributable to the climate, distinctive landforms, and largely well-preserved ecosystems. Its large rain forests play host to a myriad of species not found elsewhere. The interior regions remain more than 70 percent pristine, but this will change if logging, mining, and deforestation are not managed appropriately.

Guyana was called British Guiana before it became independent in 1966. The region was home to Carib and Arawak groups before the arrival of the Dutch, who established the first colony there in 1616. By 1831, the British had taken control and declared the colony British Guiana. African slaves were the main source of plantation labor until slavery was abolished in 1834. At that point, indentured servants from many countries were targeted for labor; most were East Indians from South Asia. There were also thousands of servants from China and other countries.

The ethnic background of Guyana's population, which is less than one million, is a mix of African (36 percent), East Indian (50 percent), and Amerindian (7 percent). The remainder are mixed and European. The religious background of the East

Indians includes both Hindus and Muslims. About 50 percent of the population has converted to Christianity. Ethnic divisions create difficulties in organizing politics and social activities. For example, one political concern is that minority groups with greater economic advantages will dominate the political arena without regard to the wishes of other ethnic groups.

Most of Guyana's people live in the coastal regions. The almost impenetrable interior has large uninhabited areas with dense tropical forests. To protect its unique biodiversity, large areas have been placed in conservation areas and national parks. More than one million acres have been placed in the Konashen Community-Owned Conservation Area (COCA) to help preserve the natural environment and provide a sustainable economic relationship with local Waiwai people. In a similar effort, the government created the Kaieteur National Park, which features the spectacular Kaieteur Falls, which is about five times higher than Niagara Falls in North America. For its size, the falls has an enormous amount of water volume flowing over its crest. Other areas of the Guyana are also being considered for preservation.

Figure 6.7: Kaieteur Falls, Guyana Kaieteur Falls has the largest single-drop waterfall by volume in the world. A national park has been created in this area to include the falls. Paul Arps – Kaietur falls from a distance (Guyana 2001) – CC BY 2.0.

Guyana's economy is based on agricultural activities and the export of extracted natural resources. Sugarcane is a main cash crop, and timber sales are increasing. Fishing is common in the coastal areas. Bauxite mining has been the main mining activity, and mining of gold and other minerals has only advanced in recent years. Guyana's economy is similar to that of many islands of the Caribbean, a relationship exemplified by its membership in **CARICOM** (the Caribbean Community). CARICOM's headquarters is located in Georgetown, the capital of Guyana.

Suriname

Suriname's colonial name was Dutch Guiana. The name was changed when independence was granted in 1975. The cultural background of the colony is as diverse as its history. Dutch Guiana was divided between British and Dutch colonists. Slavery and plantation agriculture were introduced in the colonial period, and the harsh conditions of slavery caused many slaves to escape the plantations and live in the forests. These runaway slaves, called **Maroons**, eventually organized into viable independent groups. Peace treaties were finally made with the Maroons to grant them legitimate rights. After slavery was abolished, indentured servants from South Asia and the Dutch East Indies (Indonesia) were conscripted into the colony's labor base.

Most of the half-million people who live in Suriname today reside along the coast. The ethnic background of the people includes East Indian (37 percent), Creole (31 percent), Indonesian (15 percent), African (10 percent), Amerindian (2 percent), Chinese (2 percent), and European (about 1 percent). All three of the Guianas are ethnically diverse, and in this setting, the diversity creates tensions related to social and economic stability. Guyana and Suriname are the least Latin regions of South America; the Creoles, for example, are mainly a mix of African and Dutch mixed with Asian or Amerindian. Establishing a stable government and conducting civil affairs are often complicated by the ethnic groups that vie for power and political control. The religious background in Suriname includes Hindu, Muslim, Christian, and a mix of African religions such as Winti.

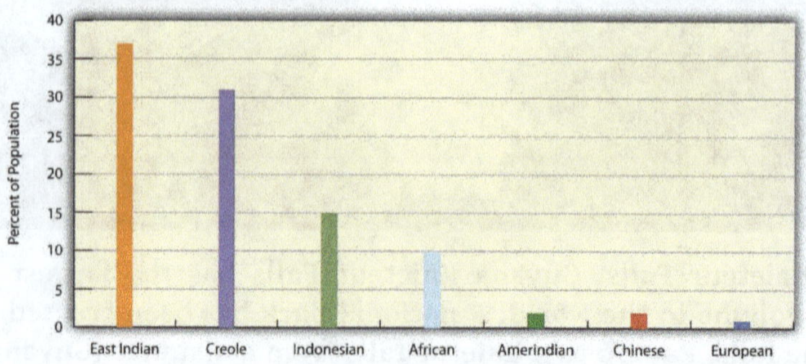

Figure 6.8: Ethnic Groups of Suriname by Percentage of the Population. Data courtesy of CIA World Factbook 2010.

Suriname depends on agriculture and natural resources as its two main economic activities. Suriname has the smallest physical area of all the independent South American states, and it has a limited ability to provide opportunities and advantages for its growing population. Bauxite is the country's main export product. In 1941, during World War II, the United States occupied the country to secure the valuable bauxite mines that were critical to the war effort. Suriname has also been expanding its gold mining, oil extraction, logging, and fishing industries. To protect its natural environment and its extensive biodiversity, the country has established national parks and created the large Central Suriname Nature Reserve, which UNESCO has named a World Heritage Site. Suriname hopes to benefit from these measures to create an ecotourism market.

Both Guyana and Suriname have had border disputes with their neighbors. Guyana has been locked in a challenge with Venezuela over its western region, which Venezuela claimed during the colonial era. A 1905 treaty ruled in Guyana's favor, but a small portion is still under dispute. The corner of Surname that borders Brazil and Guyana has been an issue of contention fueled by the potential for mining of gold and other minerals in the area. The maritime boundary between Suriname and Guyana is also a point of contention. Suriname's border with French Guiana has a disputed area along the southern corner next to Brazil. These disputes emerged out of colonial agreements and poorly defined treaty boundaries.

French Guiana

French Guiana is still a colony (department) of France. With fewer than 250,000 people in an area slightly smaller than Kentucky, this is a sparsely inhabited area. Half the population lives in the capital city of Cayenne. Most of the population is Creole, mainly a combination of African and French with various Asian groups mixed in. The people work in subsistence agriculture or for the French government. A noted feature of the colony's heritage is the former off-shore prison on **Devil's Island**, which France used to secure its most undesirable prisoners. More than fifty thousand French prisoners were delivered to the Devil's Island facility during the colonial era, but less than 10 percent survived. The prison was closed in 1951.

Figure 6.9: The Ariane 42P Carrying the TOPEX/Poseidon Spacecraft This Ariane 42P was launched from the European Space Agency's Guiana Space Center in Kourou, French Guiana, on August 10, 1992. Photo courtesy of NASA – public domain.

The European Space Agency maintains a launch center in French Guiana because of its favorable climate and launch conditions. France maintains the facility and has highly subsidized the department's economy. This European support provides the population with a higher standard of living than in many other South American countries. Economic activities have included fishing, agriculture (bananas and sugar), and logging.

Deforestation from the sale of hardwood lumber has become a problem throughout the Guianas as it leaves the soil open for erosion. Logging endangers the area's fragile but extensive biodiversity. Roads, dams, and gold mining have also contributed to the erosive destruction of the environment.

Cultural Regions of South America

It is impossible to understand the current conditions in South America without first understanding what occurred to create those conditions. This is why studying European colonialism is so important. Colonialism changed the ethnicity, religion, language, and economic activities of the people in South America. The past five hundred years have tempered, stretched, and molded the current states and regions of the South American continent. To identify standards of living, ethnic majorities, and economic conditions, it is helpful to map out South America's various **cultural regions.**

In South America, five main cultural regions indicate the majority ethnic groups and the main economic activities:

1. Tropical Plantation Region
2. Rural Amerindian Region
3. Amazon Basin
4. Mixed Mestizo Region
5. European Commercial Region (Southern Cone)

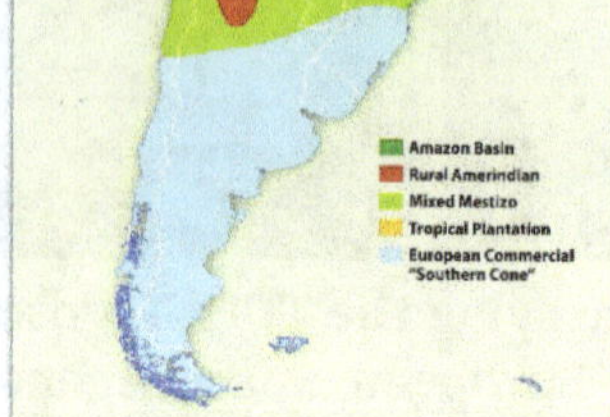

Figure 6.10: Cultural Spheres of South America.

These are generalized regions that provide a basic understanding of the whole continent. Technological advancements and globalization have increased the integration of the continent to the point that these regions are not as delineated as they once were, but they still provide a context in which to comprehend the ethnic and cultural differences that exist within the realm.

Tropical Plantation Region

Located along the north and east coast of South America, the Tropical Plantation Region resembles the Caribbean rimland in its culture and economic activity. The region, which extends as far south as the Tropic of Capricorn, has a tropical climate and an agricultural economy. Europeans opened up this area for plantation agriculture because of coastal access for ships and trade. The local people were forced into slavery, but when the local people died off or escaped, millions of African slaves were brought in to replace them. After slavery was abolished, indentured servants from Asia were brought to the Guianas to work the plantations. The Tropical Plantation Region has a high percentage of people of African or Asian descent.

Figure 6.11: Young Women in Salvador, Brazil Salvador, Brazil, is located along the coastal region of South America where the Tropical

Plantation Region was prominent. Most people in this region are of African descent. Carnal.com Studios - Salvador Carnaval Comanches 03 - CC BY 2.0.

Rural Amerindian Region

The Rural Amerindian Region includes the countries of Ecuador, Peru, and Bolivia. The ruling Mestizo class that inherited control from the European conquistadors mainly lives in urban areas. Most of the rural Amerindian population lives in mountainous areas with type H climates and ekes out a hard living in subsistence agriculture. This is one of the poorest regions of South America, and land and politics are controlled by powerful elites. The extraction of gold and silver has not benefited the local Amerindian majority, which holds to local customs and speaks local languages.

Figure 6.12: Amerindians The Amerindian woman and child in this photo live in the Sacred Valley of the Andes in Peru. Thomas Quine - Mother and Child - CC BY 2.0.

Amazon Basin

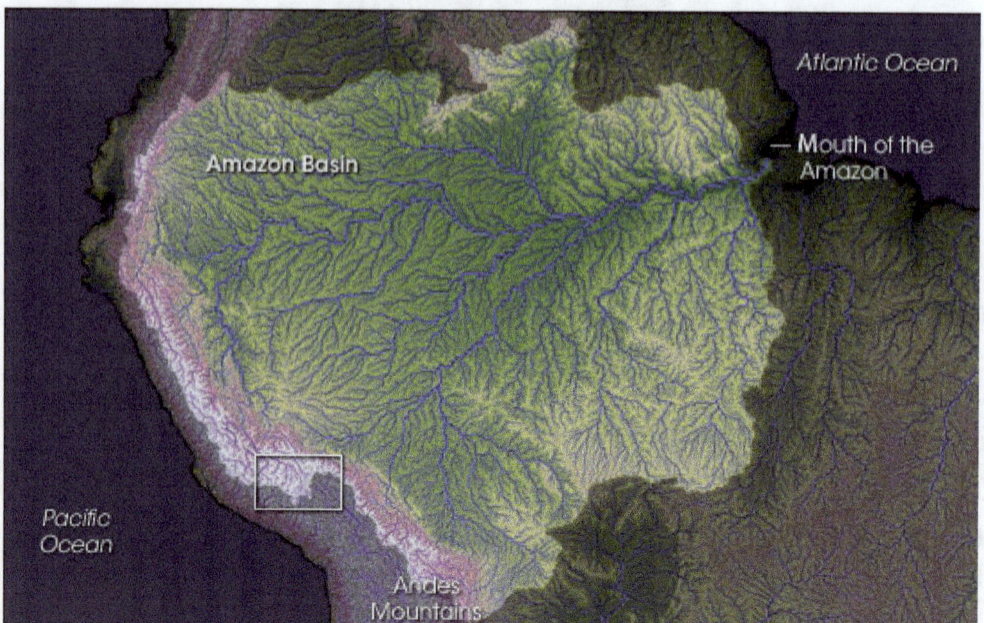

Figure 6.13: Amazon River Drainage Basin The Amazon has more than 1,100 tributaries. Image courtesy of NASA - public domain.

The Amazon Basin, which is characterized by a type A climate, is the least-densely populated region of South America and is home to isolated Amerindian groups. Development has encroached upon the region in the forms of deforestation, mining, and cattle ranching. Large deposits of iron ore, along with gold and other minerals, have been found in the Amazon Basin. Preservation of the tropical rain forest of this remote region has been hampered by the destructive pattern of development that has pushed into the region. The future of the basin is unclear because of development patterns that are expected to continue as Brazil seeks to exploit its interior peripheral region. Conflicts over land claims and the autonomy of Amerindian groups are on the rise.

Mixed Mestizo Region

The Mixed Mestizo Region includes the coastal area of the west and the interior highlands of the north and east. This region between the Tropical Plantation Region and the Rural Amerindian Region includes a majority of people who share a mixed European and Amerindian ethnicity. It is not as poor as the Rural Amerindian Region and yet not as wealthy as the European-dominated region to the south. Paraguay falls into the Mixed Mestizo Region, as do other portions of other South American countries such as parts of Brazil, Colombia, and Venezuela. Paraguay is mainly Mestizo, but its economic qualities resemble that of the Rural Amerindian Region to the north, even though Paraguay is not located in the mountains.

European Commercial Region (Southern Cone)

The southern part of South America, called the European Commercial Region or the Southern Cone, includes Chile, Argentina, Uruguay, and parts of Brazil. European ethnic groups dominate this region and include not only Spanish and Portuguese but also German, Austrian, Italian, and other European ethnic heritages. Fertile soils and European trade provided early economic growth, and the region attracted industry and manufacturing in the later decades of the twentieth century. There are not many Amerindians or people of African descent here. More than 90 percent of all the people in Argentina, Chile, and Uruguay are of European descent and live in urban areas. With a highly urbanized population and with trade connections to a globalized economy, it is no surprise that the Southern Cone is home to South America's most developed economies.

Globalization and Trade

South America has been fragmented by European colonialism, which established colonies and economic dependence on its European masters. The colonial economic patterns did not encourage the South American countries to work together to create an integrated continental trade network. Countries outside the continent have promoted trade partnerships to benefit from South America's natural resources and agricultural exports. The establishment of the European Union and the North American Free Trade Agreement (NAFTA) created globalized trading blocs that challenged the South American countries to consider how to take advantage of trading opportunities within their realm to protect and support their own economic interests.

Since the 1990s, cooperation and business ventures have started to form within the realm to create a more integrated network of trade and commerce to benefit the countries of South America. Transportation and communication systems are being developed through joint ventures by internal investment groups. River and road systems continue to be managed and developed for improved transport of people and goods throughout the continent. Free-trade agreements have been implemented to support the integration of internal economic networks and competition in the global marketplace.

In 2008, the South American countries formed the **Union of South American Nations (UNASUR)** to oversee the customs unions and trade agreements within the realm. One of the more established trade associations is **Mercosur (the Southern Cone Common Market)**, created in 1995 by the southern countries. It has evolved to include most countries in South America and is the most dominant trade agreement in the realm. Full members of Mercosur include Argentina, Uruguay, Paraguay, and Brazil. Chile, Bolivia, Peru, Ecuador, and Colombia are associate members. As of 2011, Venezuela's membership was under review. The **Andean Community** (Colombia, Ecuador, Peru, and Bolivia) was established in 1969 but did not gain ground until 1995, when it established stronger trade measures. Multinational corporations have supported the creation of a **Free Trade Area of the Americas (FTAA)** to include all of the Western Hemisphere in one unified trade association. It has not been approved and has received strong opposition from Mercosur and economic forces that support a more localized economy controlled by local people.

South America faced division and competition during the colonial era between the Spanish and the Portuguese. Today's new era of corporate colonialism has created similar fragmentation and divisions. The level of trade between the countries of South America and the United States and Europe varies widely. Countries such as Colombia and Chile have well-established trade relationships with the United States and are unwilling to jeopardize those trade connections to strengthen ties with their neighbors that have less-supportive political relationships with the United States. External global trade arrangements often provide financial benefits to individual countries that might not be shared by the bordering countries in the same region. South America's historical fragmentation has not made it easy to unify the continent under a singular trade agreement to complete against the European Union or NAFTA.

Key Takeaways

- The extensive Andes Mountain chain and the massive Amazon River dominate the realm's physical geography.
- The Spanish and the Portuguese were the two main colonial powers that dominated South America. The Guianas were the only part of the continent not dominated by these two European powers.
- Britain, Holland, and France formed colonies in the Guianas. The slave trade brought many people of African descent to the Guiana colonies. After slavery was abolished, indentured servants from Asia were brought to the Guianas to support the labor base.
- Identifying the majority ethnic groups in South America can be helpful in classifying the various cultural regions of the realm. Colonial activities and ethnic backgrounds are consistent enough to formulate regions with similar characteristics.
- Globalization and the creation of economic or political units such as the European Union and NAFTA have prompted the South American countries to work together to implement cooperative trade agreements and create the Union of South American Nations.

Discussion and Study Questions

1. What are the two main physical features of South America? Where are they located?
2. What two main European colonial powers dominated South America?
3. What are the majority ethnic groups in each of the Guianas?
4. How do most people in the Guianas make a living?
5. Where are the five main cultural regions of South America?
6. What are the majority ethnic groups in each of the cultural regions of the continent?
7. Why would the Southern Cone have a stronger economic position than other regions?
8. What environmental concerns are prominent in the Guianas?
9. What are the main economic trade agreements for the South American countries?
10. Why has it been so difficult to unify the countries in this realm into a single trade zone?

Geography Exercise

Identify the following key places on a map:

- Altiplano
- Amazon Basin
- Amazon River
- Andes Mountains
- Atacama Desert
- Brazilian Highlands
- Cape Horn
- Cerrado
- Devil's Island
- Easter Island
- European Commercial Region
- Falkland Islands
- Galapagos Islands
- Guiana Highlands
- Guianas
- Lake Maracaibo
- Lake Titicaca
- Llanos
- Madeira River
- Mato Grosso Plateau
- Mixed Mestizo Region
- Orinoco River
- Pampas
- Paraguay River
- Paraná River
- Patagonia
- Rio de la Plata
- Rio Negro
- Robinson Crusoe Island
- Rural Amerindian Region
- Sao Francisco River
- Southern Cone
- Tierra del Fuego
- Tordesillas Line
- Tropical Plantation Region
- Xingu River

6.2. URBAN NORTH AND ANDEAN WEST

The Urban North and Andean West represent two distinct cultural and geographical landscapes within the broader context of a country's diversity. The Urban North, characterized by its bustling cities, modern infrastructure, and vibrant cultural scene, reflects the rapid pace of urbanization and economic development. Here, technology and innovation thrive, shaping a lifestyle that is often fast-paced and interconnected. In contrast, the Andean West, with its majestic mountains and rich indigenous heritage, offers a stark juxtaposition. This region is steeped in tradition, where age-old practices coexist with contemporary challenges. The Andean communities often emphasize sustainable living and a deep connection to the land, showcasing resilience and cultural richness. Together, these regions highlight the dynamic interplay between modernity and tradition, urban life and rural values, creating a unique tapestry of human experience and expression.

Venezuela: Oil, Politics, and Globalization

Bordering the Caribbean is the large urban country of Venezuela. The Andes Mountains reach into the northern part of the country and make up the terrain of the northern coastal region all the way to the capital city of Caracas. The large grassland plains of the Llanos extend farther south from the Colombian border to the Orinoco River delta. The Llanos is a large, sparsely populated region that makes up about one-third of the country. It is remote, susceptible to flooding, and used mainly for raising cattle. In the southeast of Venezuela are the Guiana Highlands, which make for a spectacular physical landscape of tropical forests and rugged mountainous terrain. The highlands include **Angel Falls**, the tallest waterfall in the world. Angel Falls drops 2,647 feet and is Venezuela's most popular tourist attraction. Lake Maracaibo, a large inland sea located in the western region of the country, is not a true lake in that it is open to the Caribbean Sea, but it is considered the largest inland body of water in South America. Lake Titicaca, located in the Andean region of the Altiplano on the border between Peru and Bolivia, is considered the continent's largest freshwater lake.

Figure 6.14: Northern South America and Venezuela Notice that most of the main cities are located along the northern coast. CIA World Factbook – public domain.

Figure 6.15: Angel Falls, Venezuela The total falls is estimated at 3,212 feet; the largest free fall is estimated to be about 2,647 feet—about half a mile. Angel Falls is the country's number one tourist attraction. Inti – El Sato Angel – CC BY 2.0.

Venezuela has an assortment of physical regions, but most of the population lives along the northern coast. About 90 percent live in urban areas, and the capital Caracas has the highest population. Less than 5 percent of the population lives south of the Orinoco River, and Amerindian groups live in the interior and along the river.

Included in the Mixed Mestizo Cultural Region, Venezuela has a heavy Spanish influence laid over an Amerindian base in a plantation region known for its African infusion. There is also a strong Caribbean cultural flavor, which is evident in the region's music and lifestyle. The official language is Spanish, but more than thirty indigenous languages are still spoken in the country.

Venezuela gained its independence from Spain in 1821 and has developed into an urban country with an economy based on oil production. A large extent of the interior is undeveloped. Venezuela does not have extensively developed agricultural production, so most food goods are imported. Lake Maracaibo has vast oil reserves beneath it that have provided substantial wealth to the country.

As much as 90 percent of Venezuela's export earnings are from the export of oil. Venezuela's national oil company, **CITGO**, has made extensive inroads into the US gas station market. The country was one of the founding members of OPEC (Organization of Petroleum Exporting Countries), which is usually associated with the oil-rich states of the Persian Gulf. In the past decade, Venezuela has been one of the top five countries exporting oil to the United States. The other four are Canada, Mexico, Saudi Arabia, and Nigeria.

As is the case with many countries, national wealth in Venezuela does not filter down to most of the population. The wealthy elite who have benefited the most from the nation's wealth often find themselves on the opposite side of the political debate from the majority, who are likely to live in poor conditions. Caracas has many upscale neighborhoods, but it also has a large number of slums on the outskirts of the city. Slums in South America go by different names, such as barrios in Venezuela or favelas in Brazil. Many of Venezuela's barrios are built on the mountainsides of the Andes.

Figure 6.16: The Two Sides of Caracas, the Capital City of Venezuela The photo on the left is of the main urban core, with upscale neighborhoods. The photo on the right is a barrio located on a steep mountainside.

Barrios are usually self-constructed slum areas and are similar to favelas in Brazil. Cristóbal Alvarado Minic - Vista de Caracas - CC BY 2.0; Danila Medvedev - A tunnel and barrios surrounding it on the hill - CC BY 2.0.

Exporting oil to the United States does not inherently lead to a friendly political relationship between the United States and Venezuela. There has been some political unrest within the country related to the current president, **Hugo Chavez**, holding continuous terms in office without term limits. President Chavez has held close ties with socialist Cuba and with the Castro regime and has made somewhat antagonistic statements about the world's core economic countries. From time to time, his rhetoric and his positions are not geared toward enhancing the country's political relationship with the United States. This situation has caused concern within the US political establishment with regard to the position that the United States takes toward Venezuela. Nevertheless, the United States remains Venezuela's number one trading partner in both imports and exports.

Politics in Venezuela often plays the wealthy elites against the poorer majority, and President Chavez has not been an exception. Chavez has supported socialist political leaders from other Latin American states and has pushed a globalization agenda along the same lines. The country's enormous oil revenues and its current political climate have increased Venezuela's visibility in the global arena, but how this will play out over the long term is unclear. In the past, Latin America has had a greater number of elected political leaders with more progressive or socialist views. These trends continue to shape the economic trade agreements between countries. Venezuela has been working to increase sales of oil to countries in Mercosur (the Southern Cone Common Market), which is the most significant trade association in South America.

Globalization is also evident in Venezuela's cultural and social dynamics. In many Latin American countries, soccer (European football) is the most popular sport, but Venezuela's biggest sport is baseball because of the influence of early US activity in the country's oil industry. Soccer is gaining attention and support, however. Orchestras and classical music performances have also gained notoriety in recent years. Concert halls from the Americas to Europe have experienced the performances of the Simón Bolívar Youth Orchestra. The Miss Venezuela pageant is a major production for the country, and Venezuela has won the top title at least five times in each of the following pageants: Miss World, Miss Universe, and Miss International.

Colombia: Drugs, Coffee, and Oil

Three ranges of the Andes Mountains run from north to south through Colombia, which is larger than the nine most southeastern US states. With a land area covering about 440,839 square miles, Colombia is more than ten times larger than the US state of Kentucky and close to twice the size of France. Colombia borders five countries, with the Caribbean to the north, the Pacific Ocean to the west, the Orinoco River to the east, and a short segment of the Amazon River to the far south. Even though agriculture has been a mainstay of the country's economic activities, because of the influence of the mountainous terrain, about 75 percent of the population lives in urban areas.

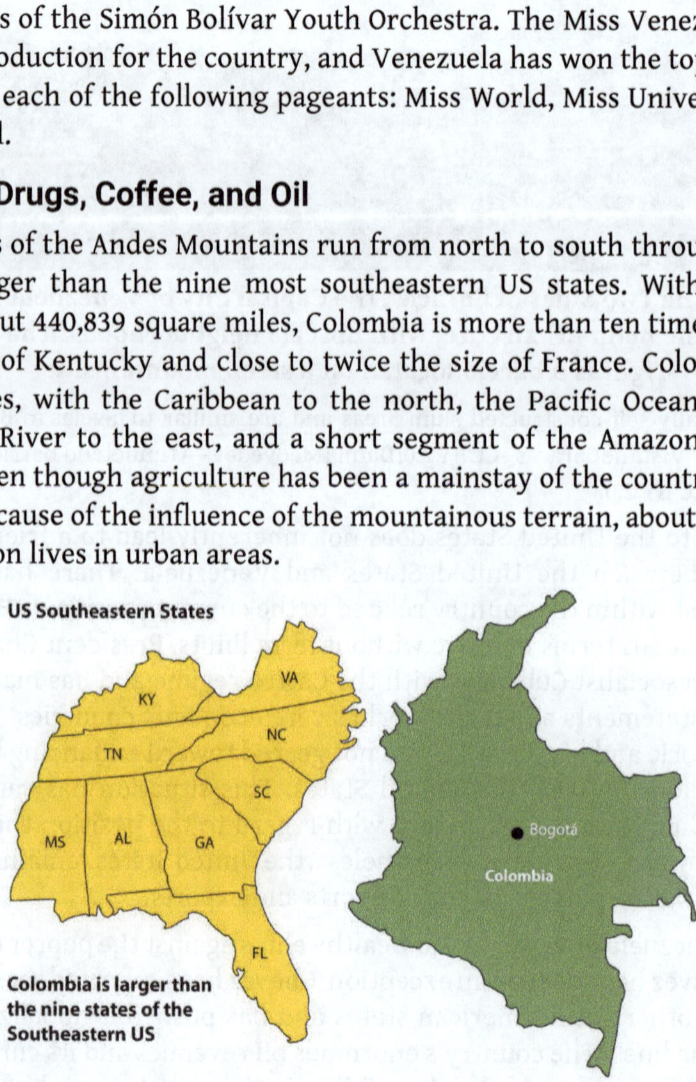

Figure 6.17: Physical Size of Colombia: 440,839 Square Miles

Colombia was a Spanish colony during the time that Spain controlled most of western South America. Colombia became independent in 1819. The region of Panama, which was first a part of Colombia, broke away in 1903 when the United States backed Panama's independence movement. After Colombia became independent of Spain, the conservatives (wealthy elite) and the liberals (poor workers) struggled to gain control of the government. Since 1948, the conflict, known as **La Violencia**, has caused more than two hundred thousand casualties. During the twentieth century the government in Colombia has not always been peaceful or stable. By the beginning of the twenty-first century the government has become more unified and the country has even witnessed an increase in tourism.

Colombia and the Drug Trade

Colombia's tropical climate and its many remote areas contributed to its development as a major coca-growing region. By the 1970s, extensive drug smuggling had developed, and powerful drug cartels became major political brokers within the country, competing against the government for control of Colombia. The largest and most organized cartels operated out of **Medellin** and Cali, the second- and third-largest cities in the country after the capital city of **Bogotá**.

The coca plant grows throughout the slopes of the Andes, from Colombia to Bolivia. Historically, locals have chewed it or brewed it into tea. Coca can alleviate elevation sickness and act as a mild stimulant. Using modern methods and strong chemicals, the coca leaves can be converted into coca paste and then into cocaine hydrochloride, a powerful narcotic. It often takes up to a ton of such chemicals as sulfuric acid, kerosene, methyl alcohol, and additional substitutes to produce a kilo of cocaine. Once the process is completed, most of the chemicals are discarded and frequently find their way into nearby rivers and streams, which are the same water supplies that local people drink, clean with, and bathe in. Birth defects have become a problem in coca-growing regions because of the high levels of chemical pollution in water supplies.

It must be noted that the short, leafy coca plant that cocaine comes from is not the same as the cacao tree that produces the beans that chocolate or cocoa comes from. They are two completely different plants with separate processes.

The United States is the largest cocaine market. Secret airfields and private boats transport the cocaine from Colombia to distribution centers in Mexico, Central America, or the Caribbean. From there, the drugs are smuggled into the United States. Colombian drugs are a multibillion-dollar industry that makes up a large portion of the Colombian economy. The effect of the drug industry on the people of Colombia is extensive—from the gunfire on the streets to the corruption of government officials. In recent years, the same drug cartels that have operated the cocaine industry have imported opium poppies, which grow well on the higher and more arid slopes of the Andes. Opium poppies are native to Asia but have been transported to South America. Opium is extracted from the seedpod and can be further refined into heroin. Colombian drug cartels, with a Mexican distribution network, have muscled into as much as 20 percent of the US heroin market. The

US government has supported the Colombian government in the fight against the drug cartels and the trafficking of illegal drugs out of Colombia.

Colombian Coffee and Oil

Colombia's two main legal exports to the United States are coffee and oil. Coffee is only grown in the tropics, since coffee trees must be grown in a frost-free environment. Coffee trees, which originally grew in Ethiopia, have since been grown throughout the world. Coffee trees can grow in elevations from sea level to six thousand feet, but most of the best specialty coffee is grown at elevations between three thousand to six thousand feet. Colombia has ideal conditions for growing coffee and was once the world's largest coffee producer; now Brazil and Vietnam each produce more.

Early coffee production in Colombia was initially promoted by Catholic priests who were influential in supporting local parishioners to grow the crop. The industry was greatly enhanced in 1927 with the formation of the National Federation of Coffee Growers of Colombia. Coffee production on the mountain ranges of the Andes in Colombia supports up to a half million small farms and local growers that make up a large portion of the coffee economy. Harvesting coffee is labor intensive and can employ large numbers of workers. The seasonal nature of the harvest also leaves workers to find employment during the rest of the growing season.

In recent years, there has been growing concern about how climate change will continue to affect the region's coffee production. In the past century, parts of Colombia experienced up to a 1 ºC average temperature increase and up to a 25 percent increase in precipitation. These climate changes have negatively affected coffee output and reduced coffee production as much as 25 percent in some areas. New agricultural methods are being developed in an attempt to counter these effects.

Oil has now become Colombia's number one legal export. Oil is found in fields in the northern and central regions of Colombia. Immense quantities of coal are also found in the same regions, but oil is more valued on the export market. Pipelines connect the interior oil fields of Colombia with the northern ports. The US market size and population make it the world's largest oil consumer. US oil companies have been investing in the development of Colombian oil for many years. Colombia has been a developing oil source even though its total extractable resources are not as vast as in other countries. For example, in 2006 the United States imported more oil from Colombia than from Kuwait, Oman, the United Arab Emirates (UAE), Bahrain, Qatar, and Yemen combined.

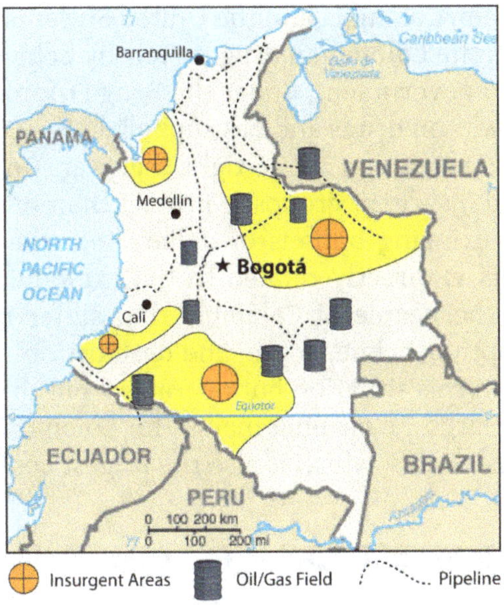

Figure 6.18: Insurgent Areas Insurgent groups control regions in Colombia as large as US states. CIA World Factbook – public domain.

Since the United States is the largest consumer of Colombian oil, it is easy to understand why the United States has a vested interest in the stability of the Colombian government. A sizable portion of Colombia is controlled not by the government but by drug cartels or other insurgent groups. Dozens of guerilla organizations also control portions of Colombia. Some insurgent groups support the government and are against the drug lords, while others fight the government and work independently or with the cartels. Drug sales, kidnappings, and extortion of legitimate businesses provide income to these groups. Thousands of children serve in these groups, and about a third of them are female. The most powerful insurgent group is **FARC (Revolutionary Armed Forces of Colombia)**, which controls entire regions the size of many US states. FARC is a recognized political entity by neighboring countries but is not given the same recognition by the United States and many external countries of the region.

Figure 6.19: Colombian Exports The three main export products of Colombia are illegal drugs, oil, and coffee. The United States is the largest consumer of all three. DEA – public domain; R. Berglee – CC BY-NC-SA; Wikimedia Commons – public domain.

The relationship between Colombia and the United States is often conflicting. The US consumer supports the Colombian drug cartels by being the largest consumer of illegal drugs. The US government, under the Drug Enforcement Administration (DEA), has declared a war on drugs and has supported the Colombian government with billions of dollars in foreign aid to fight that war. On another front, US oil corporations have paid insurgent groups to protect their oil assets. Oil is exported to the United States, bringing billions of dollars into the Colombian economy. The chaos in Colombia is directly related to the exploitation and marketing of their resources. It is the people of Colombia that suffer in the cross fire from this civil war of corruption, crime, death, and destruction. The United States is a counterforce partner in this situation but operates from the consumer end of the resource pipeline. The largest consumer market for Colombia's export of oil, drugs, and coffee is the United States, which is also the largest contributor of foreign aid to Colombia.

Rural Amerindian States of Ecuador, Peru, and Bolivia

The Central Andes, which includes Ecuador, Peru, and Bolivia, were home to the Inca Empire. The empire had gone through some internal divisions and was working on unifying the region when Francisco Pizarro's small army defeated the Incan warriors and brought about colonial rule beginning in the 1530s. Many cultures lived in the Central Andes before the Inca, and their legacy continues in the customs and the ways of the Amerindian people who still live there today. Spanish is the official language, or the lingua franca, but indigenous languages are widely spoken and dominate in the rural areas and remote villages. Ecuador, Peru, and Bolivia make up the core of the Rural Amerindian Region of South America. There have been border disputes among the three countries, and also with their neighbors. Nevertheless, they all share the Andes and have many things in common.

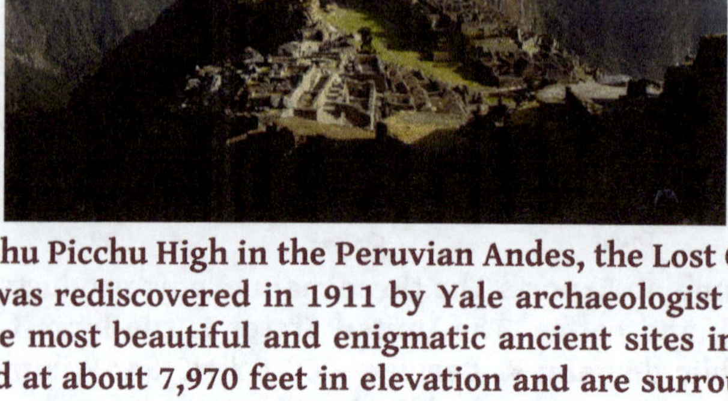

Figure 6.20: Machu Picchu High in the Peruvian Andes, the Lost City of the Incas, Machu Picchu, was rediscovered in 1911 by Yale archaeologist Hiram Bingham and is one of the most beautiful and enigmatic ancient sites in the world. The ruins are located at about 7,970 feet in elevation and are surrounded by higher peaks of the Andes. Funkz – Machu Picchu 12 – CC BY 2.0.

Physical Geography

The physical geography of the Central Andes includes more than just the high Andes Mountains, although they dominate the landscape. The coastal region to the west of the Andes is generally warmer than the cooler climate of the mountains. The equatorial region is rather humid. The coastal region in southern Peru is dry and arid because of the ocean currents and the rain shadow effect of the Andes, which creates the Atacama Desert that extends up from northern Chile. Southwest Bolivia has some of the world's largest salt flats in this dry and barren region. In the interior, on the eastern side of the mountain ranges, is the huge expanse of the Amazon Basin. Tropical and humid with heavy precipitation is generally the climate rule. Rain forests and jungle fauna can be found on the eastern slopes. The Altiplano region has the high-elevation Lake Titicaca. The variations in physical terrain provide extensive biodiversity in animal and plant species. It also supports a variety of economic activities to exploit the bountiful natural resources.

Even though the Altiplano region borders the Pacific Ocean, it also links directly to the Atlantic Ocean. The headwaters that create the Amazon River start in Peru, and by the time the water reaches the Peruvian city of Iquitos, the river is large enough to accommodate large shipping vessels. Iquitos is a port city for the Atlantic Ocean with access to Europe, Africa, or eastern North America. The port also links the region with Brazil's free-trade zone in Manaus, which has access to large oceangoing shipping and an international market.

Economic Geography

The region's main income comes from exports of minerals, fossil fuels, and agricultural products. Oil is the number one means of gaining national wealth in Ecuador and Peru; natural gas is the number one export of Bolivia. Gold, silver, tin, and other minerals are also abundant and are being exploited as conditions allow. The Spanish opened up a large silver-mining operation in **Potosí**, Bolivia, which continues to be exploited with modern mining methods. Potosí, one of the earth's highest-elevation cities at 13,420 feet above sea level, was once the largest silver mine in the world.

The city of Lima, Peru, was built on wealth from gold and silver extracted from the Inca Empire and the Andes Mountains. As is the case in many peripheral regions, the economic state of Peru, Ecuador, and Bolivia is dependent on global prices for their export products. There has been considerable conflict and political wrangling about who controls the wealth from extractive economic activities. Originally, the Spanish conquistadors took the materials and the wealth. They were replaced by Mestizo land owners and wealthy elites who struck deals with international corporations. The corporations exploited the countries' natural resources, with little profit actually ending up in the hands of most of the people. These issues remain at the top of the political agendas in all three countries.

Figure 6.21: Quito, Ecuador Quito, Ecuador, is an urban center high in the Andes, with a population of more than 1.5 million. This photo shows the large buildings of the central business district with the mountains in the background surrounding the city. Quito is at about 9,200 feet in elevation and is considered the second-highest-elevation capital in the world after La Paz in Bolivia. David Berkowitz – Panoramic View – Quito, Ecuador – South America – CC BY 2.0.

Poverty and the exploitation of natural resources usually result in environmental degradation unless proper measures are taken to prevent it. The area's heavy reliance on oil and gas extraction to gain national wealth has come at a great cost to their environment. Many oil spills have caused oil to enter the freshwater supplies of local residents and pollute the rivers and streams of the Amazon Basin. Mining has traditionally devastated the land because large portions of earth are removed to extract the ore or mineral. Pollution is causing a loss of habitats and destroying ecosystems, and few measures are being taken to prevent it.

Deforestation is being caused by the timber industry and by clearing for agriculture. Overgrazing and the removal of the trees leave the soil open to erosion.

Tourism

Tourism is expanding to connect travelers with opportunities to explore Incan and pre-Incan sites, which are the main attractions. One of the main tourist attractions in Peru is the ancient city of Machu Picchu in the Andes not far from Cuzco. In 2010, Peru gained over two billion dollars from the tourist activities of about two million foreign tourists (Andina). Ecuador's major tourist attraction is the Galapagos Islands, which aided Charles Darwin in understanding natural selection and the evolutionary process. Bolivia has a number of ancient sites that predate the Inca and have become major tourism destinations. The ancient city of Tiahuanaco and the enigmatic Lake Titicaca are good examples.

Tourism can be a great source of economic income but it can come at a cost to the environment. There is always concern that high-traffic tourism sites like Machu Picchu can be degraded by the sheer mass of people visiting the site. The environmental imprint may be extensive. The term ecotourism has been used to indicate the activity of people traveling to experience and enjoy the natural world with an aim not to damage the environment in the process. The main objective was to make the tourism activity sustainable, which promoted stewardship of the land and respect for its attractions. Jonathan Tourtellot, director of the National Geographic Society's Center for Sustainable Destinations, coined and prefers the term geotourism, which can be translated as the stewardship of place and the preservation of its essential character (Miller-McCune). These concepts are becoming more integrated with the tourism industry to promote a sustainable model for high traffic sites like the Galapagos Islands with fragile ecosystems.

Political Issues

Economics usually drive politics. Accordingly, Peru, Ecuador, and Bolivia have endured some serious ups and downs in their political environment. Corruption, authoritarianism, and human rights violations have been common accusations toward the political leadership of the countries. To address the economic condition of his country in 2000, Ecuador's president announced that the country would adopt the US dollar as its main medium of exchange.

Figure 6.22: Indigenous Women on Their Way Home from the MAS congress in Bolivia, January 2009 MAS is the Movement for Socialism, which has been active in Bolivian politics. Norsk Folkehjelp Norwegian – Urfolk i Bolivia – CC BY 2.0.

This may have brought some economic stability, but it did not address the problems of a high national debt and a fluctuation in commodity prices. All three countries have undergone political turmoil. Large percentages of the populations live in poverty. Bolivia is considered one of the poorest countries in South America. In 2006, Bolivia elected a socialist president from the MAS (Movement for Socialism) party who was from a minority Amerindian group rather than a member of the wealthy elite. In Peru, a number of presidents have been forced to resign, and military coups have also produced leadership changes.

Population and Culture

Population growth is a major factor in the future of Peru, Ecuador, and Bolivia. In 2010, Bolivia had more than ten million people, Ecuador had more than fourteen million, and Peru had about twenty-nine million. More than 30 percent of the population of Ecuador and Bolivia resides in rural areas and make a living from subsistence agriculture. All three countries have large populations in relation to the production of adequate food. Peru and Bolivia are large countries in physical area but do not have a high percentage of arable land.

Rural-to-urban shift is increasing and the major cities are continuing to expand, overtaxing public works and social services.

The culture of the Central Andes is heavily influenced by its rural Amerindian heritage. The foundation of the traditional agrarian society has been subsistence agriculture. One-third of the population in Ecuador and Bolivia and up to one-fourth of the population in Peru continue to live a traditional way of life. Local cuisine reflects the connection to the land. Potatoes, maize, guinea pigs, and fish are common fare in rural areas. The cities are encountering international influences that are changing the demands in local cuisine and culture. Traditional food, arts, and local crafts still thrive in the local districts and for the tourism market.

Figure 6.23: The Andean West and Paraguay This map also displays a part of the Amazon Basin, the Atacama Desert, Altiplano, and the Mato Grosso Plateau. University of Texas Libraries.

This region's location on the Pacific Rim of South America has contributed to an Asian influence, which has integrated itself with the local culture. Former president of Peru Alberto Fujimori had Japanese ancestry and held dual citizenship in both Peru and Japan. Similar to Havana, Cuba, in Middle America, Lima, Peru, is also home to one of the earliest Chinatowns in the Americas, where the Chinese culture has mixed with the Latino culture to create a unique cultural blend. Bolivia is landlocked and does not have the advantage of a west-coast port city to interact with the Asian marketplace. However, its culture is still impacted by globalization and is evolving from within.

Paraguay

Paraguay is located in the Mixed Mestizo Cultural Region between Brazil, Argentina, and Bolivia. This landlocked country is not located in the Andes. The country's poor economic characteristics and troublesome political dynamics are similar to those of its neighbors in the Central Andes. Like other Latin American countries, most of the population practices Catholicism. Spanish is one of the main languages along with **Guarani**, the local language of the people. The country's name comes from the Paraguay River, which flows through the region and provides hydroelectric dams that provide all the electricity for the country. Paraguay suffers from a lack of infrastructure. The government has not been able to provide for the needs of a growing population with a fertility rate of more than three children per family.

As much as 40 percent of the population makes its living from agriculture. However, conditions in the rural areas are poor: less than 10 percent of the land is arable. There is not much agricultural growth that could boost the economy. Paraguay is the sixth-largest producer of soybeans in the world, and cattle ranching is its other strong commodity. A large portion of the marshlands have been transformed for agricultural purposes, but this has caused a loss of wetlands as a habitat.

The economy is afflicted by poverty and an absence of opportunities and advantages. Connections to global markets are hindered by the lack of an international port. Corruption and unstable governments are the political system's legacy. There is a wide disparity between the social elites, who own a high percentage of the land, and most of the population, which remains in poverty with poor living conditions. Large family size and rural-to-urban shift has caused rapid growth of the urban areas. Sixty percent of the population lives in the cities, and they often suffer from insanitary conditions because of water pollution. The largest slums in the country are found in the expanse of the capital city of Asuncion, which has almost two million people. In the urban areas, there is a strong informal market economy that thrives on imported goods being redistributed to other countries, but there is no formal record-keeping system. An enormous number of black-market goods are brought in and resold to neighboring countries such as Brazil and Argentina, its two largest trading partners. In such conditions, crime and illegal activities thrive, and the rule of law is difficult to enforce.

Key Takeaways

- Most of the people in Venezuela live in cities along the Caribbean coast. The country has always been urban and has not developed its agricultural production. Venezuela depends on oil exports to gain most of its national wealth.
- The United States is the largest consumer of products exported from Colombia. Illegal drugs, oil, and coffee and are the three main export products of Colombia. The United States directly or indirectly supports the three main factions vying for power within Colombia.
- The Andean countries of Ecuador, Peru, and Bolivia share similar developmental dynamics, including a large percentage of Amerindian people and reliance on the export of natural resources to gain national wealth. The wealth gained from exports is concentrated in the hands of elites rather than filtering down to most of the people.
- Paraguay is a landlocked country between the Andes and the Atlantic coast. It is a poor nation with few opportunities to advance its standard of living. The country has been a main staging ground for unregulated goods to be redistributed throughout the region.

Discussion and Study Questions

1. How intense is the current rural-to-urban shift in Venezuela? What accounts for this unique situation?
2. What are the three main export products of Colombia? How is the United States involved in each of them?
3. Why is coffee not grown in the continental United States? Where are the main types of coffee grown?
4. What three groups vie for control of the various regions of Colombia? How does the United States directly or indirectly support all three groups?
5. What environmental problems result from the cocaine production process?
6. How did colonial activity assist in determining which ethnic group or groups would be at the top or bottom of the socioeconomic hierarchy in the Andean West region of South America?
7. What are some of the common difficulties with the political systems of the countries in the Andean West?

8. What does Paraguay have in common with Bolivia? How have the two countries each addressed this common situation?
9. Where has socialism been promoted or supported in the countries of South America?
10. What types of agriculture are found in the north and western regions of South America?

Geography Exercise
Identify the following key places on a map:

- Angel Falls
- Machu Picchu
- Pacific Rim

References

Andina, "Peru's Inbound Tourism Revenue to Reach $2.2 Billion This Year," www.andina.com.pe/ingles/Noticia.aspx?id=nfbX5vudSxw=.

Miller-McCune, "Can Tourism Be Sustainable?," http://www.miller-mccune.com/environment/can-tourism-be-sustainable-16362/.

6.3. BRAZIL

A Portuguese Colony

Brazil, the largest country in South America, is similar in physical area to the continental United States (i.e., the United States without Alaska or Hawaii). Catholicism is the dominant religion and Portuguese is the primary language. Once a Portuguese colony, the country's culture was built on European immigration and African slave labor, making for a rich mixture of ethnic backgrounds.

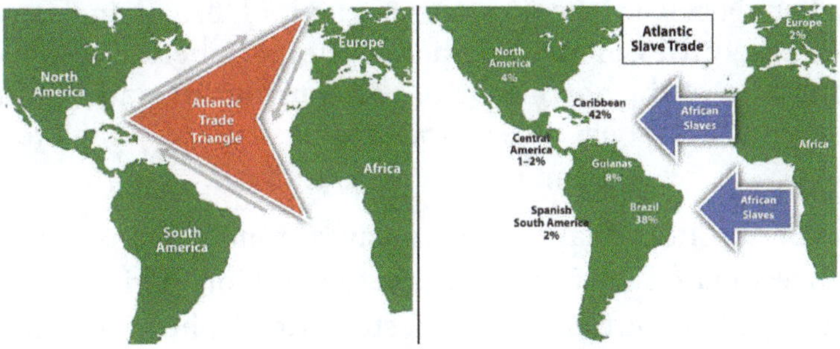

Figure 6.24 Atlantic Trade Triangle

In colonial times, Brazil was a part of the Atlantic Trade Triangle, which functioned as a transportation conveyor, moving goods and people around the regions bordering the Atlantic Ocean. Colonial merchant ships financed by Europe's wealthy elite brought goods and trinkets to the African coast to trade for slaves, who were shipped to the Americas and the Caribbean to diminish the labor shortage for the colonies. The last leg of the Atlantic Trade Triangle moved food crops, sugar, tobacco, and rum from the colonies back to the European ports. The merchant ships never sailed with an empty hold, and their successful voyages provided enormous profits to the European financiers.

Manufactured trinkets were sent to Africa from Europe, slaves were sent to the Americas, and plantation products and rum were sent to Europe. The Atlantic slave trade was responsible for bringing more than ten million African slaves to the Americas. Brazil received the largest number of slaves.

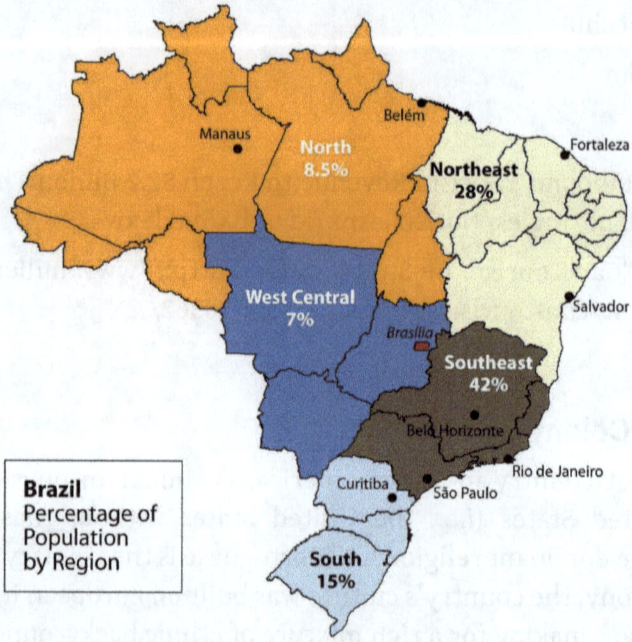

Figure 6.25: Population of the Various Regions of Brazil Indicating the Core Region around the Cities of the Southeast The peripheral region includes the large Amazon Basin of the north. Notice the many small states in the Northeast from Fortaleza to Salvador; these states are densely populated. Wikimedia Commons – CC BY-SA 3.0.

The total number of individuals taken as slaves from Africa is unclear and often debated. It is estimated that more than ten million African slaves survived the Middle Passage from Africa to the Western Hemisphere, which is more than the current population of Bolivia. Slavery supplied cheap labor for the plantations and agricultural operations in the New World. Brazil took in more African slaves than any other single country—at least three million. Colonial Brazil thrived on early plantation agriculture. When slavery was abolished in Brazil in 1888, the freed slaves found themselves on the lower end of the socioeconomic hierarchy. People of mixed African descent now make up more than one-third of Brazil's population. The Afro-Brazilian heritage remains strong and dominates the country's east coast. The African influence is evident in everything from the samba schools of the Brazilian carnival to the music and traditions of the people. In spite of Brazil being a culturally diverse country, Africans still have not found themselves on an even playing field in terms of economic or political opportunities in positions of power in the country.

Southeast Core: Urbanize and Industrialize

Brazil's human development patterns are an example of the core-periphery spatial relationship. The main economic core area is located in the **southeast** region of Brazil, an area that is home to the largest cities of the realm and acts as the hub for industrial and economic activities. Political and economic power is held by elites residing in the urban core areas. The rural northern Amazon Basin is the heart of the periphery, providing raw materials and resources needed in the core. The periphery has a small population density, and most are Amerindian groups that make a living from subsistence agriculture, mining, and forestry. Rural-to-urban shift has prompted many of the rural poor to migrate to the large cities.

Founded in 1554 as a Catholic mission, the city of **São Paulo** rests at the heart of the core region. Its pattern of development is similar to that of Mexico City. Coffee production was the early basis of the local economy. São Paulo is located about thirty miles inland from the coast. It has grown to be the center of trade and industry for Brazil, but **Rio de Janeiro** receives greater attention for tourism. With more than twenty million people, São Paulo is the largest industrial metropolis south of the Equator, the largest city in South America, and among the largest cities on the planet.

Rio de Janeiro, Brazil's best-known city, is a travel and international business center with a population of more than ten million. The city is renowned for its carnival festivities and famous coastline. Tourists are attracted by its cultural attractions and coastal setting, with beautiful sandy beaches and the landmark Sugarloaf Mountain located in an open bay. **Salvador**, located on the coast, was Brazil's first capital. Rio de Janeiro became the capital in 1763, but to further develop Brazil's interior, in 1960 the capital was moved from Rio to the *forward capital* of Brasilia. Forward capitals are created to either shift development or to safeguard a geographical region. Brazil has an enormous interior region that it wishes to continue to develop for economic gain and the creation of the forward capital of Brasilia is in line with this objective.

Figure 6.26: Christ the Redeemer This giant statue overlooks the immense urban landscape of Rio de Janeiro, with Sugarloaf Mountain in the background. Kirilos – Rio de Janeiro – CC BY 2.0.

The three cities of São Paulo, Rio de Janeiro, and Brasilia, along with their urban neighbors, anchor the core region in the south. Brazil's internal migration to these cities follows a rural-to-urban or periphery-to-core pattern, so it is no mystery why these three cities continue to grow at an unprecedented rate. São Paulo is more than sixty miles across. As migrant workers from the countryside and from the rural northeast migrate to the cities looking for work, they expand the city through self-construction. Slums, called **favelas**, extend out from the central city for miles.

The cultural fabric of Brazil has been built by immigrants from many countries. Brazil's diverse population has linked the country to the homelands of its immigrants and established trade and cultural connections that benefit the country in the global marketplace. The more than two hundred million people that make up Brazil's population are diverse and yet integrated, even if many still hold to the traditions or cultural heritage of their country of origin. After World War II, a large number of Japanese migrated to São Paulo. Today, Brazil boasts the largest Japanese population outside Japan. Many people from both Western and Eastern Europe have made Brazil their home, and large numbers of people from Lebanon, Syria, and the Middle East have immigrated here.

In spite its cultural diversity, Brazil has two overarching cultural forces that have helped hold the country together: the Portuguese language and Catholicism. These two centripetal forces help establish a sense of nationalism and identity. The Portuguese language has been adapted to the Brazilian society to reflect a slightly different dialect than the Portuguese spoken in Portugal. The Portuguese language has more of a unifying effect than religion. Though about 70 percent of the population claims to be Roman Catholic, additional religious affiliations in Brazil range from the African influence of Umbanda to the Muslim minority. Protestant denominations are the second-largest religious affiliations in Brazil. Secularism is on the rise, and many do not actively practice their specified religion.

In both population and physical area, Brazil ranks fifth in the world. Brazil is as urban as the United States or countries in the European Union. The index of economic development can be applied to Brazil: as a region urbanizes and industrializes, the family size usually decreases and incomes usually increase. Family size has dropped significantly, from 4.4 children in 1980 to 2.4 children in 2000, and continues to decline. Religious traditions give way to the urban culture and secular attitudes of modernity when it comes to family size.

Brazil is an urban country. Approximately 87 percent of Brazil's population lives in urban areas. Urbanization seldom eliminates poverty. Though the index of economic development indicates that urbanization will increase incomes for the population as a whole, poverty is a standard component of any large urban area in most places of the world, and the favelas of Brazil are similar to slums elsewhere.

Figure 6.27: Favela La Rocinha, Rio de Janeiro, Brazil Favelas are usually self-constructed and start out as slums. Many lack ownership rights, police protection, or public services. With time, some become established neighborhoods. Pedro Lopez – Favela, Rio de Janeiro – CC BY 2.0.

The favelas of Brazil's urban centers are dynamic places, where land ownership, law and order, and public services are questionable. Millions of people live in the favelas of Brazil's large cities. Poverty and the search for opportunities and advantages are common elements of urban life. The core cities of Brazil suffer from the same problems as other developing megacities: overcrowding, pollution, congestion, traffic jams, crime, and increased social tensions. The energy generated by the sheer magnitude of people, industry, and commerce continues to fuel Brazil's vibrant growth and its many large, dynamic cities.

The pattern of wealth developing in Brazil is quite common: wealth and the ability to obtain wealth are held in the hands of those at the higher end of the socioeconomic ladder. The pyramid-shaped graphic used to illustrate Mexico's social layers in Chapter 5 "Middle America", Section 5.2 "Mexico" can also be applied to Brazil (see Figure 6.28 "Socioeconomic Classes in Brazil, Also Common throughout Middle and South America"). The minority wealthy elite own most of the land and businesses and control more than half the wealth.

Corporate colonialism is quite active in Brazil. Multinational corporations take advantage of the country's development activities and muscle in on the profits, which seldom reach the hands of the majority at the lower end of society.

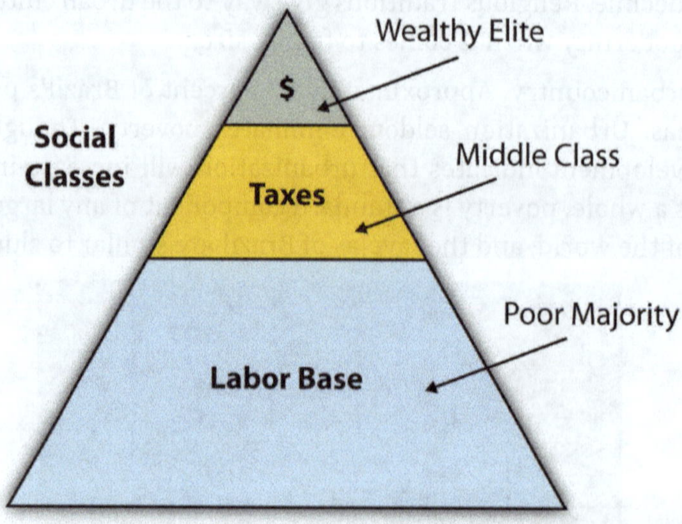

Figure 6.28: Socioeconomic Classes in Brazil, Also Common throughout Middle and South America

Brazil is an emerging power in the world marketplace. Brazil has the largest economy in South America and is a parallel force with the United States in the Northern Hemisphere. The country has urbanized and industrialized to compete with the global economic core areas in many ways. Brazil is among the ten largest economies in the world.

Brazil has favorable resources and labor to complete in the global marketplace. Its agricultural output has grown immensely over the past few decades. Brazil is a major exporter of soybeans, coffee, orange juice, beef, and other agricultural products. Brazil is the largest coffee producer in South America, but coffee only constitutes about 5 percent of its current annual exports. Coffee production is extensive in the Brazilian Highlands just inland from the coast. In addition, Brazil exports more orange juice than any other country and is second in the world in soybean production. The vast central interior regions such as the cerrado continue to be developed for industrial farming of massive food crops. However, exports of industrial manufactured products have surpassed agricultural exports in 2010. The availability of abundant minerals and iron ore has supported an expanding steel industry and automotive manufacturing, and its industrial activity continues to develop. Competitive high-tech companies continue to emerge, and production has increased in semiconductors, computers, petrochemicals, aircraft, and a host of other consumer-based products that provide economic growth.

Brazil economically dominates and anchors South America. The economic integration being developed throughout the South American continent has great potential because it combines the energy region of the north, the commercial centers in the south, and the agricultural sectors as complementary forces within Brazil's industrial base. Brazil is at both the physical center and the economic center of the continent.

Brazil is not only emerging on the world's stage on the economic front, but it also has a strong social and cultural presence in the world. The large metropolitan areas of the country are a Mecca for fashion and the arts. Carnival is the most well known of Brazil's cultural festivals, but the country is host to a number of other international events. Brazil is active in world sports competition and hosted the 2007 Pan American Games. Rio de Janeiro has been scheduled to host the 2016 Summer Olympics and the 2014 FIFA World Cup. The Fédération Internationale de Football Association (International Federation of Association Football) is commonly known by the acronym FIFA and is the international governing body of soccer (football). The 2010 FIFA World Cup was held in South Africa.

Northern Periphery: The Amazon Basin

Just as the megacities of southern Brazil anchor the core of the country, it is the rural expanse of the Amazon North that makes up the periphery. A region the size of the US Midwest, the Amazon River basin is a frontier development area that has been exploited for its natural resources. Rubber barons of early years traveled up the Amazon River and established the port city of **Manaus** to organize rubber plantations for automobile tires. The Amazon River is large enough for oceangoing vessels to travel to Manaus. Today, Manaus has a free-trade zone with an entire industrial complex for the production of electronic goods and an ultramodern airport facility. Smaller ships can continue up the Amazon River all the way to Iquitos in Peru, which makes Manaus an ideal core city for economic trade; smuggling; and transshipment of illegal goods, including exotic animals from the region, such as monkeys, beautifully colored parrots, and other birds.

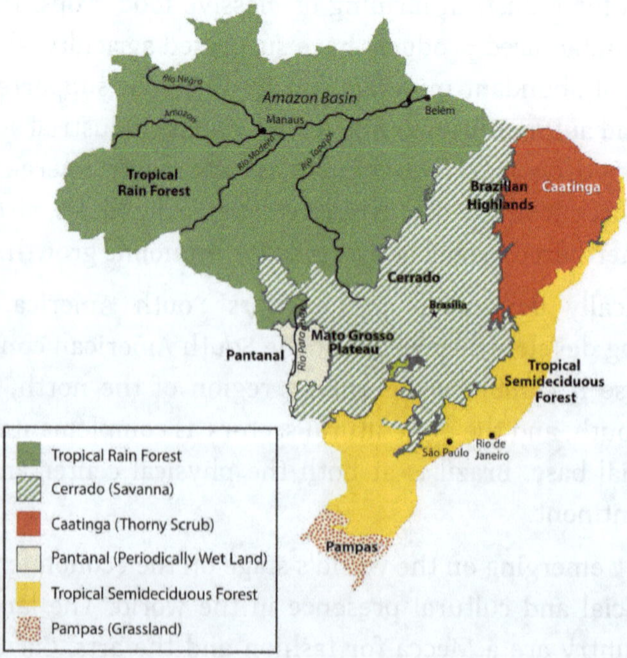

Figure 6.29: Natural Vegetation of Brazil The immense area of the tropical rain forest can be seen on this map. The extensive cerrado region is noted on this map between the tropical rain forests of the Amazon Basin and the deciduous forests of the southeast. The soils of the thorny scrub region along the northeast are not as conducive to large-scale agriculture as is the cerrado. Source: Based on map courtesy of University of Texas Libraries.

The other regions of the Amazon Basin have not been as fortunate as Manaus. Deforestation from cattle ranching, logging, and mining have devastated parts of the tropical rain forests of the Amazon Basin. The Amerindian populations have also suffered from encroachment into their lands. Only about two hundred thousand Amerindians are estimated to remain in Brazil, and most reside in the Amazon interior. This region boasts one of the world's leading reserves of iron ore; as much as one-third of Europe's iron ore demands are met through extensive mining southwest of **Belém**. In this area, a development scheme known as the **Grande Carajás Project** was implemented to create the infrastructure to support the mining complex and aid in the exploitation of the iron ore and other minerals. A large dam was built to provide hydroelectric power, and a railroad line was laid down to transport the ore to the Atlantic coast. The availability of transportation and energy attract other economic activities. Deforestation for cattle ranching and farming has increased. Many poor settlers have arrived seeking employment opportunities or a chance to obtain land. The rapidly expanding development activities in the Amazon basin have boosted the region's economic situation, but at the same time there is growing concern about the preservation of the natural environment.

Gold mining, an activity that has been widespread in the Amazon region for many decades, also presents environmental problems. Toxic substances such as cyanide and mercury used in the collection of gold from rivers and streams have entered the waters and tributaries of the Amazon River. The pollutants then enter the food chain and harm the ecosystem. The mining of ores and minerals has enticed people to immigrate to the area, creating frontier boomtowns with few public utilities or social services and poorly organized law and order. The exploitation of natural resources exacts a cost from the environment and the local indigenous Amerindian people. Amerindian groups have used the land as hunting grounds for centuries, but the rapid influx of gold miners and land speculators has caused conflicts with local groups who claim the land. Violence in the Amazon and battles for control over land and resources have been well documented.

Figure 6.30. Deforestation in the Amazon Deforestation in the Amazon is caused mainly by logging, agriculture, and mining. Agricultural practices can include slash-and-burn farming and cattle ranching. Matt Zimmerman – Slash and burn agriculture in the Amazon – CC BY 2.0.

Deforestation has reduced the habitat critical to the survival of native species. An estimated 50 percent of the earth's species live in tropical rain forests, which only cover about 5 percent of the earth's surface. Tropical rain forests in the Amazon Basin are being cut down at an unsustainable rate, creating serious environmental problems. Loggers cut down the large trees, and the rest are usually burned to allow the ash to provide nutrients for other plants. The cleared areas are most often used by cattle ranchers until the soil is no longer viable. Then more forest is cut down and the process continues. Speculation that land prices will increase as the region becomes more developed encourages this process.

The forest has many layers of habitat. Soils in the tropics are extremely low in nutrients, which have been leached out by the abundant rainfall. The nutrients are on the surface layer of the ground built up from falling leaves, branches, and debris decomposing on the forest floor. The removal of the forests removes these nutrients and results in serious soil erosion. Tropical forests usually expand along the edges where falling debris from the trees collects and provides nutrients for young plants.

There is much discussion among environmentalists, scientists, and other concerned people about deforestation in the Amazon region, usually with a focus on the devastation that results from the loss of forested areas. Forest studies have indicated that tropical rain forests are actually quite resilient and can recover with proper forest management. However, clear-cutting large, wide areas for timber leaves an area devastated for an extended period. Clear-cutting could be replaced by strip cutting, which would harvest trees in narrower strips, leaving rows of trees standing. Strip cutting allows for more edges to be available for young plants to get their start to replenish the forest.

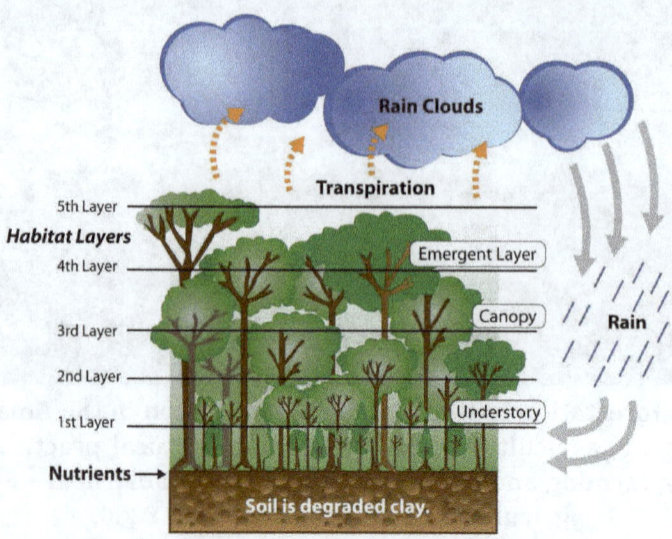

Figure 6.31: Dynamics of a Tropical Rain Forest

This image illustrates that the nutrients of the tropical rain forest ecosystem rests on the forest floor because tropical soils are degraded. The forest canopy has many habitat layers providing homes to a diversity of organisms. Dense tropical rain forests also contribute to the process of transpiration, which cycles precipitation from the ground back into the atmospheres where it can collect and return in the form of rain.

Additional forest habitat is lost when rivers are dammed up for hydroelectric power and thousands of acres of tropical rain forests are flooded. Decomposing plant material in the reservoirs behind the dams increases the water's acidity, causing erosion to the hydroelectric turbines. The turbines have a short life span, so the dams are a short-term energy solution that creates long-term problems. The immense flooded areas reduce the habitat for tropical organisms and restrict the use of the land for other purposes. Flooding can extend into traditional hunting grounds for indigenous groups, who have little political power to restrict this type of development.

The Amazon Basin is in the midst of a development pattern in which peripheral regions are exploited for natural resources or agricultural production. Much of the wealth generated ends up in the hands of investors who live in the core cities or in other countries rather than local people, a common and recurring pattern of economic exploitation of the periphery. The great frontier of the Amazon Basin continues to attract thousands of new immigrants seeking out opportunities or employment. As roads and highways are built to open up new paths through the region, even more of the forest gets cut down to accommodate the development pattern. This region will continue to be developed and altered, and the government of Brazil will have to continue to control its evolving dynamics in order to preserve the natural environment.

Regional Brazil

The core-peripheral spatial relationship can help us understand the power of Brazil's southeast core and the natural resource base of the country's peripheral north. This can explain the basic layout of Brazil's economic geography, but the northeast, the west central, and the south are three additional regions to consider that contribute to the geographic understanding of the country. Each has specific qualities that make it unique in Brazil's diversity of landscapes.

The **northeast** region is anchored along the coast, where plantation agriculture thrived during the colonial era. African slaves became the main labor base, and once freed, they made up most of the population. This agricultural region still grows sugarcane along the coast and other crops inland where the rainfall is reduced. The inland region includes parts of the Brazilian Highland, which runs parallel to the coast. Agriculture has traditionally required a large labor base, and family sizes in rural areas have been larger than their urban counterparts. This trend has given rise to a regional population of more than fifty million. However, the agriculture-based economy has not been able to supply the number of jobs and opportunities required for this large population. Poverty and unemployment in the northeast are high and have a devastating effect on the people. Some farms are not large enough for families to make a living on. Individuals in the northeast have developed a pattern of migration to the larger cities of the core area, looking for opportunities and employment—a pattern that fuels the self-constructed housing in the favelas of Rio de Janeiro and São Paulo.

The **west central** region has been opening up for development since the last part of the twentieth century and has experienced enormous advancements in industrial agriculture. In this region, vast cerrado grasslands are being plowed and converted into colossal fields of soybeans, grains, and cotton. The northwest portion of the cerrado is the huge Motto Grosso Plateau, which almost covers the largest state of the region. The landscape of this plateau is mainly scrub forest and savanna. Traditionally, the savanna portions have been used for grazing livestock, but in recent years more of the land is being plowed for growing agricultural crops. The entire west central region of Brazil is a giant breadbasket for the country. Its full agricultural potential has not been reached. Industrial agriculture requires infrastructure and transportation networks to transport the harvests to market, but the government has not kept up with the development of infrastructure at the same rate as the private sector has been developing industrial agricultural production. This region has enormous potential for agricultural expansion in the future.

Figure 6.32: Metropolitan Cathedral of Brasilia The forward capital of Brasilia, built in 1960, showcases many architectural styles. Carla Salgueiro – Brasilia – CC BY 2.0.

Located on the eastern edge of the west central region is the forward capital of Brasilia. Its creation was prompted by the need to further develop the country's interior. Built in 1960 and now a metropolitan center with more than 3.8 million people, the city was planned and designed to be a capital city. Because of its rapid growth and development, the city faces issues similar to those found in Rio de Janeiro or São Paulo. Rural-to-urban shift has brought many rural people to Brasilia in search of opportunities and employment. Just as Rio de Janeiro and São Paulo have favelas, Brasilia has slums and self-constructed districts, too. Moving the capital to Brasilia moved the focus from the coastal region of the south to the interior. However, a large percentage of the interior lies beyond the city of Brasilia. Highways that reach the Amazon can be accessed from the capital, but large areas of Brazil's interior remain a frontier unexploited by development. Developing these areas may bring great opportunities and benefits to the country; however, recent development activities have also brought devastation to the tropical rain forests in many parts of the frontier.

The three states of the **south** region are located well south of the Tropic of Capricorn and extend south to the border with Uruguay. This extrusion is often called South Brazil. Brazil was colonized by the Portuguese, but many of the immigrants to the south were from other parts of Europe, including Germany and Italy. The Italian immigrants developed a growing wine industry, and the German immigrants raised cattle and grew other crops. The region's good soils and moderate climate support many types of agriculture, which have dominated the early development of the region. Success in growing rice and tobacco and raising livestock has brought prosperity to the people who live here. Brazil has been one of the world's main producers of tobacco for many years, and the tobacco from the south is highly regarded for its nicotine content.

Figure 6.33: Miolo Vineyard The Miolo Vineyard is located in the interior of the Brazil's affluent south region. Many Italian immigrants settled here and developed a wine industry that continues today. Jeff Belmonte – Vinicola Miolo | Miolo Vineyard – CC BY 2.0.

The south is one of Brazil's most affluent regions, and it has gained much wealth from agricultural activities. Farming is only one of the region's highly developed economic activities. The area is also blessed with natural resources such as coal that is shipped north to the main cities where steel is produced. The wealth of the region has provided support for high-tech industries, which are attracted to the region because of the supply of skilled labor, access to quality transportation, and communication links. Computer companies have established software firms that have in turn attracted other new companies. A technology center similar to California's Silicon Valley has combined with the manufacturing complex that has sprouted up along the coast of the south. The government and the business community have both provided economic incentives for these and other emerging enterprises.

The attractiveness of the region in terms of both physical geography and economic stability has prompted some who live there to suggest that the region should separate from Brazil and become an independent country. Along the same vein, residents of the region have discouraged immigrants from poorer regions of other Brazilian states from moving here. Poor migrants seeking opportunities or employment are often rejected and provided transportation back to where they came from just to keep them from becoming residents. Nevertheless, many people have migrated to the south. The region is host to immigrants from Japan and the Middle East. The progressive cities and striking environmental conditions will continue to attract people to live here.

The Brazilian City of Curitiba in the SouthBrazil's "green city" of Curitiba in the south region has been a model of urban planning that has been used in the development of cities around the world. Curitiba is an attractive city in its urban design, environmental friendly attitudes and its green spaces. The city even served as Brazil's capital for a few days in 1969 during a transition period before the forward capital of Brasilia was built.

The city is an industrial hub for southern Brazil and has diverse commercial activities—from manufacturing to financial services and tourism. Automobile manufacturing is a major part of the industrial sector and is the second largest car maker in the country. The modern airport, bus system, and planned green spaces create a favorable environment for commercial ventures, which are promoted via business incubators to attract new enterprises.

In 2010, Curitiba had an overall metropolitan population of about 2.3 million but has used mass transit and environmental planning with great efficiency. Urban planning has helped develop a combination of streets and districts that encourage walking, bicycling, and mass transit as the main means of access. Pedestrian malls with shopping, restaurants, and services are common. More than sixty miles of bike routes allow commuters to bicycle to work. The rapid transit system, regularly used by approximately 85 percent of the people in the city, is rated as one of the best in the world. As much as 70 percent of the city's trash is recycled, providing additional employment opportunities. At the same time that the population doubled, the use of automobiles declined by 30 percent. However, the city still has the highest rate of automobile ownership per person in Brazil. The decline in automobile use has had a positive effect on air quality by reducing the amount of harmful exhaust emissions. Serious urban planning has gone into creating Curitiba as an efficient and yet progressive city. In 2010 Curitiba was awarded the Globe Sustainable City Award, which was set up to recognize cities and municipalities that excel in sustainable urban development around the world (EkonomiNyheter).

Figure 6.34: Rua XV de Novembro The mayor of Curitiba developed a master plan for the city in 1972. One weekend, the mayor changed one of the main downtown avenues into a pedestrian-only street without announcing it to the public. At first the local businesses objected to it, but once the concept caught on, the street became one of the most actively visited parts of the downtown area. Lu – Canon-proimage – CC BY-NC 2.0.

Key Takeaways

- Brazil was a Portuguese colony during the colonial era. This is why the Portuguese language and the Catholic religion are dominant components of the Brazilian culture. The strong African heritage comes from the many

African slaves brought in during the colonial era to work the plantations. Immigrants from many other countries have settled in Brazil as well.

- The urban southeast, with its large cities of São Paulo and Rio de Janeiro, represents the core economic area of the country. The large cities of Brazil continue to expand to accommodate the influx of new immigrants. The outer edges of the cities expand through self-construction in slums called favelas.
- The northern regions of the Amazon Basin represent the periphery of the country. The north is being exploited for its natural resources. Development patterns have caused extensive deforestation of the tropical rain forest of the region. The main causes are agriculture, logging, and mining.
- Other regions of Brazil include the west central, with extensive agricultural activities; the northeast, heavily populated with serious poverty; and the south in the protruded area bordering Uruguay, which is an affluent region with high standards of living.

Discussion and Study Questions

1. Explain the Atlantic Trade Triangle. How did it operate? How was Brazil involved?
2. Where is Brazil's core region? Where is its peripheral region? What are its core cities?
3. What three cities make up Brazil's political capital, cultural capital, and largest city, respectively?
4. What is a forward capital and why does Brazil have one? Are favelas located in the forward capital?
5. Why is Manaus such a core city for its region? What makes it so attractive to economic development?
6. What are the main causes of deforestation in the Amazon Basin? How can deforestation be reduced?
7. What is the economic situation in the northeast? How are people attempting to remedy the economic situation?
8. What is promising about the west central region of Brazil? What is needed to develop the area?
9. Why is the south region so affluent? What is this region doing to attract economic development?
10. How has rural-to-urban shift been evident in the population of Brazil? Where has it been occurring?

Geography Exercise
Identify the following key places on a map:

Cities of Brazil

- Belém
- Brasilia
- Curitiba
- Manaus
- Rio de Janeiro
- Salvador
- São Paulo

Regions of Brazil

- North
- Northeast
- South
- Southeast
- West central

Physical Regions

- Amazon Basin
- Brazilian Highlands
- Cerrado
- Mato Grosso Plateau

References

EkonomiNyheter, "Globe Award 2010: The Brazilian city Curitiba awarded the Globe Sustainable City Award 2010," www.ekonominyheter.se/pressmeddelanden/globe-award-2010-the-brazilian-city-curitiba-awarded-the-globe-sustainable-city -award-2010,16486.

6.4. THE SOUTHERN CONE

The Southern Cone region of South America includes the countries of Uruguay, Argentina, and Chile. The name is an indication of the physical shape of the southern portion of the continent. The region is located south of the tropics. The Tropic of Capricorn runs just north of Uruguay and cuts across the northern regions of Argentina and Chile. The Southern Cone has more moderate temperatures than the tropics. Type C climates dominate in Uruguay, the Pampas region of Argentina, and central Chile. The region has extremes in weather and climate. The Atacama Desert and Patagonia both have type B climates because of a lack of precipitation, which stems from the rain shadow effect of the Andes Mountains. Highland type H climates follow the Andes chain through the region and exhibit their typical pattern of warmer temperatures at lower elevations and colder temperatures at higher elevations.

Figure 6.4.356.4.35: The Three Main Regions of Chile (*left*); Argentina and Uruguay with the Regions of Argentina Outlined and Labeled (*right*). CIA World Factbook – public domain.

The countries of the Southern Cone share similar economic and ethnic patterns. Agriculture has been a major focus of the region's early development pattern, but today a large percentage of the population is urbanized. The European heritage of most of the population ties this region to Europe as an early trading partner. The global economy has given cause for these countries to form trade relationships with many countries. The physical geography has provided many opportunities for human activities. The mountains, plains, and coastal areas provide a diversity of natural resources that have been exploited for national wealth. All three countries have primate cities that hold a high percent of the country's population. Primate cities are usually twice as large as the next largest city and usually are exceptionally expressive of the national feeling and culture. In this case, all three primate cities are also the capital cities of each country. The Southern Cone is an urban region with higher incomes and higher standards of living than many other parts of South America.

Uruguay

Uruguay is located along the South Atlantic coast bordering Argentina and Brazil. The only South American country smaller in physical area is Suriname. French Guiana is also smaller than Uruguay but remains a department of France. Uruguay has the location, natural resources, and global trade connections to provide ample opportunities and advantages for its people. The agricultural lands of the Pampas extend into Uruguay and provide for an extensive agricultural industry with livestock, small grains, vegetables, and dairy. This agricultural base has been the foundation of Uruguay's growing economy.

Uruguay has been able to integrate itself into the global trade arena and has entered into a postindustrial development status. Postindustrial activities have included computer software development and advancements in information technologies. Uruguay is a modern and well-developed country. About 88 percent of the population is of European descent, and more than 92 percent of the population is urban. The capital city of Montevideo, a cosmopolitan city with a population of about 3.5 million, is home to about 40 percent of the population. The coastal region is an attractive place for tourists and locals who enjoy the beautiful beaches along the shores of the Atlantic.

Figure 6.36: Beachfront in Punta del Este, Uruguay Notice the high-rise apartments and commercial buildings. Uruguay is highly urbanized and has been an attractive place for tourism. Ernesto Jorysz – Punta del Este (Uruguay) – CC BY 2.0.

Montevideo is also the location of the headquarters for Mercosur (the Southern Cone Common Market). Mercosur was created in 1991 to open trade of goods and services among South American countries and promote development within the realm. Current members of Mercosur include Uruguay, Paraguay, Brazil, and Argentina. Venezuela's application for full membership is pending. Additional associate members of Mercosur include Peru, Ecuador, Colombia, Chile, and Bolivia. Mercosur is on a parallel path to compete with the North American Free Trade Agreement (NAFTA), the Central America Free Trade Agreement (CAFTA), the European Union, and other global trade associations.

Argentina

South America's second-largest country is Argentina. In physical area, Argentina is ranked eighth in the world. The Andes Mountains span its entire western border with Chile. At the southern end of the continent is Tierra del Fuego. Argentina is a land of extremes. **Mt. Aconcagua** is the highest mountain in the Western Hemisphere at 22,841 feet in elevation, and **Laguna del Carbon** is the lowest point in the Western Hemisphere at -344 feet below sea level. Parts of the northern region have a tropical climate; the southern region extends into tundra-like conditions with treeless plains. Argentina also claims the Falkland Islands, which are now controlled by Great Britain. In 1982, Argentina's ruling military junta attempted to retake control of the Falklands, but the British navy halted the action in what became known as the Falklands War.

Physical Regions

Argentina can be categorized into a number of regions that correlate roughly with the varied physical and cultural landscapes of the country. The main regions include Chaco, Northern, Mesopotamia, Cuyo, Pampas, and Patagonia. The Northern region of Argentina has one of the highest average elevations because of the Andes Mountain Ranges. The Andes ranges widen as they proceed northward to the west of Chaco and are home to fertile river valleys. The northern ends of the ranges extend into Bolivia and enter the Altiplano of the Central Andes.

The Chaco region, which is formally called the Gran Chaco, extends from northern Argentina into western Paraguay. Scrublands and subtropical forests dominate the landscape. There is a wet season as well as a dry season suitable for raising livestock and some farming. Western Chaco, which is closer to the Andes, is drier with less vegetation and is known for its high temperatures during the summer months. To the east, the Chaco region receives more rainfall and has better soils for agriculture. The agrarian lifestyle dominates the cultural heritage of this region. In the 1920s and 1930s, the Chaco region attracted a large number of Mennonite immigrants from Canada and Russia who established successful farming operations mainly on the Paraguay side of the border and also extending into Argentina.

To the east of the northern region—on the other side of the Paraná River and reaching to the banks of the Uruguay River—is the region called Mesopotamia, whose name means "between rivers." This unique region has a variety of features, from flatlands for grazing livestock to subtropical rain forests. The most noteworthy feature is the expansive **Iguazú Falls** on the Iguazú River, located on the border of Brazil and Argentina. It is a series of 275 parallel waterfalls that are just short of two miles across. It has the greatest average annual flow of any waterfall in the world. Most of the falls are more than 210 feet high; the tallest is 269 feet. The spectacular Iguazú Falls is a major tourist attraction, drawing people from all over the world.

Figure 6.37: The Iguazú Falls on the Iguazú River on the Border between Brazil and Argentina The headwaters of the Iguazú River are near Curitiba in Brazil. The river converges with the Paraná River about 14 miles downstream from the Iguazú Falls at the point where Brazil, Paraguay, and Argentina meet—called the Triple Frontier. The falls is approximately 1.7 miles along its edge, which is divided by rocky islands on both the Argentinean and Brazilian sides. The Iguazú Falls are a major tourist attraction for the region. Jeffrey Bary - Iguazu Falls - CC BY 2.0.

The Cuyo region is located along the Andes Mountains in the west central part of the country. Mt. Aconcagua is located here, along with other high mountain peaks. This arid region gets most of its moisture from melting snow off the mountains, which irrigate the rich agricultural lands that produce fruits and vegetables. The Cuyo is a major wine-producing region; it accounts for up to three-quarters of the country's wine production. Picturesque vineyards and farms make the Cuyo a favorite tourist destination in Argentina. **Mendoza** is the country's fourth-largest city. Low mountain ranges form the eastern border between the Cuyo and the Pampas.

The Pampas is a large agricultural region that extends beyond Argentina and includes a large portion of Uruguay and the southern tip of Brazil. With adequate precipitation and a mild type C climate, the Pampas is well suited for both agriculture and human habitation. The rich agricultural lands of the Pampas include the largest city and the country's capital, **Buenos Aires**, which is home to up to a third of the nation's population. The Pampas provides some of the most abundant agricultural production on the planet. The western grasslands host large haciendas (prestigious agricultural units) with cattle ranching and livestock production. This area has elevated Argentina to its status as a major exporter of beef around the world. Agricultural production has been a major part of the nation's economy. One hundred years ago, the export of food products made Argentina one of the wealthiest countries in the world. In today's global economy, the profit margins in agricultural products are not as lucrative, and industrialized countries have turned to manufacturing for national wealth. Argentina continues to have a strong agricultural sector but has been increasing its industrial production in order to secure a strong economy.

Figure 6.38: Rain Shadow Effect The rain shadow effect in southern South America creates the Atacama Desert and an arid Patagonia. University of Texas Librariers.

Patagonia is a large expanse of the south that is semiarid because of the rain shadow effect. This area possesses enormous natural resources, including large amounts of oil and natural gas. Deposits of gold, silver, copper, and coal can be found here. Raising livestock has been the main livelihood in Patagonia, which is otherwise sparsely populated. Patagonia includes the southern region of Tierra del Fuego and the rugged Southern Andes, which have some of the largest ice sheets outside Antarctica and many large glaciers that provide fresh water that feeds the region's streams and rivers. Patagonia also has a number of scenic lakes. Abundant wildlife can be found along the Atlantic coast, including elephant seals, penguins, albatrosses, and a host of other species. This region demonstrates that Argentina is blessed with a variety of physical landscapes and natural resources that can help support its economy and its people.

Population and Culture

Argentina, with a population of about forty million, is a country of immigrants and a product of the colonial transfer of European culture to the Western Hemisphere. During the colonial era, millions of people immigrated to Argentina from Western European countries such as France, Germany, Switzerland, Portugal, Greece, the British Isles, and Scandinavia. Additional immigrants came from Eastern Europe and Russia. Eighty-five percent of the population is of European descent; the largest ethnic groups are Spanish and Italian. The Mestizo population is only at about 8 percent. A small number of people from the Middle East or East Asia have immigrated and make up about 4 percent of the population. Less than 2 percent of Argentines declare themselves to be Amerindians.

Old World European customs mix with New World Latin American traditions to form a cultural heritage unique to Argentina. This cultural heritage can be experienced in the metropolitan city of Buenos Aires, where all facets of society and culture can be found. With a population of about thirteen million—one-third of Argentina's total population—Buenos Aires is a world-class city. Argentina is an urban country: more than 90 percent of the population lives in cities. The rural side of the culture has often been characterized as the traditional gaucho (cowboy) image of the self-reliant rancher who herds cattle and lives off the land. Beef is a mainstay of the cuisine in much of the country. The urban culture includes the traditional Argentine tango with music and camaraderie in upscale night clubs. These traditional images may be stereotypes, but the cultural scene in Argentina is heavily invested in the international trends of the modern world. The cultural landscape has become integrated with fashions and trends from across the globe.

Figure 6.39: Palermo District, Buenos Aires This photo illustrates the enormous expanse of the most populous area of Buenos Aires, Argentina. Blatant World – Palermo, the most populous district in Beunos Aires – CC BY 2.0.

People and Politics

The people of Argentina have a good standard of living, and the country is up-and-coming on a world scale. The main problem has been the instability of the government during the twentieth century. The Argentine constitution provides for a balance of powers between the judicial, legislative, and executive divisions, similar to that of the United States. For many decades in the twentieth century, the state vacillated between a democratically elected government and military rule. The swings back and forth have been consistent and regular, which has wreaked havoc on the economy, resulting in everything from hyperinflation to brutal authoritarian rule.

An example of the national swings in Argentina occurred between 1946 and 1955, when **General Juan Perón** was elected president. His wife, **Eva** (popularly known as Evita), became a public sensation. "Peronism" started out with populous support and a shift toward improving working conditions and increasing government spending. At the same time, censorship, isolationism, and repression of civil rights were elevated to a point of social unrest. Opposition members were imprisoned or killed. Eva Perón died of cancer in 1952, and Juan Perón was eventually ousted from office and fled to Spain. He later returned from exile to gain prominence in politics once again but died in office in 1974. Other presidents came to power only to be deposed or ousted by military coups. The instability in the political arena created problems for the economic sector, which had to deal with inflated currency and an unattractive environment for foreign investments. More recently, there has been some progress in stabilizing both the government and the economy, but political instability remains a factor. Argentina has abundant natural resources, adequate infrastructure, and an educated work force. The country has all the necessary means to launch into the future with a strong economy—as long as it is able to establish a stable government and a sound economic agenda.

Chile

Chile is a long, narrow country on the western edge of southern South America. Chile is 2,500 miles long and only 90 miles wide on average. This country borders the Pacific Ocean on one side and the Andes Mountains on the other. Chile has a variety of environmental zones, administrative districts, and climate patterns. Temperatures are cooler as one moves south toward Tierra del Fuego, which is split between Chile and Argentina. Rain has never fallen in select areas of northern Chile, which includes the Atacama Desert, one of the driest places on Earth and home to one of the world's greatest copper and nitrate reserves. The sodium nitrates found in the Atacama Desert are used in plant fertilizers, pottery enamels, and solid rocket fuel.

The Rain Shadow Effect

The climate is due to the rain shadow effect. In northern Chile and the Atacama region, prevailing winds reach northern Chile from the east and hit the Andes Mountain chain, which are some of the highest mountains on the continent.

Figure 6.40: Mt. Fitz Roy's Rugged Landscape Mt. Fitz Roy is in the Andes on the border between Chile and Argentina. The mountain range borders Patagonia on the Argentinean side in the remote region of the south. The village of El Chaltén, which is the main access to the mountain, can be seen in the lower right of the photo. The unique terrain of the mountain is often photographed but seldom climbed. Josh and Erica Silverstein – Patagonia – CC BY 2.0.

The height of the Andes causes any moisture from rain clouds to precipitate on the eastern slopes. The western side of the Andes Mountains at that latitude receives little or absolutely no precipitation, causing extreme desert conditions in the Atacama region of Chile. Southern Chile receives a large amount of rainfall because the prevailing winds at that latitude come from the west. Here the winds, which have picked up moisture over the South Pacific Ocean, hit the western side of the Andes. The air then precipitates out its moisture as it rises up the mountainsides of the western slopes of the Andes. Less moisture reaches the eastern side of the mountains, creating a rain shadow with arid and dry conditions for the region called Patagonia in southern Argentina. The Andes are not as high in elevation in the south, which allows some precipitation to fall on the rain shadow side.

Chile can be divided into three regions:

1. **Northern Chile**, with the dry Atacama Desert
2. **Central Chile**, with a mild type C climate, adequate rainfall, and good farmland
3. **Southern Chile**, with lots of rainfall, rural, isolated islands, and mountains

Central Chile is the core region because it has a valuable port in **Valparaiso** and the country's capital city, **Santiago**, which is also Chile's most populous city. Central Chile is also home to more than 90 percent of the country's population.

Socioeconomic Conditions

The people of Chile are 95 percent European and Mestizo. They have worked to establish a good education system and an increasing standard of living. The country had a relatively stable government until the 1970s when a short revolution brought a military dictatorship to power until 1989. The 1990s brought about a surge in economic development that has continued into the twenty-first century. The political situation has stabilized, and the first female president, Michelle Bachelet, was elected in 2006. The political system is faced with the unequal distribution of wealth that is common in Latin America and many other countries of the world. Half the country's wealth is concentrated in the hands of about 10 percent of the population. About 50 percent of the population is on the lower end of socioeconomic scale. Dire poverty exists in Chile, but it is not as prevalent here as it is in the Central Andes, Paraguay, or Northeast Brazil. Chile has a thriving middle class that has made good use of the opportunities and education that Chile has offered them.

Chile is blessed with natural resources that include the minerals of the Atacama Desert, extensive fishing along the coast, timber products from the south, and agricultural products from central Chile. All these factors have brought about an emerging development boom and have attracted international trading partners.

The stable government and the growing economy have successfully kept inflation low, kept employment high, reduced poverty, and brought in foreign investment. In the globalized economy, Chile has managed to work with various trading partners to increase its advantages and opportunities in the international marketplace.

Chile and NAFTA

Chile has been increasing its trading activities with its counterparts in the Pacific Rim in Europe and North America. Chile has an abundance of fish in its coastal waters and copper and minerals in the Atacama Desert and has been exporting timber products from its southern region. The United States is one of Chile's main trading partners. Chile's main commodity exports to the United States include paper, minerals, metals, and copper. Major agricultural products that are shipped to the United States include processed fruits, tomatoes, grapes, vegetables, and wines. There are several good reasons why the United States trades with Chile even though it is the farthest country from the United States in South America. International trade and the seasonal patterns of agriculture have much to do with the benefits. The seasonal variations between the northern and southern hemisphere create a balance between the two. When it is winter in the north, it is summer in the south. Each has an opposing growing season for fruits and vegetables that can complement the other. Chile is an excellent counter to the opposite growing season in the Northern Hemisphere.

If you recall Section 5.2 "Mexico" on Mexico and NAFTA, you will remember that after NAFTA was implemented, Chile was considered a "silent" partner. In 1994 during a summit meeting, the president of Mexico, Canada's prime minister, and the US president agreed to allow Chile to join the North American Free Trade Agreement (NAFTA). Chile's president declared his support for his country to join NAFTA. The agreement could not be completed because of negotiation approval by the US Congress. At this time Chile entered into independent free trade agreements with both Canada and Mexico. A free trade agreement was later signed between the US and Chile in 2003. Thus, Chile has free trade agreements with all three North American countries and can benefit by the mutual agreements that parallel NAFTA. In this situation, Chile is an associate trading partner and enjoys the privileges of NAFTA membership without technically being listed as a member. The United States was really looking to counter Chile's trade with Japan and Europe. Increased trade with North America diminished Chile's trade with the European Union or the East Asian community.

As copper and natural resource prices fluctuate, creating unstable economic conditions, Chile has been expanding into the manufacturing sector. To retain the highest value-added profits and expand its economy, Chile will have to enlarge its manufacturing sector. Today Chile produces modest quantities of basic chemicals and electronic goods. Chile's expansion of business and commerce has been endorsed as a model of successful economic development to be emulated by other Latin American countries.

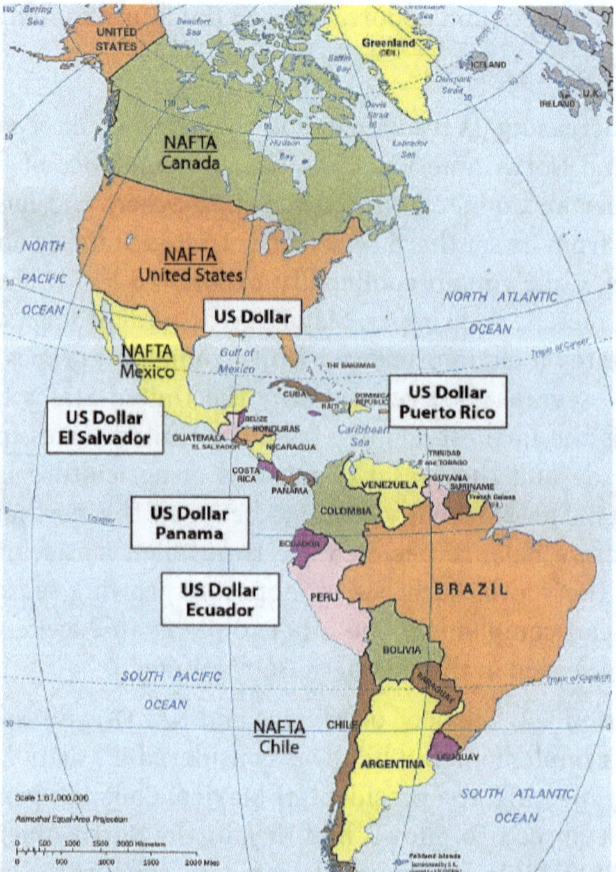

Figure 6.41: NAFTA Members and US Dollar Users as of 2010 Chile is not a full member of NAFTA but has separate free-trade agreements with all three NAFTA countries and is considered an unofficial member. University of Texas Librariers.

Chile has strong ties to the economies of North America, but in spite of close ties with the north, Chile has retained its unique status in the Southern Cone. Chile still has its own currency even though countries with smaller economies, such as Ecuador and El Salvador, have adopted the US dollar as their medium of exchange. As of 2010, the NAFTA members of Canada, the United States, and Mexico, along with their trading partner of Chile, all used different currencies. The US dollar is the most widely used currency in the world and is also used in other Latin American countries. There has been talk of creating a similar currency within NAFTA called the Amero that would replicate the European Union's solution, which was to create a single currency, the euro. World currencies fluctuate in value, and a stable currency would increase the capacity for creating larger trading blocs that could do business on a more even economic playing field. Chile has individual free-trade agreements with all three members of NAFTA, so any change in currency with the NAFTA countries would also affect Chile.

Key Takeaways

- The Southern Cone possesses large, diverse physical landscapes. Variations in terrain include tall Andean mountain peaks, desert conditions, prairie grasslands, and semitropical regions to the north.
- The Southern Cone is an urban realm with high percentages of the population living in large cities. Primate cities dominate each country. Large sections of each country's interior make up the extensive rural periphery with activities based on natural resources.
- The Southern Cone is a European commercial area, and more than 85 percent of the population is of European ancestry. There are few Amerindian minorities or immigrants from Africa or Asia in the Southern Cone.
- Argentina is a large country in physical area and has a number of important regions that provide opportunities for economic prosperity. The country, however, has lacked a stable government and a consistent business climate to take advantage of Argentina's favorable geographic situation.
- Chile has emerged on the global stage as a model for economic growth and stability. Its people have a higher standard of living, and the country has abundant natural resources. Chile has been a trading partner with NAFTA countries.

Discussion and Study Questions

1. What are the main physical regions of Argentina and Chile?
2. What role does agriculture play in the human activities of Uruguay?
3. How did Uruguay manage to build up its high standard of living?
4. How does the core-periphery spatial relationship apply to Uruguay, Argentina, and Chile?
5. Why is there a desert in Chile that is partially located in the tropics? Why is this desert so important?
6. How does the rain shadow effect apply to both Argentina and Chile?
7. What are the main attributes and contributions of the Pampas and Patagonia?
8. What are the three regions of Chile and how do they complement each other?

9. How is Chile associated with NAFTA? Why does Chile have free-trade agreements with every NAFTA member?
10. How does Chile complement the agricultural production of the United States and Canada?

Geography Exercise

Identify the following key places on a map:

- Atacama Desert
- Cape Horn
- Chaco
- Cuyo
- Falkland Islands
- Iguazú Falls
- Iguazú River
- Mendoza
- Mesopotamia
- Pampas
- Paraná River
- Patagonia
- Tierra del Fuego
- Uruguay River
- Valparaiso

6.5 CONCISE SUMMARY

The equator traverses the center of South America at its broadest expanse, aligning with the Amazon River. The primary physical features are the formidable Amazon River and the vast Andes Mountains. The Andes, the world's longest mountain range, extend from Venezuela to southern Chile. The Amazon River possesses the greatest water discharge of any river globally, and the Amazon Basin encompasses the largest tropical rainforest on the planet. Rainforests include less than 5 percent of the Earth's surface yet possess the highest biodiversity of any biome, including up to 50 percent of the planet's creatures.

All of South America, with the exception of the Guianas, was colonized by Spain and Portugal. The Guianas were situated near the coastline, where plantation agriculture thrived. African slaves constituted the primary labor force on the plantations. Upon the abolition of slavery, indentured workers from the Asian colonies supplanted enslaved individuals. The majority of the people in Guyana and Suriname is of Asian descent. The cultural geography of South America can be comprehended by delineating the cultural zones established by predominant ethnic groups and the human activities in which they participate.

Venezuela and Colombia are the predominant nations in northern South America. Venezuela is an urban nation, with a significant portion of its revenue derived from oil exports. Colombia is characterized by rugged terrain and extensive tropical forests adjacent to the Amazon. Colombia's primary export commodities are illegal narcotics, oil, and coffee. The United States is the primary purchaser of these products. The governance of Colombia's physical territory is partitioned among the government, drug cartels, and guerrilla factions.

The Andean West region of South America was the site of the old Inca Empire, which was subjugated by Spanish adventurer Francisco Pizarro in 1533. Ecuador, Peru, and Bolivia possess high-elevation areas in the Andes, notably the Altiplano region, which encompasses Lake Titicaca. The majority of the population is Amerindian and adheres to the customs and traditions of their ancestry. Oil and mineral resources have been a primary source of wealth for the region; nevertheless, the gains have not consistently benefited the majority population, since they are frequently concentrated among a minority at the apex of the socioeconomic hierarchy.Brazil was colonized by Portugal, which imparted the Portuguese language and the Catholic religion to the nation. During the colonial period, a significant influx of African slaves was brought to the country, contributing to its heterogeneous society. A significant number of Europeans and Japanese also relocated to Brazil. The country's vast expanse facilitates a variety of human endeavors. The principal urban region is situated along the southeastern coastline, encompassing the major cities of São Paulo and Rio de Janeiro. Brasilia, the future capital, is situated inland at the outside of the central area.The rural outskirts of Brazil encompass the huge Amazon Basin, characterized by tropical rainforests and extensive underdeveloped areas. The Amazon Basin, first occupied by Amerindian populations, is undergoing development for agriculture, mining, and timber extraction. This form of development is ravaging extensive areas of rainforest and exerting escalating strain on the dwindling indigenous populations. Extensive mining activities have driven development, resulting in significant environmental issues related to habitat destruction and the devastation of tropical rainforests. The Southern Cone of South America features diverse physical landscapes, ranging from the semitropical regions in the north to the arid grasslands of Patagonia in the south. The Atacama Desert, located on the western coast, results from the rain shadow effect created by the towering Andes Mountains. The fertile agricultural expanse of the Pampas serves as the region's breadbasket, yielding substantial quantities of grain and livestock. Coastal regions are advantageous for fishing and facilitate access to international markets. The southern regions of South America also possess significant European populations. There are few Indigenous Americans or individuals of African descent. The majority of individuals reside in urban regions, with one-third of Argentinians inhabiting Buenos Aires. Chile has been a significant trading partner with the nations of the North American Free Trade Agreement (NAFTA) and possesses a stable and expanding economy. Argentina is a vast nation with significant potential for economic advancement.

www.ingramcontent.com/pod-product-compliance
Lightning Source LLC
LaVergne TN
LVHW062219200926
848484LV00006B/1429